T0335397

MONSOON RAINS, GREAT RIVERS, AND THE DEVELOPMENT OF FARMING CIVILIZATIONS IN ASIA

The Asian monsoon and associated river systems supply the water that sustains a large portion of humanity and have enabled Asia to become home to some of the oldest and most productive farming systems on Earth. This book uses climate data and environmental models to provide a detailed review of variations in the Asian monsoon since the mid-Holocene and their impact on farming systems and human settlement. Future changes to the monsoon due to anthropogenically driven global warming are also discussed. Faced with greater rainfall and more cyclones in South Asia, as well as drying in North China and regional rising sea levels, understanding how humans developed resilient strategies to past climate variations is critical. Containing important implications for the large populations and booming economies in the Indo-Pacific region, this book is an important resource for researchers and graduate students studying the climate, environmental history, agronomy, and archeology of Asia.

Peter D. Clift is the Charles T. McCord Chair in Petroleum Geology at Louisiana State University. He is an affiliate faculty at the Massachusetts Institute of Technology and Yunnan University, China, and a fellow of the Geological Society of America. In 2008, he coauthored *The Asian Monsoon: Causes, History and Effects* (Cambridge University Press).

Jade d'Alpoim Guedes is an associate professor at the Department of Anthropology, Scripps Institution of Oceanography, University of California, San Diego. She is an environmental archeologist and ethnobiologist who employs an interdisciplinary research program to understand how humans adapted their foraging practices and agricultural strategies to new environments and have developed resilience in the face of climatic and social changes.

MONSOON RAINS, GREAT RIVERS, AND THE DEVELOPMENT OF FARMING CIVILIZATIONS IN ASIA

PETER D. CLIFT

Louisiana State University

JADE D'ALPOIM GUEDES

University of California, San Diego

CAMBRIDGE
UNIVERSITY PRESS

CAMBRIDGE
UNIVERSITY PRESS

University Printing House, Cambridge CB2 8BS, United Kingdom

One Liberty Plaza, 20th Floor, New York, NY 10006, USA

477 Williamstown Road, Port Melbourne, VIC 3207, Australia

314–321, 3rd Floor, Plot 3, Splendor Forum, Jasola District Centre, New Delhi – 110025, India

79 Anson Road, #06–04/06, Singapore 079906

Cambridge University Press is part of the University of Cambridge.

It furthers the University's mission by disseminating knowledge in the pursuit of education, learning, and research at the highest international levels of excellence.

www.cambridge.org
Information on this title: www.cambridge.org/9781107030084
DOI: 10.1017/9781139342889

First published 2021

A catalogue record for this publication is available from the British Library.

Library of Congress Cataloging-in-Publication Data
Names: Clift, P. D. (Peter D.), author.
Title: Monsoon rains, great rivers and the development of farming civilisations in Asia / Peter D. Clift, Louisiana State University, Jade d'Alpoim Guedes, Scripps Institution of Oceanography, University of California, San Diego.
Description: Cambridge, UK ; New York, NY : Cambridge University Press, 2021. | Includes bibliographical references and index.
Identifiers: LCCN 2020028436 (print) | LCCN 2020028437 (ebook) | ISBN 9781107030084 (hardback) | ISBN 9781139342889 (ebook)
Subjects: LCSH: Meteorology, Agricultural – Asia. | Rain and rainfall – Asia. | Monsoons – Asia. | Crops and climate – Asia.
Classification: LCC S600.7.R35 C55 2021 (print) | LCC S600.7.R35 (ebook) | DDC 630.2/515095–dc23
LC record available at https://lccn.loc.gov/2020028436
LC ebook record available at https://lccn.loc.gov/2020028437

ISBN 978-1-107-03008-4 Hardback

Contents

Acknowledgments

Peter Clift (PC) wishes to thank support from Louisiana State University during the writing of this book. Much of the material was put together during a fellowship at the Hanse-Wissenschaftskolleg (HWK) in Delmenhorst, Germany. PC especially thanks Doris Meyerdierks at the HWK and Heiner Villinger at the University of Bremen for their help. The final phases of the work were made possible by time spent at Yunnan University, the University of Texas at Austin, and the University of Hong Kong. PC also wishes to thank Sophie Vincent and Erinn Buhyoff for their help with various manuscript preparation tasks. He dedicates his work on this book to the memory of his mother, Margaret Teresa Clift, without whose help and love none of this would have been possible, as well as that of his father Donald William Clift, who first introduced him to Asia.

Jade d'Alpoim Guedes is grateful to the University of California, San Diego, for support during the writing of this book. She is grateful to Kyle R. Bocinsky and to Jacqueline Austermann, who assisted with making figures. She dedicates this book to her daughter, Anwen Else d'Alpoim-Zhang, who was patient while she worked on this book during her very first year on earth and to her family, most notably her husband, Chunbai Zhang who supported her throughout the writing process.

We wish to thank Sarah Lambert and Susan Francis at Cambridge University Press for their encouragement and patience during the writing of this work and to the comments of reviewers who helped improve this manuscript.

1

Introduction

1.1 Climate and Farming Societies in Asia

Today roughly 60% of the world's population resides in Asia under the influence of that continent's monsoon system, which dominates the regional climate and is, in turn, linked to surrounding climate systems in the Indian and Pacific Oceans, as well as Australasia. Asia is also the home of some of the earliest and longest-lived farming societies in the world, as well as some of the earliest urban cultures. Variations in the intensity and timing of rainfall, changes in temperature, and switches in the courses of rivers that rely on monsoonal rains have been tied to changes in the location and density of ancient settlements, and to the types of subsistence strategies employed by these sites' inhabitants.

In this book, we explore how variations in the intensity of the Asian monsoon have influenced the development of different farming strategies throughout Asia and how, in turn, humans have developed solutions to deal with periodic fluctuations in processes that are driven by the Asian monsoon. Variability in the monsoon will continue to test the resilience of farming strategies in the future. It is generally, if not universally, accepted that we now live in an age of continued climate change that seems likely to have been caused by human activities rather than being part of a natural cycle. The vast majority of climate scientists now agree that human-induced emissions of CO_2 and methane to the atmosphere are one of the most important causes of the changing climate we are now experiencing (Solomon et al., 2009; Zeebe et al., 2016). We use the past as the key to the future in showing how past variations in the temperature, precipitation, and changes in river systems, which are influenced by the monsoon, may have impacted past societies and examine the solutions that these societies came to in order to address these changes. This process is expected to hold importance in predicting what may happen in the future in the context of a warming global environment.

Although global climate has been relatively stable since the start of the Holocene, around 10,000 years ago, it is also apparent that even during this relatively stable phase that there have been significant perturbations. The period of maximum warmth in the Early-Mid Holocene was followed by a cooling phase after around 2000 BCE (Stager and Mayewski, 1997; Wanner et al., 2008). Since that time a number of climatic cycles have been observed, including the Roman Warm Period (250 BCE to 400 CE), the Medieval Warm Period (950–1250 CE) and the Little Ice Age (1550–1850 CE). These have found their expression through the intensity of summer monsoons in Asia because of coupling between the climate systems of the high latitudes and those in equatorial regions. Conversely, there is evidence for feedback between processes controlling the monsoon in Asia and climate systems elsewhere in the world (Ivanochko et al., 2005). In particular, we focus on the relationships between intensity of Northern Hemispheric Glaciation and the strength of summer monsoon rains because on longer timescales these appear to be closely correlated.

In the biophysical sciences, humans are sometimes seen as alien to or even parasitical on the physical environment. In an extreme representation, the role of humans is not to function as stewards but to conquer tame and subdue the environment (Beinart and Coates, 1995). Conversely, some historians and ethnographers have looked at the environment as a mere background and at climatic change as being so slow, so vast in scale, as to be virtually irrelevant (McIntosh et al., 2000). In archaeology a similar tension is present, with some approaches (particularly those highlighted in popular media) being deterministic about the role that climate played (Diamond, 2005) to others that ignore the role it could have played and critiquing approaches that invoke climate as attributing it to great or too simplistic a role, or ignoring the role played by human agency (Brumfiel, 1992; Erickson, 1999). Indeed, the majority of approaches trying to link ancient climate to changes in ancient subsistence strategies have been largely correlative: that is to say, they take changes in one variable, normally a paleoclimatic proxy and correlate this with known changes in the archaeological record. The paleoclimate community is often guilty of the latter when trying to ascribe greater significance to their findings. Archaeologists have also had difficulties in employing the paleoclimatic record in a way that allows them to accord it an appropriate role and evaluating its impact has also often been either to correlate (Contreras, 2015; d'Alpoim Guedes, 2016; d'Alpoim Guedes et al., 2016c) or conveniently ignore its potential role.

Inherent in any approach that takes into account the role of climate are ideas about "collapse" (Diamond, 2005; Fagan, 2004). While the term "collapse" is often applied in the media, what exactly is meant by this process is often unclear. Tainter (1988) argues that collapse takes place when we see a rapid loss of an established level of sociopolitical complexity. Tainter (1988) was careful not to imply that a loss of centralized control or hierarchy was somehow negative. We argue, alongside most archaeologists, that new ways of operating and organizing society are often simply adaptive: while these might be intolerable to those in positions of power, they may give greater agency and benefits to those that may have been oppressed by entrenched systems.

Middleton (2017) reviewed how ideas about "collapse" have been presented in both popular media and the archaeological literature, often in an oversimplifying manner. He highlighted, for instance, how authors like Diamond have used a Neo-Malthusian perspective to make collapse synonymous with "declines in population" that have causes inscribed in humans' role within ecosystems: ones to which few archaeologists subscribe. We aim to avoid these simplistic notions of social change that these authors critique and rather focus on highlighting the adaptive strategies employed by humans when faced with changing climatic conditions. We seek in this book to achieve a middle ground: one that neither ascribes a deterministic role for climate, nor ignores its presence, but one that seeks to provide a higher resolution picture of the intensity, nature and type of changes in the mean state of climate and more specifically how the monsoon, may have impacted the agricultural strategies they employed. Throughout this book, we also recognize that the relationship between humans and the environment is not uni-directional: humans themselves impact and change their surrounding environment.

Although we touch on the impact changes in climate may have had on different societal units, when relevant, we focus specifically on the effects that changing variables had on one aspect of human enterprise: their subsistence regimes and farming systems. Rather than purely drawing correlations between changes in rainfall or temperature with documented shifts in human subsistence, we strive to understand the mechanism via which changes in these variables might have impacted field or settlement locations, crop growth or the ability to access certain resources.

A growing literature in resilience theory (de Vries, 2006; Minnis, 1999; Redman, 2005; Redman and Kinzig, 2003; Van de Noort, 2011) and "human ecodynamics" and "socioecological aspects" of the human past (Barton et al., 2012; Kohler et al., 2007; McGlade, 1995; Van der Leeuw and Redman, 2002), which sees humans as agents that actively shape their own environment through niche construction, but also as being embedded within their ecosystems and coevolving with them is more in line with the approach we seek to adopt in this book. Like other organisms, humans have modified their subsistence strategies to ensure their ultimate success, often through thousands of years of trial and error. We thus chart changes in adaptive subsistence strategies used by humans without assuming that any of these shifts involve a type of "collapse": whatever that term might mean. Here, we prefer the term "mode shifts" to describe abrupt changes toward new and discrete climate states. These are shifts in climate that go beyond the normal range of variation and that create a new stable state that is characterized by a new suite of climatic conditions. It is often these fundamental shifts in biophysical systems that have precipitated new responses from humans.

Rather than focus on any particular theoretical framework, we highlight the traditional systems of ecosystem and plant management employed by humans inhabiting the monsoon region in the past. We argue that Asian farming systems are the products of thousands of years of human, plant, and ecosystem coevolution and thus form unique adaptations to the ecology of monsoon Asia. This wide range of strategies may prove to be our greatest asset as we face ongoing and future climatic change. These systems have been developed through trial and error over the course of thousands of years: periods of time that have encompassed a wide range of different states of the monsoon.

1.2 The Monsoon in Asia

In this book we consider how the development of climate, and particularly the monsoon system, has affected the development of agricultural strategies across Asia and, in turn, how decisions made by humans resulted in changes in the landscape. The development of the monsoon itself will be addressed in much greater detail in the following and subsequent chapters, but here we introduce the term and explain what is meant when we talk about this. The monsoon is a seasonal climate system with a summer and winter component. Most continental masses have a monsoon system (Kim et al., 2008), but the one affecting Asia is much stronger than most because Asia is much more extensive and has higher elevations than most other continents. Despite this, the Asian monsoon can be thought of as the single largest component of the global monsoon that enhances regional precipitation by drawing the Intertropical Convergence Zone (ITCZ) deep into continental interiors. The latitude and geographic features of Asia increase its ability to experience strong seasonal temperature variations. In its most basic form, the monsoon is the product of the pressure differences between the continent and the surrounding oceans, with summer heating resulting in a continental low-pressure system, while in the winter a high pressure area develops over Siberia (Manabe and Terpstra, 1974; Wang, 2006; Webster, 1987). This has the effect of driving onshore winds from the Indian and Pacific Ocean toward the Asian interior in the summertime, while in the winter winds blow out of the continent toward the south and east, often carrying large volumes of eolian dust (Arimoto, 2001). In Asia, this results in a reversal of the atmospheric circulation seen elsewhere in the world where summer heating is greatest at the equator, not in the mid latitudes. The net result of this is that mid latitude Asia is not desert-like, as seen in Africa or the Americas, but is well supplied by summer rains, making it habitable and fertile for human settlement.

The Asian monsoon is known to vary on a variety of timescales, some of them linked to the tectonics of the Asian continent and especially to the uplift the Tibetan Plateau (Molnar et al., 1993; Prell and Kutzbach, 1992) and Himalaya (Boos and Kuang, 2010; Clift and Webb, 2018), while others are modulated by other climate systems, especially the El Niño-Southern Oscillation (ENSO)(Lau and Wang, 2005). Over longer time periods there is a link between monsoon strength and the onset of glaciation in the northern hemisphere because it is often argued that summer monsoons are linked to the intensity of solar insolation (Clemens et al., 2010; Gupta et al., 2005), although this simple model has recently been called into question (Clemens et al., 2018). Strictly speaking, the monsoon represents an atmospheric pressure system and, therefore, it is experienced in the form of winds. Our association with the monsoon and rainfall comes about because around the Indian Ocean, and especially in southern parts of China and Indochina, the winds bring moisture from the ocean, which is then precipitated over the land and especially against the southern topographic front of the Himalaya, as well as in the east against the eastern flank of the Tibetan Plateau. In Arabia, however, summer monsoon winds are strong but mostly dry, with an important exception in southern Oman.

Because rainfall is crucial to farming systems in South and East Asia, and the vast majority of this is brought by the monsoon, there is a link between the strength of summer monsoon rains and the productivity of the agricultural systems, which sustain civilizations across the continent. That is not to say that the link is simple or linear but agriculture without sufficient rain at least season-ally is not practical. A quick look at the Earth's climatic belts shows that typically the mid latitude regions are generally arid regions. Deserts dominate in such areas as Afghanistan. However, regions at similar latitudes in India and southern China are humid, disrupting the global climatic pattern. The underlying reason for this distortion is the strength of the Asian monsoon system that draws moisture deep into the continental interior. Without the Asian monsoon, the population densities that we have in these regions would be unsustainable (378 people/km^2 in India compared to 43 people/km^2 in Afghanistan) and more akin to those known in North Africa and Arabia. The monsoon is, therefore, funda-mental to the maintenance of farming society in Asia, as it has been since humans first began to farm on the continent.

1.3 Environmental Impact of the Monsoon

The greatest dangers to human societies are rapid changes in environment that do not give sufficient time for people to respond and adjust. It thus helps us to understand the potential risks from changes in the monsoon if we can understand how the modern monsoon influences environment and human settlement. Figure 1.1 shows the time averaged annual rainfall across Southeast Asia based on data from NASA's Tropical Rainfall Measuring Mission (TRMM) covering the period 1998 to 2007. The satellite data is in accord with rain gauge data in showing the strongest rainfall across the Eastern Bay of Bengal, northeast India, and parts of Indochina. Rain is also strong in southern China and along the Ganges River valley adjacent to the Himalaya, as well as along the western coast of India. The vast majority of this precipitation is monsoonal and can be compared with the environment of Asia to assess what impact this rainfall has. Figure 1.2 shows ecological zones for the same area. It should be remembered that at these same latitudes in Africa we see widespread desert conditions in the Sahara, and even slightly further to the west deserts exist in Afghanistan and through much of the Middle East. Without, the Tigris and Euphrates rivers much of Mesopotamia would be uninhabitable. Indeed, there is even a significant desert in western India and southern Pakistan, the Thar Desert, which lies on the western edge of the monsoon rainfall. This desert is significantly

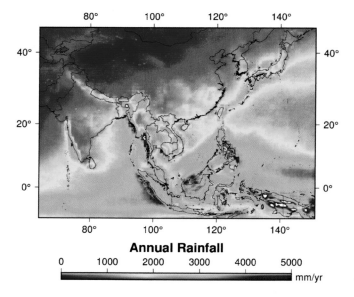

Figure 1.1 Map showing continent-wide mean annual rainfall in mm/year. Data from NASA TRMM spanning 1998 to 2014 (http://trmm.gsfc.nasa.gov/). Figure courtesy of Bodo Bookhagen.

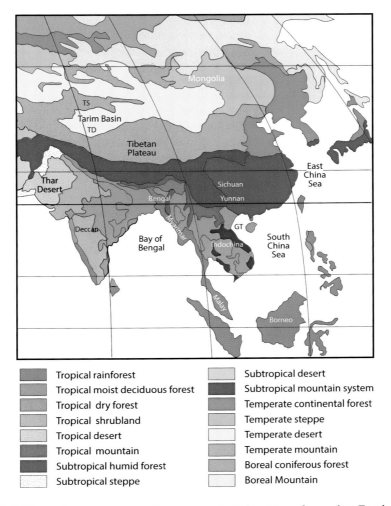

Figure 1.2 Vegetation ecozones of monsoonal Asia. Map from the Food and Agricultural Organization of the United Nations (www.fao.org/home/en/). GT = Gulf of Tonkin, TS = Tian Shan, TD = Taklimakan Desert.

influenced and supplied with sediment by summer monsoon winds (East et al., 2015; Singhvi et al., 2010) but, as Figure 1.1 shows, the precipitation is weak in the western part of the subcontinent. Figure 1.2 reveals significant zonation of vegetation that is at least partly linked with latitude, as it is in other parts of the world. As we go further north in Asia there is a gradation into temperate deserts and steppe, before the sparse, high-latitude coniferous forests of the Arctic Taiga.

The high areas of the Tibetan Plateau have their own special flora that are uniquely adapted to high altitude conditions (i.e., cold and dry; Figure 1.2), but also to the relative aridity of the northern and western plateau (Figure 1.1). The pattern we see now is one of tropical rainforest or moist deciduous forests throughout Southeast Asia and the islands of Indonesia, as well as in Bengal and other parts of northeast India. There is a clear gradient from east to west across India from tropical forests and swamps near the Ganges–Brahmaputra delta, toward the west and the deserts around the Indus Delta in southern Pakistan. Tropical forests exist along the west coast of India, immediately adjacent to the coast on the western flank of the elevated Deccan Plateau. As might be expected, there is a close parallel between the intensity of monsoon precipitation and the abundance of tropical rainforests in low latitudes across South and Southeast Asia. Where monsoon rain is weak, we instead see deserts developed, as in the Middle East.

The only exception to this general pattern is along the topographic front of the Himalaya where altitude and orographic rainfall effects dominate (Figure 1.1). In all other cases, those lands lying south of latitude 20° N are well covered by dense forest. Likewise, in China, especially in eastern China, we see the growth of some subtropical forests across much of the central and southern part of the country, transitioning north into temperate forests. These lie to the east of the Tibetan Plateau and extend as far west as the Sichuan Basin. The vegetation in China mostly differs from that seen in Indochina and Bengal because it lies further north. Much of the moisture delivered to southwest China is, moreover, coming across Indochina from the Bay of Bengal (Drumond et al., 2011), so that some of the precipitation is lost on the hills in Myanmar and Yunnan Province. It is this topographic barrier that accounts for the tongue of temperate forest that extends toward the Gulf of Tonkin from the more extensive regions of such flora in the southern Tibetan Plateau. It is noteworthy that some parts of western Indochina also have a less dense forest cover. This partly reflects slightly higher elevations and also the topographic barrier of the coastal mountains, that extend through Myanmar and into the Malay Peninsula. These are a focus for orographic precipitation for the moisture moving to the northeast from the Bay of Bengal. Biomes become more arid moving into central Asia away from the East China Sea, because as the East Asian Summer Monsoon front moves away from the coast it drops its moisture across eastern China, aided by the progressive increase in altitude going west. Further west in the region of the Tarim Basin, north of the Tibetan Plateau and south of the Tian Shan, the precipitation is negligible, allowing the Taklimakan Desert to form in this rain shadow.

We demonstrate that the pattern of floral and ecological zones in Asia is largely a function of altitude and latitude, but that rainfall intensity, largely governed by the strength of the summer monsoon is also important. Monsoon intensity, often focused by topography, accounts for the gradient in precipitation from east to west across the Indian subcontinent, for the strong rainfall in western coastal India, and for the south to north decrease in rainfall across eastern China. We might reasonably expect, therefore, that when the intensity of the summer monsoon changes, the environments in these parts of Asia would be subject to potentially significant change. This is particularly important for human settlement because these also happen to be the regions that are most heavily settled in the present day, but also regions that in the past played an important role in the development of the different systems of cultivation that sustain these areas today (Hosner et al., 2016; Roberts et al., 2016).

Figure 1.3A shows the population density within South and East Asia in 2000 and particularly highlights the high concentrations found in the Indian subcontinent and in Eastern China and more specifically within the Valley of the Ganges–Brahmaputra River, in the Sichuan Basin and in Eastern China especially in the lower reaches of the Yangtze basin and north, all the way to the Bohai Sea. Other population concentrations observed include that on the Red River delta plain in northern Vietnam, around the Mekong Delta in southern Vietnam and Cambodia, as well as a smaller concentration in the Chao Phraya delta in Thailand. Those population densities located close to the coast are at less risk from losing sufficient water supplies in the form of rainfall because precipitation in those areas is not entirely driven by the strength of the Asian monsoon, although a decrease in the summer monsoon would affect the volume of rainfall even in those places. Coastal regions are, however, more vulnerable to the effects of sea-level rise. The particularly extensive continental shelves of Southeast Asia, especially the Sunda Shelf, and the East China Sea were fully exposed at the Last Glacial Maximum (LGM) and were progressively inundated during the deglaciation. Times of rapid rise would have been potentially damaging to early settlements, for example, the 8.2-thousand-year-ago (ka) Event but especially Meltwater Pulse 1-A (14.7–13.5 ka) during which sea level rose around 20 m in only 500 years (d'Alpoim Guedes et al., 2016a; Weaver et al., 2003). Regions within the continental interior that only receive significant rainfall because of the summer monsoon winds would be most strongly affected by a change in intensity, with the exception of regions in Southwest Asia that also receive winter rainfall brought by the westerly jet, mostly particularly in the form of heavy rain associated with atmospheric vortices known as "Western Disturbances" (Hunt et al., 2018). Further east the winter is a drier season, away from coastal zones.

1.4 Past Monsoon Environments

As noted earlier, we would expect the intensity of the summer monsoon to be affected by changes in temperature and solar insolation. Monsoon intensity has been linked to the extent of Northern Hemispheric Glaciation and the climate of the

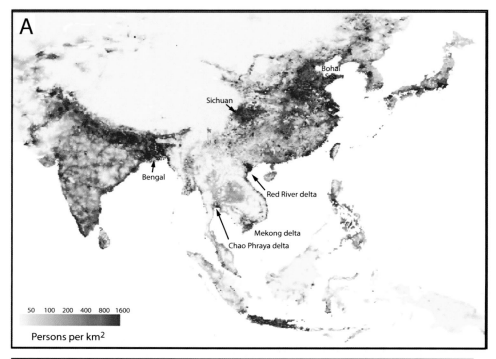

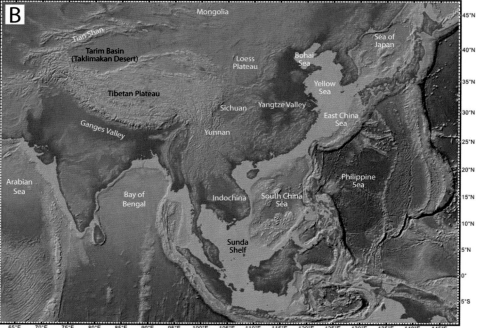

Figure 1.3 (A) Population density maps of monsoonal Asia in 2000. (B) Shaded topographic map of monsoonal Asia emphasizing the high ground in the Tibetan Plateau and to the north into the Tian Shan and Mongolia. Note the correspondence of heavily settled regions and low altitude plains in South and East Asia. Data from United Nations Environment Program/Global Resource Information Database (UNEP/GRID).

North Atlantic region, which is also controlled by solar insolation. Recent work offshore the Yangtze River mouth that monitored the discharge from that river showed a dominant variation on 41 and 100 thousand year (k.y.) time scales, linked to orbital eccentricity and obliquity and implying a strong ice volume influence on East Asian summer monsoon strength (Clemens et al., 2018). Because we know ice volumes have changed substantially in the past 20,000 years (Lambeck et al., 2014), we have good reason to believe that the monsoon has changed in its intensity during the settlement of the Asian continent by humans. How past monsoon strength has been reconstructed will be discussed in greater length in Chapter 2, but here we note that there is both modeling and observational data that confirms significant changes in climate during the period of settlement and that must have affected where settlement occurred and the subsistence choices of communities in these areas.

Figure 1.4 shows an example of how environments across Asia have changed during the period of human settlement. Modern pollen assemblages (Figure 1.4A) are broadly consistent with the ecological zones shown in Figure 1.2 (Yu et al., 2000), confirming that they can act as reliable proxies of past environment. There is a dominance of subtropical broadleaf flora in southern China, with a more temperate forest in northern and eastern China. In the far west of the country there is some evidence for a cool, mixed forest, although the Taklimakan desert of the Tarim Basin does not appear to have much influence on the data collected within China. By contrast, in northeast China the transition to more cold-weather environments going further north is apparent, with floral pollen assemblages indicating cold-weather forests and even Taiga. Although this does not give us a perfect image of what the modern vegetation is like the pollen is at least a reasonable proxy for the present day.

Figures 1.4B and C show the pollen types identified from 6,000 and 18,000 years ago, respectively, in the Early Holocene, which is typically associated with strong summer monsoon, and just after the LGM, which is subject to weak summer rains. Yu et al. (2000) argue that at these times the vegetation was somewhat different than the present day. Around 6,000 years ago, the reconstructed ecozones are not so different from the modern pollen data points, although there is evidence for desert and steppe-like conditions, in particular, across much of northern China, whereas these are restricted to the west in the present day. More recent palynology data from northwest China supports the idea of reduced desert/arid region during the Early Holocene (Herzschuh et al., 2004).

The Holocene 6,000 years reconstruction does at least show evidence for a distribution of vegetation dissemination in eastern China that is similar to today, but this is not apparent at the LGM. In the Last Glacial reconstruction,

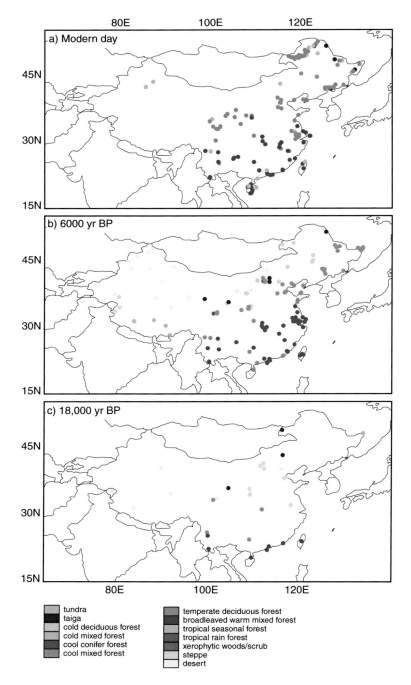

Figure 1.4 Modern Day Biomes (A) reconstructed from surface pollen data compared with those reconstructed from fossil pollen data at 6.0 ka (B) and 18.0 ka (C). Figure modified from Yu et al. (2000).

we see desert conditions in the west of the country and steppe and/or desert extending all the way to the eastern coast of China, and as far south as the modern Yangtze Delta (Yu et al., 2000). What little subtropical forest existed was restricted to the southern coast, whereas inland even southern and central China appear to have been forested by assemblages only now found in the extreme northeast. Areas that are now wet and warm enough to be important producers of rice would have been unable to do so at the LGM, and possibly even during the Early Holocene, when wild rice was restricted to areas much further south than at present (Huang and Schaal, 2012). If changes in environment of this magnitude were to occur in the future, as a result of global climate change, then the potential impact on the dense populations in this part of Asia could potentially be catastrophic.

That the environmental conditions in Asia were substantially different during the recent geological past is also borne out by constraints from numerical modeling. Although long-term climate modeling is prone to significant uncertainties, the climatic and geologic conditions during the cold Younger Dryas (11–12 ka), warm Bølling-Allerød (13–14 ka), and Early Holocene (8–10 ka) are similar to the modern, and so it is possible to make relatively realistic climate models for those time periods. These models can be used to understand what was controlling the climate at those times. Bush (2004) used numerical modeling methods to predict changes in summer monsoon precipitation between the LGM, the Early-Mid Holocene, and the present day (Figure 1.5). In that study a modification of the numerical General Circulation Model (GCM) was used that gave a spatial resolution of 3.75° in longitude and 2.25° in latitude. Temperature, wind, and precipitation predictions were made for the different stages of deglaciation based on what was known about changing atmospheric CO_2 concentrations, the exposure of continental shelves caused by lower sea levels (Fairbanks, 1989) and the extent of ice sheets (Peltier, 1994). These latter two factors are important because they affect the albedo of the continents and thus the temperature differences between land and ocean that are the root cause of the Asian monsoon.

The Bush (2004) GCM model, in agreement with several others (Braconnot et al., 2007; Kutzbach, 1981), predicts that the Western Pacific region would have been much drier during the LGM, as a result of the exposure of the Sunda Shelf to subaerial conditions. The Mid-Holocene model predicts lower rainfall in Central Asia, similar to that seen in the Yu et al. (2000) reconstruction, and confirmed by later work (Chen et al., 2015a; Feng et al., 2006). The model does predict heavier precipitation than presently experienced along the Himalayan mountain front that are driven by changes in the direction of monsoon winds. This prediction is also consistent with data from lakes in northern India (Dixit et al., 2014; Enzel et al.,

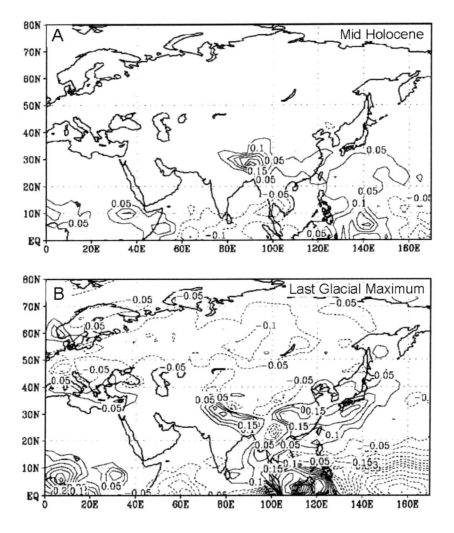

Figure 1.5 Difference in annual mean precipitation (in cm/day) over Asia compared to modern values calculated by modeling methods by Bush (2004) during (A) the mid-Holocene and (B) at the LGM. The contour interval in both panels is 0.05 cm/day. Images reproduced with permission of Taylor & Francis Journals.

1999; Prasad and Enzel, 2006). Stronger summer monsoon winds also enhanced evaporation over the Arabian Sea bringing that moisture to the Himalaya (Figure 1.5A). In the Mid-Holocene simulation, annual mean surface temperatures across northern Asia are projected to have been warmer than present and to have exhibited a stronger seasonal cycle, consistent with a stronger obliquity-driven forcing. By contrast, temperatures were cooler across South Asia, because of a warmer Western Pacific Warm Pool. The dry conditions in Central Asia are linked strongly to the extent of the Fennoscandian ice sheet (Zhao et al., 2017a). More generally, the Asian monsoon since 20 ka has been controlled by orbitally modulated solar insolation, but also by the changing temperature in the Western Pacific Warm Pool and by exposure of the Sunda Shelf (Yang et al., 2015a). In the LGM simulation, much of continental Asia is drier than present, particularly downwind of the Fennoscandian Ice Sheet (Figure. 1.5B). Predicted increased precipitation along the Himalaya is a result of slight changes in the direction of the monsoon winds. If desert margins in the vicinity of the Chinese Loess Plateau are taken to have a particular value of soil moisture in the model then there is a significant advance of the desert margin toward the southeast in the LGM simulation and a retreat toward the northwest in the mid-Holocene simulation (Bush, 2004).

1.5 Niche Construction and Plant–Human Coevolution

Human behavior is not simply imposed on the biophysical environment but is rather nested within it (McIntosh et al., 2000). "Humans, other biological systems and physical systems form complex nests of hierarchically organized ecosystems" (McIntosh et al., 2000). Human farming systems in the Asian monsoon region have coevolved with and within other biological and physical landscapes.

Because changes in the monsoon have affected the physical environment, they have also affected the distributions of plant and animal species and the actions of humans within them. Conversely, humans, perhaps more than other animal, have modified their surrounding physical environment throughout the course of history, through the process of niche construction (Smith, 2015). Beginning roughly 10,000 years ago, humans in a wide range of regions across the globe began to domesticate the first plants and animals, that is to say: engaged in activities that modified the phenotype and genotype of these species. Zeder (2015) defines domestication as

a sustained multigenerational, mutualistic relationship in which one organism assumes a significant degree of influence over the reproduction and care of another organism in order to secure a more predictable supply of a resource of interest, and through which the partner organism gains advantage over individuals that remain outside this relationship, thereby benefitting and often increasing the fitness of both the domesticator and the target domesticate.

Research on rice domestication has shown, for instance, that likely unconscious selection for grains that remain attached to the plant at maturity took place as humans harvested wild stands of rice. Humans also modified the environment in which this plant grew to increase its productivity: they moved it from deeper water stands where it grew as a perennial, to shallower environments like rice paddies, where water deficiencies precipitated higher seed production at precise moments: stimulating its development as an annual (Fuller, 2007; Fuller et al., 2009).

However, archaeologists are only able to view the end product of domestication, in other words, the microscopic traces of the already selected tough rachis (the part of the rice plant that keeps the seeds attached throughout maturity). In order for domestication to take place, sometimes millennia of patterns of mutualistic behavior between human foragers, plants and animals had to have taken place. Often that process involved human modification of the landscape. The beginnings of the "Anthropocene" or the most recent geological time interval have been strongly debated and is broadly speaking understood to represent a time when human activities became a major geological force in shaping the surface of the planet and affecting the geological record (Crutzen and Stoermer, 2000). Some people have suggested that it first became a relevant term during the late eighteenth-century Industrial Revolution, or just before (Crutzen, 2002), especially since this was the time when worldwide populations began to increase rather sharply and the output of greenhouse gases, especially CO_2 accelerated. By contrast, a pulse of sediment flux to the ocean, linked to deforestation and the onset of agriculture has been noted in many Eurasian delta systems after 3,000 years ago and most strongly after 1,000 years ago (Jenny et al., 2019; Syvitski and Kettner, 2011).

An alternative point of view argues that most of the Holocene, in other words the period that corresponds to plant and animal domestication, represents the Anthropocene. This theory was propounded by William Ruddiman (2003) and is one that he based on the observation that most interglacial periods were relatively short-lived and swiftly followed by progressive cooling toward a new glacial maximum (e.g., Figure 1.6). By contrast, the Holocene has had a relatively stable climate lasting around 10,000 years with no sign of renewed cooling, indeed quite the opposite. Ruddiman suggested that the reason for this stability was the steady impact of *Homo sapiens* on the environment through the start of agriculture and animal husbandry. The clearing of forests and especially the spread of rice agriculture being the sources of significant methane, which is a strong greenhouse gas (Ehhalt, 1974; Fuller et al., 2011; Li et al., 2009; Tyndall, 1861). Although however, it has been pointed out that the Holocene is not the only long-lived interglacial period, it is, undeniable that such periods are rare during the Pleistocene and that the Holocene represents an unusual phase of global climate. In Ruddiman's view, it is our activities themselves that have stabilized the climate and prevented the onset of a new glaciation.

As the home of much of the world's human population in the present day, Asia has had a fundamental role to play in the development of the Anthropocene hypothesis. The human–climate–environment interactions described in this book may play a fundamental part in our wider understanding of the Anthropocene, especially

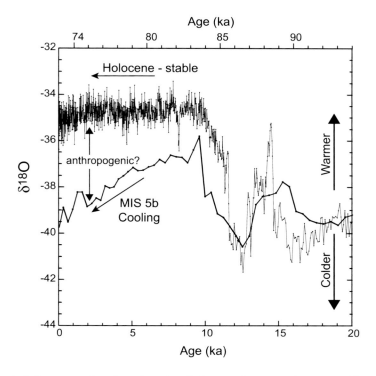

Figure 1.6 Diagram showing that the Holocene is unusual as a stable period following rapid warming. MIS 5b is more typical in showing immediate cooling after the maximum warming. Data from Stuiver and Grootes (2000).

since Asia has been the home to several of the earliest urban societies and might be expected to show the impacts of human settlement before other parts of the world.

1.6 Connecting Changes in the Monsoon to Changes in Human Society

Many attempts to link changes in climate to changes in human strategies have faced a number of challenges. Many of these arise from the need to select variables and scale of measurements that are relevant to humans (McIntosh et al., 2000). Humans have developed a wide range of solutions to adapt to changing river courses, temperature, and precipitation: solutions that are highly local in nature. Paleoclimate reconstructions tend, however, to operate at broader sometimes regional scales.

Simply correlating changes in climate with changes in human civilization characterize many of the approaches that attempt to connect changes in ancient climate to human action (Contreras, 2015). A first challenge comes from translating data from distant paleoclimate proxies, or regional reconstructions of paleoclimate, into data that is relevant at the local spatial scale at which humans operate. A second disconnect often takes place on a temporal scale. Human strategies for dealing with changing climate operate on the decadal or centennial scales, whereas paleoclimate proxies are often much lower resolution. A challenge related to this issue is that sometimes archaeologists and those working in Quaternary sciences do not always use the same dates or methods of calibrating radiocarbon dates, sometimes creating a further disconnect in timing (Madella and Fuller, 2006). A third challenge arises through the nature of the proxy data itself: an issue we discuss in more depth in Chapter 2. For instance, oxygen isotopes, while potentially useful as a proxy for monsoonal intensity, cannot be readily translated into variables such as millimeters of rainfall or temperature that mattered to humans and the crops that they grew. We thus argue that explicit process models are necessary to couple these datasets (Contreras, 2015; d'Alpoim Guedes et al., 2016c) linking changes in the mean state of precipitation, temperature and changing river courses to human's ability to cultivate plants. Climate itself does not impact humans, but rather it impacts the things that humans do. In other words: where they are able to grow crops and where they can place their settlements. In order to clarify these relationships, we aim throughout this book to create a roadmap of areas and periods of time where humans may have faced challenges in carrying out certain types of farming systems and identify solutions that they employed to meet these challenges.

The agricultural strategies developed by humans can be impacted by changes in the mean state of climate in several different ways. There are a number of key abiotic factors involved in plant growth and development. These include nutrients, soil, air (especially CO_2 content), sunlight for photosynthesis, and sufficient temperatures and water for maintaining life.

1.6.1 Adapting to Changes in Temperature

The first way that changes in climate could have impacted agricultural strategies was through changes in temperature. Temperature can affect plant growth at several stages in the plant life cycle. Different species have different temperature ranges within which germination will occur. Trying to germinate seeds in conditions below or above this range will not lead to germination. In the spring, cool soil temperatures can be a limiting factor for germination rates (Raven, 2009). Temperature can also play a critical role in flowering, and, therefore, also in seed and fruit setting and development. Temperatures that are too high or too low can prevent fruit setting and development (Raven, 2009). Flowering and grain filling are important phenological stages to consider in plant development (Raven, 2009): ones that are essential for the production of crops employed by man. It is essential to consider temperature differences between ecological settings when thinking about choice of crop, but also these crops' vulnerability and potential for success. The length of the growing season is determined by the dates of arrival and end of satisfactory conditions for growth. This is generally determined by the start of temperatures necessary for plant germination, and for maintaining the growth of delicate seedlings. Once the plant has reached maturity it is essential that satisfactory temperatures persist for long enough for the plant to set seed or fruit and for that seed or fruit to reach maturity. Many crops are not frost tolerant and, for the majority of crop plants, the dates of last and first frost establish the limits of growing season length.

However, more than growing season, it is important that a sufficient number of heat units are present in order for a crop to achieve maturity. Different crops require different numbers of GDD in order to complete their lifecycle. Compared to length of the growing season and frost-free days, consideration of GDDs allows a wider range of thermal factors affecting plant growth to be captured. While in some areas, estimates of frost-free days or of growing season length may provide similar estimates, GDDs are more accurate as a measure of calculating where crops can complete their life cycle for several reasons. Many plants require not only the presence of frost-free days in order to complete their lifecycle, but also units of heat that may be well in excess of temperatures above 0°C. For instance, rice is not only killed by frosts but also has optimal germination temperatures of 18–35° C and as a result simply calculating the beginning and the end of the arrival of frost-free days would likely overestimate the area in which rice was able to grow. Too much heat can also be problematic for crop growth. In certain cases, plants require a number of days of low temperature to grow properly. This vernalization requirement is true of crops growing in cold regions. For instance, apples and peaches require 700–1,000 hours below 7° C before they break their rest period. If this cold requirement is not met then fruit will not be set properly (Butler and Huybers, 2013). Killing degree days measure temperatures above which crops can suffer from heat stress. For instance, global maize yields are forecasted to decline in response to increasing temperatures (Butler and Huybers, 2013).

Changes in temperature are difficult for humans to control. While humans can make minor adjustments to the heat units available to plants through adjusting planting dates, and more recently the use of greenhouses, the limits of temperature have been something that humans have had to accommodate through either shifting the repertoire of crops grown or shifting the location of fields.

Figure 1.7 shows a map of available GDDs for a series of key crops that have been either domesticated or introduced to the Asian monsoon region. It is clear from these figures that high latitude Asia, the Tibetan Plateau, and its margins are areas where crops are temperature limited today and may have been in the past. For crops such as rice, temperatures in northern China can also be problematic. In deep southern China (with the exception of the higher altitude Yunnan-Guizhou Plateau), South Asia, and Southeast Asia, temperatures largely do not govern the ultimate range of crops distribution, except perhaps when temperatures that are too warm are reached and where future killing degree days may impact their distribution. In Chapters 4 and 5, we will examine the impact that changing temperatures over the course of the Holocene had on the distribution of GDDs, particularly in East and Northern Asia and the impact that this may have had the types of adaptive farming regimes that were employed by humans in the area.

1.6.2 Adapting to Changes in Rainfall for Rain-fed Reliant Regimes

Crops also require sufficient water to complete their lifecycles. Unlike temperature, humans have developed a wide range of irrigation and landscape infrastructure strategies to deal with water shortages or excess that we review subsequently. A wide range of crops that were domesticated in Asia, such as rice and often wheat and barley are grown under irrigated conditions, however, others such as millets have traditionally been rain-fed and rely on the periodicity of monsoon rains (Gadgil and Rupa Kumar, 2006; Lu et al., 2009). Figure 1.8 shows a map of precipitation-limited regions for several key rain-fed crops (rice, wheat and barley, millet) and across monsoonal Asia. Both upland (rain-fed) and irrigated types of rice exist. It is clear from these maps that wide parts of Asia present real difficulties for cultivating these rain-fed varieties. These include large parts of northern and southwest China, but also crucially, large areas of South Asia and Southeast Asia, where the cultivation of upland rice has been archaeologically and historically important (Figure 1.8A). More arid rain-fed crops, such as wheat and barley, fare much better (Figure 1.8B and C), although in some areas such as northwestern India, Pakistan, parts of the Tibetan Plateau, Xinjiang, and Mongolia, the development of irrigation systems were either necessary or rain-fed systems may have been possible or impossible with minor fluctuations away from contemporary levels.

1.6.3 Changes in River Load and Channel

Rivers are crucial to farming systems that rely on irrigation. Disruptions to the load and course of rivers can be devastating for the farmers that rely on them. Changes in summer monsoon rains are critical in governing the discharge in Asian river

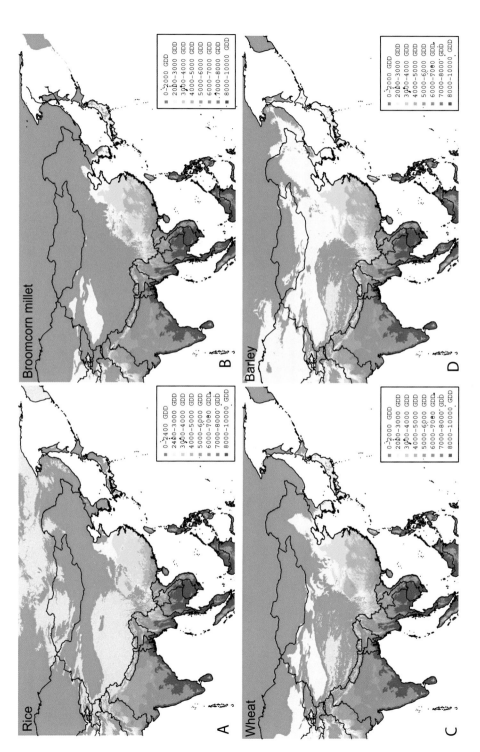

Figure 1.7 Map of available growing degree days (GDD) for a series of key crops that have been either domesticated or introduced to the Asian monsoon region. Areas in blue represent areas that are too cold for these crops cultivation (A) rice, (B) broomcorn millet, (C) wheat, and (D) barley. Data derived from Worldclim.org. (Fick and Hijmans, 2017).

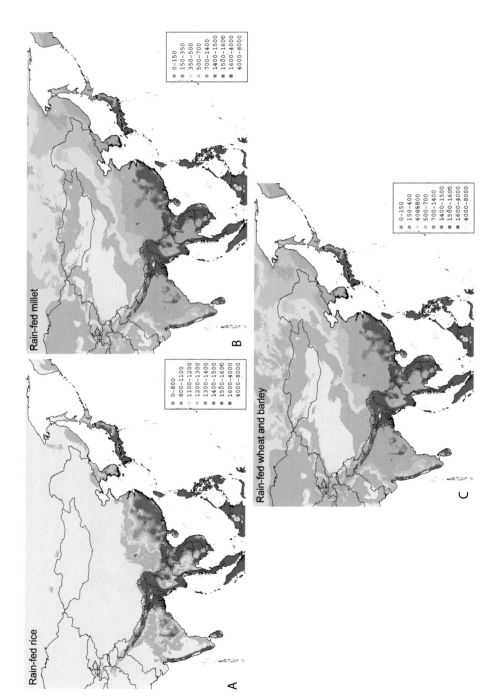

Figure 1.8 Viability limits for rain-fed crops. Areas in blue represent the zones below the minimal water requirements for these rain-fed crops. (A) Rain-fed rice; (B) rain-fed millets, (C) rain-fed wheat and barley. Data derived from Bioclim (Fick and Hijmans, 2017).

systems and influencing the success of irrigation-fed agriculture, or alternatively seasonal inundation-based methods that are strongly dependent on summer rainfall intensity. As well as the total discharge, the seasonality of rivers water supply is critical in limiting the type of agriculture that can be developed in any given region because of variable drought tolerance of different species. Rising regional temperatures might lead to higher snowmelt and spring discharge, at least temporarily (Immerzeel et al., 2010). At the same time, greater storminess can cause more frequent catastrophic flows, which can destroy fields fed by these rivers. Rivers themselves are subject to autocyclic avulsion and changes in their flow directions. While some rivers are entrenched in valleys or meander across wide floodplains many in South Asia, especially those draining directly from the Himalaya, construct large (50–100 km across) alluvial fans across whose surface the river system migrates for tens of kilometers over time periods of decades to hundreds of years (Chakraborty et al., 2010). Rapid changes in the position of the channel in such a system can have devastating effects both for the region abandoned by the river course and for those areas where large-scale flow suddenly commences (Jain et al., 2018; Wells and Dorr, 1987). In some instances, large-scale changes in the direction of flow have been recognized, such as the reversal of flow in the Yamuna River from west to east prior to the Holocene (Clift et al., 2012), as well as the flip-flopping of the Yellow River Delta in north China, north and south of the Shandong Peninsula (Slingerland and Smith, 2004). Figure 1.9 shows a map of areas across Asia with the major river basin catchments highlighted.

1.6.4 Areas Susceptible to Sea-Level Rise

Changing temperatures and the mean state of global climate has led to major changes in the sea level since the end of the LGM (around 18 ka) and throughout the Holocene (d'Alpoim Guedes et al., 2016a). This has particularly been the case throughout vast areas of eastern China, Bangladesh, and Thailand since the LGM. During Meltwater Pulse 1A (13,500–14,700 years before present (BP)), a massive continental area off the coast of East and Southeast Asia was lost to rapidly rising water (Hanebuth et al., 2000). The Sunda and East China Shelf regions are noteworthy in being particularly extensive compared to other continental margins worldwide. Sea-level transgression reached its highest point during the Holocene climatic optimum at around 6,500 years BP (Zong, 2004), although some argue for generally steady sea level since around 6,000 years BP (Lambeck et al., 2014; Peltier, 2002). In the case of China, this placed much of northern Zhejiang province, adjacent to the Yangtze delta underwater, creating substantial impacts for early

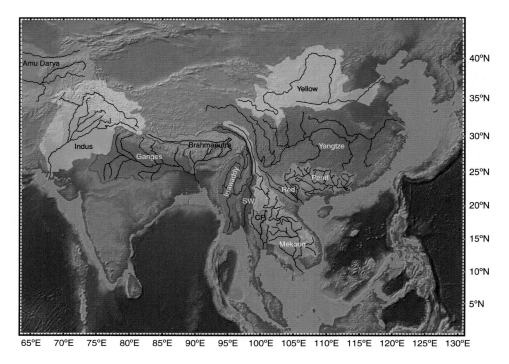

Figure 1.9 A map of Asia showing the catchments of the major rivers discussed in the text.

humans in the area (d'Alpoim Guedes et al., 2016a; Linwang, 2012; Zhu et al., 2003). Figure 1.10 shows a map depicting the effect of sea-level rise during the LGM and following Meltwater Pulse 1A. It also shows the Holocene thermal maximum high stand at around 6,500 years BP. Comparison between the latter and the modern highlights areas that may be affected by rising sea level if global temperatures reach levels comparable to those of the Holocene thermal maximum. Rising sea level has been extremely difficult for humans to deal with in the past and likely will be as well in the future. While humans have historically constructed barriers, such as locks or gates, or adopted strategies of coastal armoring such sea walls or levees, as well as investing in elevated or floating housing, finding solutions to twenty-first-century sea-level rise will likely not be without considerable disruption to current urban planning. For farmers coping with sea-level rise, fields are often the first elements of the human modified landscape to be abandoned, although we review below some of the adaptive strategies that humans have employed to protect fields from rising sea level.

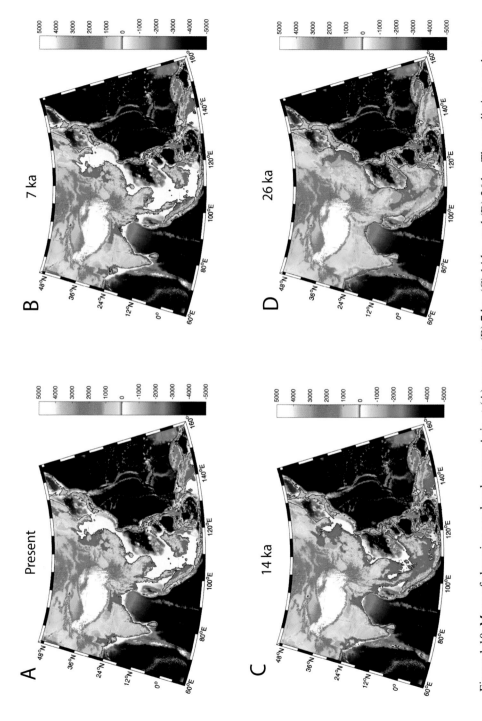

Figure 1.10 Maps of changing sea level across Asia at (A) present, (B) 7 ka, (C) 14 ka, and (D) 26 ka. The prediction employs an earth model with a 96-km-thick lithosphere, lower mantle viscosity of 5×10^{21} Pa s and an upper mantle viscosity of 5×10^{20} Pa. s. Figure courtesy of Jacqueline Austermann.

1.7 Human Farming Strategies

Here, we describe the different types of strategies that humans across Asia have employed to deal with environmental variability in their ability to forage or farm sufficient food. Although often multiple factors intersect to reduce returns in available resources, we discuss human adaptive strategies by the ways that they seek to respond to the impact of different environmental factors.

1.7.1 Strategies to Address Challenges Relating to Temperature

Adapting to cooling or warming temperatures that push crops or wild resources outside of their thermal niche can be challenging for humans to deal with. While varying the timing of planting can sometimes help avoid unseasonable frosts, this strategy often does not help with more durable, long term shifts in temperature.

Marston (2011) argued that diversification had been a key strategy used by both human foragers and farmers to reduce the chance of subsistence failure. He distinguished between two different types of diversification: diversification in the range of plants grown or foraged, and spatial diversification. Foragers can diversify the range of plants and animals they forage to include more cold- or heat-tolerant species, shifting away from species that are no longer within their range. For agricultural societies, diversification can mean the adoption of new frost or cold-tolerant species. For instance, an ancient agricultural text of the Northern Wei Dynasty (completed between 533 and 544 CE) lists a number of varieties of crops, such as different varieties rice that express traits of cold tolerance. An important example of the latter is the introduction of double cropping *Champa* rice, which was introduced around 1000 CE (Barker, 2011; Guiyun et al., 2015): a variety we describe in Chapter 4.

Humans can also exploit a range of temperature gradients by investing in different forms of spatial diversification. In the Andes, for instance, a single social unit can exploit a vertical archipelago, which allows the group to utilize the resources across a wide range of different altitudinal niches. Heat adapted crops, such as maize, can be farmed by members of the group located in warmer areas, whereas colder adapted crops such as potatoes can be farmed in higher altitudes and exchanged with group members residing in both areas (Murra, 1985; Van Buren, 1996). Food sharing, storage, and surplus production in these and other systems have played a crucial role in strengthening humans ability to respond to climatic uncertainty.

Increasing investment in other diversification strategies, such as pastoralism and trade, have been mainstays of groups located along China's northern and south-western border throughout history (d'Alpoim Guedes and Bocinsky, 2018).

Examples of trade are too numerous to list, however, we touch upon some of these topics throughout our later chapters as we examine the development of the Silk Road (Chapter 6), the role played by pastoralists in expanding trade networks throughout Asia, long-distance trade for resources in the Indus Valley (Harappan) Civilization that may have been carried on Monsoon winds (Chapter 3) and finally the role played by the Grand Canal of China, which allowed the Chinese state to maintain control over its northern margins by shipping agricultural products from the South to the North (Chapter 6). Throughout history, both farmers and foragers throughout Asia have incorporated another key form of diversification into their diet: wild plant and animal foods, allowing them to tap into resources outside the range of cultivation. Anyone who has ever attended a banquet in China can testify to the wide range of prized wild greens and tubers, as well as animals and insects that have featured on menus and still do today.

Finally, when the pressure from changing temperatures becomes too great and when culture does not allow for flexibility in changing foraged resources of crop repertoires, migration (or mapping onto available resources) can be a key strategy used by humans to cope with changing temperatures.

1.7.2 Changes in Water Availability Due to Precipitation

Compared to temperature, humans have been much more adept at developing in situ adaptive strategies to cope with fluctuations in water availability. Humans across monsoonal Asia have developed a wide range and scale of different systems of irrigation. For the Harappan civilization, Miller (2006b) lists a range of different water supply systems: (1) riverine inundation, (2) rain-fed, (3) small-scale canal irrigation to extend the inundation, and (4) well or lift irrigation. Large-scale irrigation canals were only introduced to the southwest Asian region during the 1900s by the British colonial government (Miller, 2006b). Smaller-scale models of water supply served farmers living along the Indus River valley for centuries. In China, particularly in southern and central China, larger-scale irrigation networks developed to fuel the growth of wet rice agriculture (Bray and Needham, 1984). Irrigation systems were crucial in rice cultivation because not only does the presence of water matter, water must arrive at the correct moment in the rice's lifecycle and be able to be drained during flower and fruit production. A wide range of complex irrigation systems arose to support these systems of cultivation, ranging from hillside tanks that fed lower lying areas through gravity to dyked or poldered fields (*wei tian*) employed in low-lying marshy regions. These are fields surrounded by a dyke or earth wall to prevent flooding. Irrigation systems in East Asia range from small adaptations such as these to much larger systems of irrigation such as the Dujiangyan weir and associated irrigation system of Sichuan (Willmott, 1989;

Sage, 1992 #673)} in southwest China, which we discuss in Chapter 4 (Bray and Needham, 1984). Sometimes farmers opportunistically employed standing water on lakes margins by creating floating fields, the frames of which were built using the roots of the aquatic plant *Zizania* (Bray and Needham, 1984). We discuss the reasons underlying the development of several of these systems in Chapter 4.

For challenges relating to both temperature and precipitation, bet hedging through *diversification of crops* has been a key strategy employed by humans around the world (Marston, 2011). This can be achieved through multi-cropping a series of different crops to hedge bets in the event of failure. An example of this constitutes "maslins" of mixed crops (wheat, barley, oats, and millets) that are sown across the Near East. This strategy allows farmers to take advantage of whichever crop survives the best under the climatic conditions of any given year (Marston, 2011). Crop diversification can also be achieved through crop rotation, which typically rotates on a yearly basis, or in areas where sufficient temperatures allow it to take place, this can even take place over the course of the same year. A key example of such a bet hedging system is the *Rabi* (winter) and *Kharif* (summer) seasons of cultivation that characterize agriculture in the Indus River valley in South Asia (Weber et al., 2010a) which we review in Chapter 3.

Finally, as discussed earlier, the spatial diversification of fields can be used to optimize use of a landscape within a given environment.

1.7.3 *Riverbank Flooding and Changes in River Course*

The morphology of the rivers on which people in monsoonal Asia rely for irrigation is heavily influenced by patterns of erosion. Indeed, the Anthropocene is much more than just a time when greenhouse gas emissions (e.g., CO_2) to the atmosphere increased substantially compared to the natural state of the planet (Steffen et al., 2018). The spread of human activities has also strongly affected the landscape, largely via agricultural activities (Hoffmann et al., 2015; Vanwalleghem et al., 2017). More recently large-scale earth movements, for example, in the containment of rivers through construction of artificial levees or damming, has also had a major impact on the landscape and human's ability to respond to natural forcing processes, such as flooding (Syvitski et al., 2005; Syvitski and Kettner, 2011). Such fluvial engineering has reduced the chance of smaller scale floods, but raised the potential impact of what will happen if and when these defenses are overwhelmed.

Throughout the history of monsoonal Asia, people have undertaken substantial projects to both control and harness the flow of rivers. Such efforts have been undertaken for a long time and at the largest scales. While the construction of the

Three Gorges Dam on the Yangtze River is a feat of modern engineering, large-scale water diversions have been undertaken for more than 2,000 years. For example, the Dujiangyan irrigation system in Sichuan Province, southwest China, was constructed after 250 BCE by the Qin State (Sage, 1992; Willmott, 1989). This control features a number of levees and weirs that were designed to keep the mainstream of the Min River open for ships, while diverting flood waters via a tunnel cut through the hillside to the agricultural land nearby (Cao et al., 2010a; Zhang and Hung, 2012). Its installation removed the flooding risks and allowed some of the most productive farmland in China to reach its potential. We review some of these innovations in Chapter 4.

Farmers in northern China developed methods of controlling erosion by building up balks at the end of the fields or by employing light ploughs that minimized soil mobilization (Bray and Needham, 1984). They also employed systems of ploughing that involved the construction of ridges and furrows that allowed farners to either drain off excess water or protect crops from drought.

Terraces in China have fulfilled multiple roles and are referred to as *San Bao*: they prevent erosion, conserve soil moisture and nutrients, as well as enable crops to be grown on steep slopes where irrigation would otherwise be impossible. Two very different traditions of terracing exist in China: dry terraces in northern China and irrigated terraces in the South. Irrigation systems associated with rice agriculture in the south of China have contributed to reducing the impact of erosion (Colinet et al., 2011; Shimpei, 2007), a marked difference with northern China, where agriculture can be a main driver of, rather than prevention against erosion.

The Yellow River in northeast China is also unique in China for the frequency of its common large-scale avulsion and changes in the course of its reaches. Figure 1.11 shows how the Yellow River has developed over the last 4,000 years on its way from its headwaters in the northern Tibetan Plateau to the Bohai Sea. After the river reaches the flatlands of eastern China downstream of the city of Zhengzhou, it has flowed both to the north and to the south of the Shandong Peninsula, which separates the Yellow Sea and the Bohai Sea. Shandong represents a topographic anomaly in this region, which is otherwise characterized by elevations of less than one hundred meters above sea level. The Yellow River flood plains have very little relief and are thus susceptible to diversion of the channel, driven by autocyclic switching as one river mouth area is subject to heavy sedimentation, with some role potentially played by the active tectonics of the area tilting the flood plains (Han et al., 2003; Zhang et al., 2018b).

The Yellow River left its route to the Bohai Sea in 602 BCE (Gernet, 1996) and shifted south of the Shandong Peninsula to the East China Sea. Although this reorganization may have been natural, sabotage of dykes, canals, and reservoirs and deliberate flooding became a standard military tactic since the Warring States period (ca. 475–221 BCE) (Allaby and Garratt, 2003) exploiting the tendency of the river to move its main channel for military advantage. The system was further used by the Mongols during conquest of the western Xia empire, when they attempted to divert the Yellow River through the fortifications of the Xia capital of Yinchuan, albeit unsuccessfully (May, 2012). The role of the Yellow River in shaping the development of the Chinese Empire may been significant, at least at times, and its changes in course must have had an influence on agriculture in largely dry northern China. Major flooding in 11 CE has been linked to the weakening and subsequent downfall of the Xin Dynasty (9–23 CE), an intermediate period between the eastern and western Han dynasties (Chen et al., 2012; Kidder et al., 2012). Another flood in 70 CE returned the river north of Shandong close to its present course with profound effects for those attempting to make a living along its banks.

1.7.4 Changes in Sea Level

Chinese farmers have also historically developed methods of dealing with incursions of saline water due to rising relative sea level. For instance, in 1026 CE, large sea walls were built along the Jiangsu coast to reclaim fields destroyed by relative

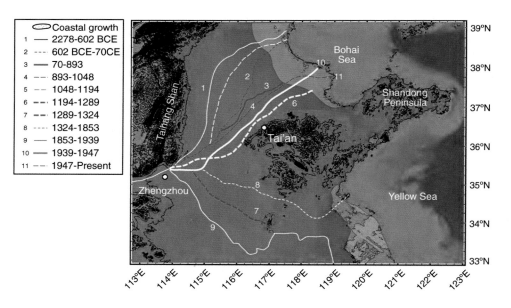

Figure 1.11 Reconstructed evolution of the Yellow River across the North China plains during the last 4,000 years. River tracks after Ye (1989). Contours are 100 m intervals above sealevels. Topographic map from GeoMapApp (Ryan et al., 2009).

sea level rise (Bray and Needham, 1984). Recent research suggests that these adaptations may be very ancient: as early as the origins of rice agriculture itself (Zong et al., 2007). Different varieties of rice that are resistant to salt have also been developed by generations of farmers and have been adapted to be employed in fields with high salt contents (Bray and Needham, 1984).

1.7.5 Maintaining Soil Fertility

Beyond salinification, farmers across Monsoonal Asia have also developed unique systems of agriculture that allow for constant replenishing of soil nutrients. Systems of shifting cultivation (once known as slash and burn), popular along East Asia's margins and throughout Southeast Asia allows the location of fields to be rotated over the course of several years to new locations, encouraging the growth of new species throughout the forest while enhancing soil fertility through burning (Conklin, 1969; 2008).

Throughout East and Southeast Asia, rice paddies constitute veritable whole ecosystems that provide both protein and carbohydrates while enriching the soil (Lansing and Kremer, 2011; Netting, 1993). Rice paddies are hosts to a wide range of fish such as carp, frogs, shrimp, crabs, and birds that farmers can harvest for protein, but also play a crucial role in fertilizing the fields through their droppings. A symbiotic fern, *Azolla* sp. is also planted in the paddies, and blooms preventing the growth of weeds. Furthermore, when it dies it sinks to the bottom of the paddies and constantly replenishes the source of nitrogen. Likewise, fish like carp play an integral role in pest control: a study showed that by bumping into rice stalks, they were able to dislodge and consume a range of insects: a form of natural pest control (Xie et al., 2011). Today, these traditional methods of cultivation dramatically reduce the need for nitrogenous fertilizers, while sustaining high and reliable grain production (Xie et al., 2011). As Lansing and Kremer (2011) point out, the overfertilization of rice with commercial nitrogenous fertilizers releases nitrous oxide, a greenhouse gas 300 times more potent than CO_2. In our rapidly warming world, we would do well to take to heart the lessons derived from traditional farming systems in Asia, as both sources of food in a time of climate change, and as mechanisms to prevent it.

1.8 The 8.2 ka Event as a Global Warming Analogue

In this book we use the past as the key to understanding the present. In addition to the intellectual interest in trying to understand why the monsoons have changed in the past and how this has impacted people, we also undertake the more practical objective of trying to predict how the monsoon might behave in the future, especially in the context of a warming global climate. Although climate change is complicated and not all parts of the planet are warming equally there is little doubt now that the polar regions, the Tibetan Plateau (Yao et al., 2012), west Antarctica, and especially the Arctic are experiencing significant change compared to the relatively recent past (IPCC, 2007; Stocker et al., 2013). Although we can use modeling methods to try and understand what impact this will have on Asian rainfall, we can also use reconstructions of past developments to assess what changes are likely and how quickly these may come into effect.

One of the most dramatic climate variations in the recent past was that which occurred around 8,200 years BP (8.2 ka), shortly after the start of the Holocene. At first sight it may seem eccentric to use the 8.2 ka event to model global warming because this was a period in which there were up to 400 years of cooler than normal conditions, even exceeding the cooling experienced during the later Little Ice Age. However, it is generally believed that the 8.2 ka event was triggered by a final demise of the Laurentide Ice Sheet and was most likely caused by the sudden draining into the North Atlantic Ocean of glacial lakes Ojibway and Agassiz, which previously dominated the geography of central North America (Barber et al., 1999; Ellison et al., 2006). This had an immediate effect on global sea level, resulting in around 2–4 m of increase throughout much of South and East Asia (Hijma and Cohen, 2010; Kendall et al., 2008), that caused flooding of coastal and delta areas and retreat of low gradient delta shorelines by tens of kilometers inland (Tjallingii et al., 2014; Wang et al., 2013).

Even more serious for the Asian environment, however, was the effect that the meltwater pulse had on global temperatures and particularly on the North Atlantic thermohaline circulation and what this in turn did to the Asian monsoon. The sudden input of a large volume of cold freshwater to the North Atlantic meant that the normal sinking of dense saline water offshore Greenland that drives the Atlantic Meridional Overturning Circulation (AMOC) was suddenly terminated (McManus et al., 2004; Thornalley et al., 2011). Because the southward-flowing North Atlantic Deep-Water flow was largely shut down, if not totally eliminated, this had a knock-on effect of reducing northward heat transport from the low latitudes in the Atlantic via the Gulf Stream, and so causing significant circum-North Atlantic cooling. Estimates of the temporary cooling caused by the discharge vary and depend on the interpretation of the paleoceanographic proxy data, but decreases ranging from 1 to 5°C have been estimated (Masson-Delmotte et al., 2013). Such a change in heat transport in the Atlantic had far, field effects. For example, some tropical paleoceanographic records in corals report a 3°C cooling from cores drilled into an ancient coral reef in Indonesia (Charles et al., 1997).

Given the likelihood of future reductions in the strength of AMOC because of progressive melting of the Greenland icecap (Enderlin et al., 2014), it is important to document how changes in the AMOC altered the monsoon climate in the past and to assess how well-coupled climate models reproduce these connections between the North Atlantic and monsoonal Asia. Recent work in eastern China supports the idea of coupling between East Asian monsoon rainfall and northern hemispheric ice (Clemens et al., 2018). The 8.2 ka event is considered particularly useful as an analog for future changes because its duration, magnitude of AMOC reduction, and background climate state are the closest to conditions expected in the near future. What is clear is that the duration of the melt-water pulse may have been as short as six months (Clarke et al., 2004). Nonetheless, based on paleoceanographic reconstructions, a freshwater perturbation of this size was insufficient to trigger a complete cessation of the AMOC. The reconstructions consistently show that the shallow and deep overturning circulation of the North Atlantic recovered completely after the cessation of melt-water discharge. However, the recovery timescale appears to have been on the order of 200 years (Bamberg et al., 2010; Ellison et al., 2006). One record points to a partial recovery on a decadal timescale (Kleiven et al., 2008), but even such a rapid reversal would still be very catastrophic to the global climate and to dense populations that inhabit coastal Asia today.

New high-resolution lake and bog records now provide expanded proxy-based reconstructions of the 8.2 ka event in monsoonal Asia (Figure 1.12A). Recent developments include more sites from the East, as opposed to the South Asian monsoon region. Furthermore, proxies of winter monsoon strength now allow the impact of the 8.2 ka event to be better quantified. Comparison of the proxy evidence with a new modeling simulation of the event using the Community Climate System Model version 3 (CCSM3)(Meehl et al., 2006; Morrill et al., 2014) prescribe North Atlantic freshwater forcing according to the latest reconstructions (Figure 1.12B and C). There is clear evidence for 8.2 ka climate anomalies at many oceanographic proxy sites distributed worldwide, emphasizing the strong and widespread impacts of the event in monsoonal Asia during both summer and winter seasons. The overall effect of rapid melting on the model is global cooling, with the strongest cooling in the North Atlantic (Figure 1.12B). However, this has an effect on Asian monsoon intensity because a cold North Atlantic causes cold Eurasian winters that intensify the Siberian High and cold summers that in turn weaken the atmospheric low during that season (Webster et al., 1998).

The model simulation generally predicts that the weakening of the summer monsoon and strengthening of the winter monsoon at that time were likely caused by a reduction of the AMOC. Rainfall anomalies are especially intense in peninsular India and in north China (Figure 1.12C), where summer rains were sharply

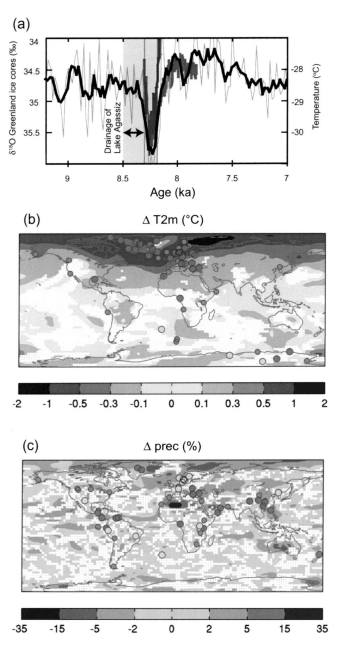

Figure 1.12 Compilation of selected paleoenvironmental and climate model data for the abrupt Holocene cold event at 8.2 ka, documenting temperature and ocean-circulation changes around the time of the event and the spatial extent of climate anomalies following the event. Published age constraints for the period of release of freshwater from glacier lakes Agassiz and Ojibway are bracketed inside the vertical blue bar. Vertical gray bar denotes the time of the main cold event as found in Greenland ice core records (Thomas et al., 2007). Thick lines in (A) denote 5-point running mean of underlying data in thin lines. (A) Black curve: North Greenland Ice Core Project (NGRIP) δ^{18}O (temperature proxy) from Greenland Summit (North Greenland Ice Core Project members, 2004). Red curve: Simulated

Caption for Figure 1.12   (cont.)

Greenland temperature in an 8.2 ka event simulation with the ECBilt-CLIO-VECODE model (Wiersma et al., 2011). Blue curve: Simulated Greenland temperature in an 8.2 ka event simulation with the CCSM3 model (Morrill et al., 2011) (B) Spatial distribution of the four-member ensemble mean annual mean surface temperature anomaly (°C) compared with the control experiment from model simulations of the effects of a freshwater release at 8.2 ka (based on Morrill et al. (2013b)). White dots indicate regions where less than three models agree on the sign of change. Colored circles show paleoclimate data from records resolving the 8.2 ka event: purple = cold anomaly, yellow = warm anomaly, gray = no significant anomaly. Data source and significance thresholds are as summarized by Morrill et al. (2013a). (C) Same as (B) but for annual mean precipitation anomalies in %. Colored circles show paleoclimate data from records resolving the 8.2 ka event: purple = dry anomaly, yellow = wet anomaly, gray = no significant anomaly. Modified from Figure 5.18 of Masson-Delmotte et al. (2013), Published by Cambridge University Press with permission of IPCC.

reduced for a period of hundreds of years. Examination of regional rainfall anomalies in East Asia reveals some important spatial heterogeneity. In the model simulation these anomalies are caused by lesser seasonal migration of the monsoon front. A connection between North Atlantic climate anomalies and those for the monsoon at 8.2 ka, both in proxy records and the model simulation (Figure 1.12C), supports the idea that the North Atlantic is tightly coupled to Asian climate, probably via its influence on central Asian conditions. A similar melt-water event in the present day would have serious environmental implications not just for countries around the North Atlantic but also for monsoonal Asia.

We emphasize that the 8.2 ka event is just one dramatic example of how Asian rainfall can be significantly impacted by other forms of climate change, especially those in the Atlantic. In Chapter 6 we discuss such interactions at a variety of timescales, noting how Asian climate responded to the Little Ice Age and the Medieval Warm Period in more recent historical times.

1.9 Summary

In this book we explore how the monsoon has changed since the start of the Holocene, how this has influenced the Asian environment and the river systems, and how these in combination have affected the development of different subsistence regimes across Asia since the Paleolithic. We also emphasize the sometimes-overlooked systems of traditional food cultivation that humans have employed to deal with these shifts. We further explore the possible impacts of future climate change and how these may influence rainfall amounts and seasonality so that we can gain some idea about how Asian environments might change in the future and what this will mean for the economies of the area, agricultural production, and for security of the various nations that live under the influence of the monsoon. As we will see, the archaeological record indicates that humans have been quite adaptable in terms of adjusting to new environmental conditions, but that rapid changes lead humans to sometimes need to drastically modify both their patterns of mobility and subsistence strategies: factors that in a modern world with nation state borders we might be less well equipped to accommodate. The sooner we understand what changes are likely to be the more likely it is that we will be able to adjust to the challenges of the next few centuries.

2

Temporal Variations in the Asian Monsoon

2.1 Introduction

The Asian monsoon evolved in its intensity since its initial strengthening, probably around 23 million years ago (Ma) (Clift et al., 2008; Sun and Wang, 2005), and possibly much earlier, before 40 Ma (Licht et al., 2014; Sorrel et al., 2017). Changes have occurred over a number of different timescales and to varying degrees. The long-term development of the monsoon appears to have been driven by the tectonically generated high topography of central Asia (Molnar et al., 1993; Tada et al., 2016) and the Himalaya (Boos and Kuang, 2010), but that influence is far too slow to affect the changes relevant to the establishment of human societies in the region. At the simplest level, it is recognized that the Asian summer monsoon has tended to be weak during periods when the Earth's climate was cold and dry, and to have intensified as the global climate ameliorated during each intervening interglacial period, most recently following the Last Glacial Maximum around 18,000 years ago (Clemens et al., 1996; Clemens and Prell, 2003; Wang et al., 2008; Wang et al., 2001). If we are to understand how the monsoon has impacted the development of human societies in Asia, we need to have a good understanding of how the monsoon has changed through the time during which Asia was first settled and throughout its subsequent history.

Correlating changes in summer monsoon intensity with other climatic or oceanographic processes allows us to better understand what processes have triggered changes in the strength of summer precipitation and to make predictions about how monsoon strength may change in the future. Because weather records – even in societies like China with long historic records – only span periods of hundreds of years, the long-term reconstruction of monsoon strength is necessarily dependent on the geological record. This means that we need to be able to measure a proxy for summer monsoon strength within sedimentary records in order to achieve a robust reconstruction. In practice, this can be a complicated issue because even the modern summer monsoon is quite variable across Asia (Figure 2.1) and is manifest in

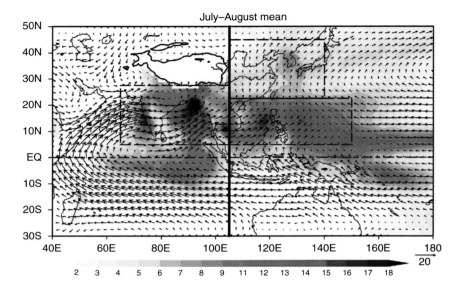

Figure 2.1 Climatological July–August mean precipitation rates (shading in mm/day) and 925 hPa wind vectors (arrows) from Wang et al. (2003). The precipitation and wind climatology are derived from CMAP (Xie and Arkin, 1997) (1979–2000) and NCEP/NCAR reanalysis (1951–2000), respectively. The three boxes define major summer precipitation areas of the Indian tropical monsoon (5°N–27.5°N, 65°E–105°E), western North Pacific (WNP) tropical monsoon (5°N–22.5°N, 105°E–150°E), and the East Asian subtropical monsoon (22.5°N–45°N, 10°E–140°E). Reprinted with permission of Elsevier B.V.

different ways in different places. For example, in the Bay of Bengal and Bangladesh the summer is characterized by extremely heavy rainfall but quite light winds. By contrast, in southwest Asia and the Arabian Peninsula the summer monsoon features strong wind and eolian dust transport rather than greater-than-average rainfall. On the periphery of the monsoon in North China, Korea, and Japan there are rains in the summer, but even more so there are powerful winter monsoon winds that transport dust from central Asia East into the Pacific Ocean (Rea, 1994).

As a result, geologists have relied on different proxies in different ocean basins to monitor the evolving strength of the monsoon. The further back in time that any given record extends, the less robust this reconstruction becomes because of additional uncertainties that are hard to predict, such as the shape and width of ocean basins, the height of mountain chains, and the direction and speed of wind systems. This is especially problematic for estimating monsoonal intensity in inland areas that are far from ocean basins, or when the monsoon reconstruction is from the deep geologic past. Nonetheless, a multi-proxy approach for the recent geological past has resulted in a reasonably coherent understanding of how monsoon strength has varied, especially since the Last Glacial Maximum (~20 ka), which is the key period for assessing the impact of monsoon temperature and precipitation on farming systems.

It is long been understood that the Asian monsoon is largely driven by temperature differences between continental Asia and the surrounding oceans that have resulted in a major reorganization of atmospheric circulation (Webster, 1987; Webster et al., 1998). While the normal "Hadley" circulation of the atmosphere involves rising hot, moist air in the tropics and descending colder, dry air in the mid-latitudes the reverse applies in the case of the Asian summer. Simply put, the heating of the land in central Asia and over the Indian subcontinent heats the lower atmosphere, resulting in this air expanding and rising, which in turn generates a low-pressure cell centered in the midcontinent. As a result of this process, warm, wet air is drawn deeper into the continental interior from the surrounding oceans, bringing rains to the Himalayan mountain front and far west into China, and even to parts of the Tibetan Plateau and Mongolia. Its great altitude and surrounding high mountain barriers keep much of the Tibetan Plateau, averaging an elevation of around 4.5 km, relatively dry compared to central and eastern China (Figure 2.1). It has also long been presumed that the presence of the Tibetan Plateau had a crucial effect on the intensity of the monsoon. This is because it has been recognized that although many continents have monsoon circulation systems, Asia's is the most intense, reflecting the especially large land area, as well as the great elevation of the Tibetan Plateau, extending over a wide area, greater than 2,500,000 km^2 (Manabe and Terpstra, 1974). Consequently, geologists have long linked the progressive uplift of the Tibetan Plateau with an intensifying monsoon since the onset of India–Asia collision around 50–60 Ma (Molnar et al., 1993; Prell and Kutzbach, 1992).

This rather simplistic model for explaining the intensity of the Asian monsoon has been challenged as additional complexities have been recognized. Ramstein et al. (1997) noted that intensification of the Asian monsoon after around 30 Ma coincided with the retreat of warm shallow seas from central Asia (Dercourt et al., 1993). This "Paratethys" Ocean would have cooled the Asian continent by preventing the land surface from heating up so intensely during the summer. Its retreat allowed atmospheric circulation to intensify as the environment became more continental (Zhang et al., 2007). Separating this effect from the climatic influence of a rising Tibetan Plateau is difficult because the two processes were interlinked, although clearly Paratethys retreat occurred during the earliest phases of uplift, dated at around the Oligocene–Eocene boundary, around 34 Ma (Bosboom et al., 2011), while modeling suggests that a plateau of the current area had to reach at least half its present height before it could have a major impact on monsoon intensity (Prell and Kutzbach,

1992). Modeling of monsoons using Eocene boundary conditions suggests that the early Himalayan mountains caused orographic rainfall in South Asia and a rain shadow to the north but that the total height of the Tibetan Plateau might not be crucial to the rainfall strength due to the advection of moist air from the Paratethys toward the southwest across the plateau at that time (Huber and Goldner, 2012).

More recently, more sophisticated climate modeling approaches have been used to try to quantify the different influences on monsoon intensity (Zhang et al., 2007). While the effect of a major plateau in central Asia was emphasized for its ability to block direct eastward flow of the westerly jet and provide heat into central Asia, climate modeling by Boos and Kuang (2010) now shows that much of the monsoon rainfall intensity in southwest Asia can instead be attributed to the blocking effect of the Himalayan ranges forming a barrier to airflow, rather than the need for a plateau (Figure 2.2). This implies that the South and East Asian monsoons are decoupled from one another. Additional differences along the strike of the Himalaya were revealed by high-resolution modeling that showed a reverse, low-level easterly jet that transports moisture from the Bay of Bengal along the flood-plains of the Ganges, with the jet constrained to the north by the blocking topo-graphy of the Himalaya and driven by the low-pressure system that forms over northern India in the summer. This reverse flow brings moisture to the plains from the east rather than from the west, as seen across much of peninsular India (Acosta and Huber, 2018).

While the Himalayan barrier may be important in the Indian subcontinent, it is less important in governing the intensity of the East Asian monsoon, whose Early Miocene (after 24 Ma) strengthening has been linked to progressive uplift of the northern Tibetan Plateau (Tada et al., 2016). Indeed, the effect of uplift is not always to cause stronger rainfall, because in western China the rising mountains caused a rain shadow and reduced rainfall at least since the Miocene (Zheng et al., 2015). Uplift of the southeastern Tibetan Plateau has been modeled to drive the progres-sive drying seen in southwest Asia, while strengthening precipitation in Southeast Asia (Molnar and Rajagopalan, 2012).

Even more surprising was the realization that not only did the monsoon not require a high Asian interior but even the presence of land itself might not be necessary. Climate modeling by Bordoni and Schneider (2008) has shown that seasonal monsoonal circulations tend to be generated even on planets that do not have any landmass on them at all. In this model of an "aqua-world," precipitation maxima in the mid-latitudes were predicted for the summer season, even in the absence of a continent, mirroring the climatic patterns observed in modern-day Asia. Precipitation rates in the simulations can exceed observed precipitation rates because the lower boundary in the simulations is entirely water covered, but this

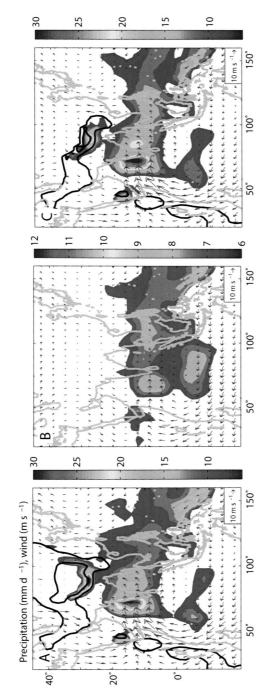

Figure 2.2 Predicted precipitation and wind from the atmospheric model of Boos and Kuang (2010). All panels are for June–August, but for the model run with (a) standard topography, (b) no elevated topography, and (c) surface elevations north of the Himalaya set to zero.

does not diminish the primary conclusion of the model. Furthermore, the results were found to be robust and do not change substantially if Southeast Asia is included in the averages. This does not necessarily mean that the presence of land or the Tibetan Plateau is not important to the evolving strength of the Asian monsoon, but only that the controls on monsoon strength are more complicated than were initially understood and that the Tibetan Plateau by itself is only part of the trigger.

2.2 Annual to Decadal Timescales

Long-term variation in monsoon intensity may be related to orbital eccentricity and variations in solar insolation, but on shorter timescales it is widely recognized that the extent of snow in central Asia, and particularly across the Himalayas and Tibet, is also an important influence (Blanford, 1884). This is because heavy snowfalls in central Asia enhance the reflection of solar energy from the region and this in turn reduces and delays the development of an atmospheric low in the following summer. The intensity of Asian snowfall is in part also controlled by processes in the North Atlantic region because these regions are connected via air transported by the westerly jet. Cold winters in the North Atlantic often correlate with cold, snowy conditions in central Asia. When Asia experiences a mild winter with little snow, the resulting warmer conditions are believed to strengthen the summer land-sea thermal contrast and thus intensify the following summer monsoon rainfall. The intensity of North Atlantic winters is controlled by atmospheric conditions over that ocean basin, and in particular to variations of the North Atlantic Oscillation, which is known to vary on timescales of 8–20 years (Hurrell, 1995). The connection between the South Asian monsoon and the North Atlantic climate is underpinned both using paleoclimate reconstructions (Shanahan et al., 2009) and climate model experiments (Zhang and Delworth, 2006). Cold winters in the North Atlantic result in colder conditions in central Asia and thus weaker summer monsoons.

The other important climatic cycle that has been linked to monsoon intensity is the El Niño-Southern Oscillation (ENSO). ENSO is a naturally occurring phenomenon that involves fluctuating ocean temperatures in the equatorial Pacific varying on timescales of around five years (Ropelewski and Halpert, 1987). During an El Niño phase, seawater in the Eastern Pacific is hotter than normal, while air pressure in the Western Pacific is high (Allan et al., 1996; Rasmusson and Carpenter, 1982). At the other end of the cycle, during a La Niña phase, the opposite conditions prevail. Although this is dominantly a Pacific phenomenon, ENSO is associated with predictable changes in the intensity of the summer monsoon, especially in India. Comparison of rainfall in India with the strength of ENSO shows that strong summer monsoon rains tend to correlate with La Niña events – that is, times when the sea surface temperature (SST) across the equatorial Eastern Central Pacific Ocean is lower than normal by 3°C–5°C (Pant and Parthasarathy, 1981). Historically, there has been a close coupling between ENSO and monsoon intensity (Figure 2.3) (Kumar et al., 1999), but this pattern has begun to break down over the last 30 years. Using "All India Rainfall" as a proxy for the strength of the South Asian summer monsoon, we see that although heavy rainfall has often been associated with La Niña that there is typically a sharp drop in summer rainfall at the end of a La Niña event (Figure 2.4). In some cases, periods of minimum rainfall coincided with La Niña – for example, between 1985 and 1987, as well as in 1979.

Kumar (1999) proposed that the decoupling of ENSO and summer rainfall was the result of the southeastward movement of the westward airflow in the tropics that is associated with ENSO. This resulted in reduced dry air subsidence over South Asia. Such a change acts to favor normal monsoon conditions rather than the weak conditions of an El Niño. It was also argued that recent increased surface temperatures over Eurasia during the winter and spring, which are a part of the mid-latitude continental warming trend (Stocker et al., 2013), may favor an enhanced land-ocean thermal gradient. This in turn is conducive to a strong monsoon, again

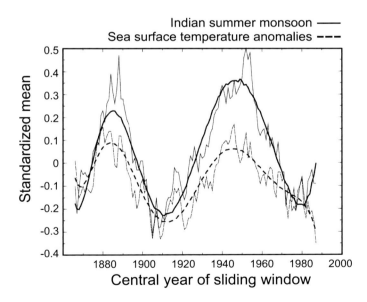

Figure 2.3 Diagram showing 21-year sliding standardized means of Indian summer monsoon rainfall (thin line) and June–August (JJA) NINO3 sea surface temperature (SST) anomalies (thin dashed line) for the period 1856–1997. The corresponding solid lines represent the smoothed values (smoothing is done by fitting a polynomial). The sign of NINO3 SST is reversed to facilitate direct comparison. Reproduced with permission from AAAS from Kumar et al. (1999).

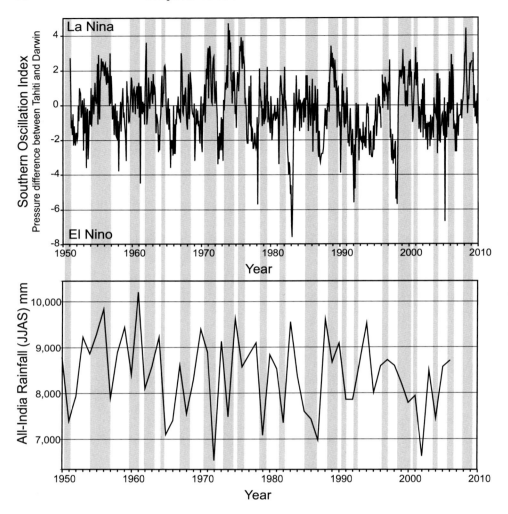

Figure 2.4 Variations in the intensity of ENSO, as measured by the Southern Oscillation Index (data from Australian Bureau of Meteorology), and the strength of South Asian summer monsoons, as represented by the All India rainfall record since 1950 (Parthasarathy et al., 1995). All India rainfall is an areal average of 29 subdivisional rainfalls. Subdivisional rainfalls are averaged district rainfalls. The district rainfall is computed from averaging all stations in the district. Rainfall amounts (in millimeters) are totals for June, July, August, and September.

breaking the former simple linkage between ENSO and monsoon intensity. The relatively short period of these ENSO cycles and their associated impact on summer monsoon rains means that this type of variability is unlikely to result in long-term drought.

2.3 Intra-seasonal Variations in the Monsoon

Rainfall during a monsoon season normally peaks in the height of the summer, albeit not synchronously across Asia, because the rain front migrates from southeast and northwest across South Asia. Rainfall then typically decreases after the peak, but in practice different influences can change the delivery of rain during a single season. Figure 2.5 shows an example of how the middle of a monsoon season can be marked by lower-than-normal precipitation, a feature that can have serious impacts on agricultural production in South Asia. Such extended periods of lower-than-normal rainfall are called "monsoon breaks." These can be severe, and in the case of the 2002 season, illustrated here, less than half the normal rainfall fell during the latter half of the monsoon season. It is noteworthy that although not much precipitation fell over peninsular India during this period, there was heavy precipitation in the tropical Indian Ocean, as well as in northeast India

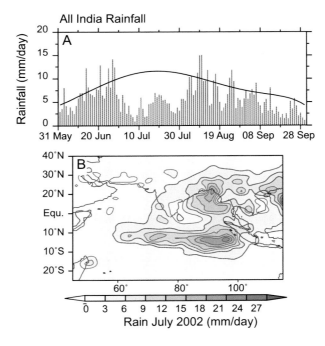

Figure 2.5 (A) Daily All India Rainfall observed during the summer of 2002. Solid line is the long-term climatological normal (source: www.tropmet.res.in). (B) Rainfall (mm/day) during July 2002 shows the striking contrast between the Indian landmass and equatorial eastern Indian Ocean. The rainfall enhancement over northeast India is typical of monsoon breaks. The rainfall data are from Climate Prediction Center (CPC) Merged Analysis of Precipitation. Figure redrawn from Krishnan et al. (2006).

(Figure 2.5B). However, central and western India were badly hit as a result of the monsoon break.

Strong rainfall in the tropical ocean associated with the monsoon break suggested that this phenomenon might be caused by ocean–atmosphere interactions (Krishnan et al., 2006). The weak precipitation onshore resulted from anomalous dry air subsidence over peninsular India, as induced by strong upward motions over the equatorial Indian Ocean. This was, at least in part, caused by the rapid intensification of an El Niño during that year. Moreover, this was not a typical El Niño because the ocean temperature maximum was located in the central, not eastern, Pacific (Gadgil, 2006). Nonetheless, at least part of the 2002 monsoon break was caused by the feedback between the atmosphere and the ocean. Prior to the onset of the summer monsoon, ocean currents normally move water to the east, and these are accompanied by the development of strong westerly wind bursts along the equator that enhance the eastward component (Saith and Slingo, 2006). It is this equatorial flow that reduces the intensity of the normal monsoon circulation and causes a monsoon break. At the same time, the restricted northward penetration of the monsoon flow resulted in an intensified equatorial trough and favored oceanic down-welling in the eastern equatorial Indian Ocean, thereby maintaining a depressed thermocline and warm subsurface temperature anomalies in the region. This process keeps the summer monsoon weak. The economic impact of the 2002 monsoon break was particularly profound because the event had been very difficult to predict (Rajeevan et al., 2010). Most normal indicators for monsoon strength had predicted that 2002 was going to be a normal, good year for summer rains. The frequency of monsoon breaks has important implications for our ability to practice farming in South Asia and may have been much more severe in the past.

2.4 Links to Australian-Indonesian Monsoon

The Asian monsoon is at least partly coupled with the Australian-Indonesian monsoon, which is the other major climatic system in the Indo-Pacific region (Wang et al., 2003). While there is general acceptance of a link on millennial timescales between the Asian monsoon and solar insolation, links between the Australian-Indonesian monsoon and that in mainland Asia are less clear (Liu et al., 2003; Magee et al., 2004). Some of the relationships between the Australian-Indonesian monsoon and the Asian monsoon can be investigated through looking at how they have covaried in the past. A sediment core collected off the southern coast of Java in Indonesia now allows that linkage to be assessed (Figure 2.6). A number of climate proxies were investigated and compared with the development of climate in mainland Asia. Figure 2.6A shows the ratio of lithogenic, eroded continental material to calcium carbonate, which is of biogenic origin and is a proxy

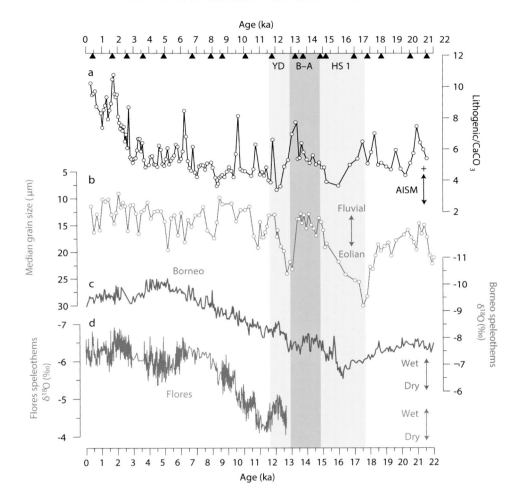

Figure 2.6 Proxy records for the Australian-Indonesian summer monsoon and other paleoclimate records over the past 22,000 years. (A) Lithogenic/CaCO$_3$ for core GeoB 10053-7 recovered offshore southern Java. (B) Median grain size of the terrigenous fraction of the sediment. (C) Stacked δ^{18}O of speleothems in Borneo (Partin et al., 2007). (D) Stacked δ^{18}O of speleothems in Flores (Griffiths et al., 2009), indicating changes in the Australian-Indonesian monsoon and sea level–related precipitation. Black triangles at the top show age control points. Figure is modified from Mohtadi et al. (2011).

for the strength of fluvial runoff. This shows a sharp increase in the eroded flux to the ocean during the warm interval of the Bølling-Allerød (ca. 12.7–14.7 ka), as well as since around 3 ka. This pattern was interpreted to reflect increased discharge from rivers draining the islands of Indonesia at those times (Mohtadi et al., 2011). This same study showed that grain size decreased at times of the fastest accumulation, which implied a dominance of muddy fluvial sediment over coarser eolian sediment. Slow lithogenic accumulation and large grain size during the Younger Dryas (11.5–12.7 ka) point to a weak Australian monsoon as the global climate cooled. Such variations are also similar to those seen in mainland Asia.

In India and China, the time of the strongest summer monsoon occurs during the early Holocene (10–8 ka), when the Australian monsoon was weak, followed by weakening in Asia, at least until around 2–3 ka. This pattern is significantly different from both the record from the Indonesian marine core and speleothem records in Borneo and from the Indonesian island of Flores. Differences between the rainfall records in Indonesia and mainland Asia may reflect the fact that precipitation there is more tropical and may combine both winter and summer precipitation (Griffiths et al., 2009). Mohtadi et al. (2011) concluded that similar long-term variations of the Australian-Indonesian winter monsoon wind and the East Asian summer monsoon rainfall probably reflected their common forcing by Northern Hemisphere summer insolation. The two systems are clearly coupled but are also operating in distinct ways to one another.

2.5 Measuring the Monsoon

Observations lie at the root of our ability to show how the monsoon has changed in the recent past and to predict how they may change in the near future. Without good observations computed general circulation models (GCMs) for the atmosphere cannot be calibrated and their accuracy remains suspect. Long-term accurate records of wind speed, rainfall, and temperature do not extend far back in time. The "All-India" rainfall and weather records extend to 1871, but before that time we have little quantifiable information about total rainfall or the seasonality of the precipitation, which is a hallmark of the monsoon (wet summer and dry winters) (Parthasarathy et al., 1994, 1995). Temperature and rainfall data at monthly resolution are collected and distributed by the India Meteorological Department (IMD). Rainfall is still measured in the traditional fashion using rain gauges, which are simple and reliable but need to be tended by human staff on a regular basis, which makes this method expensive. Measurements are typically done with an inner cylinder that can be filled by 25 mm of rain, with overflow spilling into an outer cylinder. Plastic gauges have markings on the inner cylinder down to 0.25 mm resolution, while metal gauges require use of a stick designed with the appropriate 0.25 mm markings because they are not transparent. After the inner cylinder is filled, the amount inside is discarded, and then it is filled with the rainfall in the outer cylinder until all the fluid in the outer cylinder is measured. The total volume is measured by adding together all the water until the outer cylinder is empty. After a precipitation measurement is made, data is typically submitted through the Internet to a central bureau, depending on the country involved, for synthesis into a regional pattern.

Systematic collection of meteorological data was first started in India in 1871, although there are examples in ancient Indian literature of scholars recording rainfall amounts, as well as the timing and directions of winds associated with the summer monsoon. Most noteworthy are entries in the *Arthashastra*, a document that records a variety of topics including statecraft, economic policy, and military strategy. It was written in Sanskrit by an author with various names, but generally identified as "Chanakya," a scholar who was active in the years 350–283 BCE (Mabbett, 1964). However, the start of systematic measurements in India in the nineteenth century followed serious agricultural losses caused by a tropical cyclone that struck Calcutta in 1864, as well as expensive failures of the monsoon rains in 1866 and 1871 (Cook et al., 2010; Kumar et al., 2006). In 1875 the Government of India established the IMD, bringing all meteorological work in the country under a central authority and with a headquarters in the then capital of Calcutta, although this was later moved to Simla, then to Pune, and finally to New Delhi. The first Director General of Observatories was Sir John Eliot who was appointed in May 1889.

More recently, rainfall gauges have been supplemented by more advanced Doppler radar systems to image areas of precipitation in real time. Regional estimates of precipitation are now best collected by satellite in order to provide the continuous area coverage that single weather stations are incapable of doing. Two satellite missions in particular have been and will be important for monsoon studies in the future, and these are known as "Precipitation Measuring Missions" (PMM). The older of the two is called the "Tropical Rainfall Measuring Mission" (TRMM), which is now supplemented by the "Global Precipitation Mission" (GPM), both of which are joint undertakings by the National Aeronautics and Space Administration (NASA) of the United States and the Japan Aerospace Exploration Agency (JAXA). The TRMM (Kummerow et al., 1998) was launched in November 1997 with data recording having started in December 1997. The GPM was launched in February 2014.

On the TRMM satellite, a 3.8 GHz Precipitation Radar provides detailed vertical distribution of radar reflectivity, related to the number and especially the magnitude of precipitation inside weather systems. The Precipitation Radar was the first such instrument in orbit. It provides detailed information on the three-dimensional structure of rain systems with a horizontal resolution of approximately 4 km and a total of 80 levels in the vertical with a resolution of 250 m. The Precipitation Radar is a cross-track scanner with a relatively narrow swath width (~215 km), although the dual frequency radar on the GPM now scans across 125 km. This limit results in moderate sampling for climate-related studies. The TRMM and GPM Microwave Imagers can provide information related to the vertically integrated ice and water path and is designed to provide quantitative rainfall information (Kummerow et al., 1998). The highest temperatures observed by the Microwave Imagers are sensitive to integrated quantities of water vapor, liquid water, and ice in the atmosphere, as well as surface temperature and wind speed over ocean regions. The horizontal resolution of the sensor is also much lower than for the Precipitation Radar and varies from around 5 km for the highest frequencies, which are sensitive to precipitation-sized ice particles, to around 40 km for the lowest frequency channels, which are sensitive to liquid water droplets. These satellites then use these data to estimate precipitation in millimeters on the land below. In contrast to the Precipitation Radar, however, the Microwave Imagers' sensor has a swath width over more than three times (~759 km), providing much better sampling of rain systems for climate applications. The Visible and Infrared Scanner (VIRS) can provide information on cloud top temperature and reflectance. At the same time, a Lightning Imaging Sensor (LIS) estimates lightning flash rates. Because the satellites are continuously measuring rainfall, it is possible to compile large data sets showing annual rainfall estimates at the scale of 5×5 km grids across all types of terrain, which is extremely useful in remote areas such as the Himalaya and Tibet (Figure 1.1)(Bookhagen and Burbank, 2006).

Satellite measurements of temperature at different levels in the atmosphere have been routinely collected for a much longer time than satellite measurements of rainfall. Weather satellites have been available to constrain SST since 1967, with the first global syntheses of SST being made during 1970. Measuring SST is much simpler than deriving land surface temperatures because land is much more physically heterogeneous than the ocean. Furthermore, satellites do not measure temperature directly but rather radiances in various wavelength bands, which are then mathematically inverted to obtain indirect constraints on temperature (Wu et al., 1999), typically at the scale of 1×1 km. The resulting temperature profiles depend on details of the methods that are used to obtain temperatures from radiances. As a result, different groups processing satellite data have generated contrasting temperature data sets that need to be normalized in order to be compared and synthesized in order to isolate long-term changes.

Syntheses of observations, largely based on rain gauge information because of the relative short duration of TRMM and GPM data, show weakening trends in South Asia summer monsoon during the second half of the twentieth century (Lau and Kim, 2010). The East Asian summer monsoon has also been documented to have decreased in intensity since the late 1970s (Gong and Ho, 2002; Yu et al., 2004b). The weakening trends in both monsoons form integral parts of the global monsoon system, so that the total global monsoon rainfall over land shows an overall weakening trend in the second half of the twentieth century (Wang, 2006). The observed weakening trend in global monsoon rainfall over land was reproduced in simulations of climate models of the twentieth century, such as CMIP3 that factor in the influence of anthropogenic forcing, as well as natural forcing, although simulated trends are weaker than those actually observed (Held and Soden, 2006; Kim et al., 2008).

Climate models (e.g., CMIP3) cannot reproduce all the observed changes in global monsoon circulations and are not yet reliable indicators of detailed future climate change (Kim et al., 2008). The observed negative trend in global terrestrial rainfall is closely related to warming trends in SST over the central eastern Pacific and the western Indian Ocean during the twentieth century (Zhou et al., 2008). While tropical SST warming is considered a primary factor in driving monsoon weakening, because it reduces the contrast between the hot continent and cooler ocean during the summer, the effect of anthropogenic aerosol emissions over land in reducing solar insolation may also be important in reducing the land–sea contrast and thus causing weakening of South and East Asian summer monsoons (Bollasina et al., 2011; Lau and Kim, 2010; Li et al., 2010). These issues are discussed in detail in Chapter 6. In order to improve these models, better ground truthing data is needed in terms of wind, temperature, and precipitation constraints, as well as atmospheric aerosol concentrations.

2.6 Reconstructing the Prehistoric Monsoon

Although we here consider the influence that the monsoon has had on the development of farming strategies throughout Asia, we recognize that the monsoon is much older than humans and their colonization of Asia. Because rainfall is hard to reconstruct in the deep geological past, many attempts to chart monsoon intensity over timescales of tens of thousands, or even millions of years, have often focused on the environmental impact of the monsoon, as preserved in the sediments stored in sedimentary basins and especially on the continental margins around Asia. Often this type of record is interpreted based on assumptions about how continental environments respond to changing rainfall amounts, its seasonality, as well as the temperatures. The same proxies have also been used in much more recent geological times to infer environmental conditions.

Monsoon intensity and environmental conditions can be measured in a number of ways. An early approach was to look at the type of vegetation growing in floodplain areas, based on the fossil record, the carbon isotope signature of soils or the presence of pollen, typically preserved in local terrestrial or offshore sediments. All these methods allow the types of plants growing at a given time to be constrained. The basic premise is that different climatic conditions allow different types of plants to prosper and that changing monsoon strength would result in a change in the array of vegetation. In particular, the relative balance between woody C_3 type plants and grass-like C_4 plants was identified as being a critical proxy. This was a useful criterion because these different types of plants have different metabolic pathways for carbon fixation in photosynthesis, which resulted in different ratios between the ^{13}C and ^{12}C carbon isotopes (Ehleringer and Rundel, 1989; Smith and Epstein, 1971). As a result, C_3 plants show higher degrees of ^{13}C depletion compared to their C_4 counterparts. On the death of the plant this character is transferred into the soil where it may be preserved in the form of concretions that can be sampled even millions of years after their formation. Quade et al. (1989) were some of the first scientists to apply this method to ancient soils exposed in Pakistan. They inferred that the transition from C_3-dominated flora to a C^4 assemblage, starting shortly after 8 Ma, represented the onset of the summer monsoon, based on the assumption that the monsoon is presently strong. Subsequently, however, it was recognized that long-term global cooling during the Cenozoic (<65 million years) may also have a role to play in driving changing flora. Moreover, it has also become clear that grassland tends to thrive in drier environments than woodlands and that the change from one type to the other need not always be interpreted in the same fashion. This is especially true when considering the monsoon, which is characterized by an intense dry season, as well as heavy summer rains. The isotopic transition noted by Quade et al. (1989) can thus be reinterpreted to reflect a drying not a wettening of the climate, or even as an increase in seasonality. Consequently, it is now more commonly accepted that the summer monsoon weakened after 8 Ma, the opposite of the original interpretation (Clift et al., 2008).

Controversy continues concerning the significance of the 8 Ma transition, which is undoubtedly a major climatic phase globally. Some of the earliest advances in understanding the Asian monsoon were made using the marine sedimentary record, and in particular the paleoceanographic history of the eastern Oman continental margin. Marine records are often superior to their continental equivalents because such records are generally more complete and are relatively easy to date using micropaleontology. Understanding how to translate these into terrestrial records can, however, be difficult. The Oman region was singled out because in modern times the summer monsoon winds blowing offshore from Arabia drive strong upwelling of cold, nutrient-rich waters along this coast, which in turn allows the proliferation of certain types of foraminifera, as well as other forms of marine life (Currie et al., 1973; Manghnani et al., 1998). In particular, a pelagic foraminifera, Globigerina bulloides, is particularly associated with the strong upwelling system and has been used as a proxy for the intensity of the summer monsoon, because stronger monsoon winds tend to favor higher abundances of G. bulloides (Curry et al., 1992).

Following the first deep marine coring in this region, Kroon et al. (1991) noted that the abundance of this organism increased sharply after 8 Ma, consistent with the findings from the Pakistani carbon isotopes. This result was further reinforced by more data from other upwelling fauna from the same cores indicating greater productivity and thus monsoon wind strength (Prell et al., 1992). More recently, evidence for monsoon-related upwelling has been pushed back to around 13 Ma onshore Oman (Gupta et al., 2015) and in the Maldives (Betzler et al., 2016). While it is certainly clear that around 8 Ma is one of significant climate transition, this does not imply that it is also a time of monsoon rainfall intensification. Recent studies in the South China Sea indicate that 8 Ma was a time when salinity in surface waters increased, suggesting reduced runoff from the continent and presumably less summer rainfall (Steinke et al., 2010). Moreover, it has been recognized that this time was one of global climatic change, with enhanced productivity across wide regions of the Indian and Pacific oceans (Peterson et al., 1992), not just in the Asian marginal seas. If this is a global event, then it is less likely that enhanced oceanic productivity is driven only by the Asian monsoon. In particular, a connection with continental rainfall is hard to demonstrate, a common predicament with monsoon proxies regardless of the timescale.

During the early stages of monsoon reconstruction work, the importance of the 8 Ma transition had been underlined by a common belief that the elevation of the Tibetan Plateau also sharply increased at this time (Molnar et al., 1993). However, more recent work using new elevation proxies now shows that the south and central Tibetan Plateau were in fact been elevated close to their modern values long before that time (Currie et al., 2005; Harris, 2006; Rowley and Currie, 2006), while the evidence for rapid uplift at 8 Ma has been replaced by a more widely accepted view that the plateau has expanded to the north and southeast progressively during the Neogene (Clark et al., 2005; Royden et al., 2008) and was probably already of some topographic significance even as early as the Oligocene (older than 35 Ma), at least in southwest China (Wang et al., 2012b; Zheng et al., 2013).

Oceanographic productivity proxies have been applied to the more recent geological past to reconstruct the intensity of the monsoon over the time periods considered in this book. For example, Gupta et al. (2003) used a number of such proxies including the abundance of G. bulloides to look at the last 12,000 years of activity in the western Arabian Sea (Figure 2.7B). Following a common pattern this study showed generally strong upwelling and, therefore, powerful summer monsoon winds during the early part of the Holocene, after 11,000 years ago (ka), with

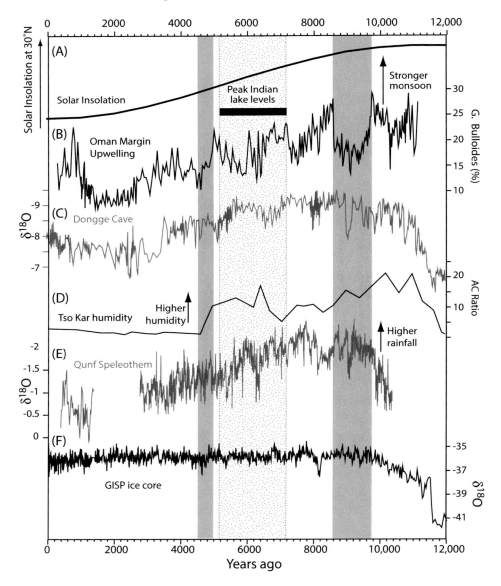

Figure 2.7 Compilation of monsoon intensity proxies showing the peak intensity in summer strength ~8000 years BP, decreasing until around 2000 years BP. (A) Solar insolation has been falling since its peak ~10,000 years BP. (B) Abundance of G. bulloides planktonic foraminifers in hemipelagic sediment on the Oman margin (Gupta et al., 2003). (C) Oxygen isotope data from stalagmites from Dongge Cave in southwest China (Dykoski et al., 2005). (D) Ratio of Artemisia/Chenopodiaceae pollen in lake sediment from Tso Kar, India (Wünnemann et al., 2010). (E) Oxygen isotope data from stalagmites from Qunf Cave, southern Oman (Fleitmann et al., 2003). (F) Oxygen isotope data from ice cores in the GISP2 site, Greenland (Stuiver and Grootes, 2000). Gray bands show periods of rapid weakening in monsoon strength across several of the study regions

a decline starting around 9 ka and lasting until around 2 ka. An increase is noted in some proxies since 1 ka. There are some departures to this general trend with a notable weaker period between 9 and 10 ka and a particularly sharp decline after 5 ka seen in the G. bulloides record. In general, the intensity of the monsoon follow the strength of solar energy reaching the mid-latitudes (Figure 2.7A), but Gupta et al. (2003) argued that there is also a link to millennial-scale climate events in the North Atlantic. In particular, they suggested that a number of intervals with weak summer monsoon coinciding with cold periods in the North Atlantic.

2.7 Chemical Weathering and the Monsoon

It has widely been assumed that chemical weathering would always intensify as the monsoon became stronger because heat and moisture are required for the break-down of silicate minerals and the formation of soils (Edmond and Huh, 1997; Meybeck, 1987; West et al., 2005). Because the summer monsoon is the primary supplier of moisture during the Asian summer, it seemed that stronger monsoons should result in faster and thus more intense weathering. However, this need not always be the case because as rainfall increases so the transport time of sediment in any given river system reduces as discharge increases, especially if there is little vegetation (Giosan et al., 2017). If transport is very rapid, then the time available for minerals to be altered to clay or other phases is strongly reduced and the degree of their total alteration thus reduces. The link between climate and weathering intensity is further complicated if the sediment is stored and then reworked during its journey between the source and the final depocenter (Hu et al., 2012). Recycling of older sediments as a result of climatically driven erosion can result in sedimentation of strongly weathered material, even if this same climate is liberating a lot of relatively fresh new material from the bedrock. Marine records showing an increase in chemical weathering in sediment from the Bay of Bengal starting around 8 Ma were interpreted to indicate stronger monsoon rains (Derry and France-Lanord, 1996), even though the rate of sediment supply, presumably linked to discharge, had reduced at this time (Burbank et al., 1993). This interpretation of a stronger monsoon thus now looks rather unlikely.

More recently, it has been shown that sediments from the Bengal submarine fan and the deep-water South China Sea show similar but reversed trends in weathering at the same time during the Miocene (Clift et al., 2008). This either means that the monsoon in the Bay of Bengal is anticorrelated with that in the rest of Asia, or more likely that the weathering response in the humid Bay of Bengal region is different from that seen in other parts of Asia. This is probably because even a weakened monsoon in this region is still strong enough to cause significant alteration of sedimentary material in the river systems.

Despite these potential complexities, proxies for chemical weathering, both geochemical and mineralogical, have been used to trace monsoon intensity over various durations of geological time. Geochemical and clay mineral weathering proxies indicate stronger chemical weathering during periods of intense monsoon in the Mekong River basin (Colin et al., 2010), as well as in the South China Sea (Wan et al., 2010b). On shorter timescales of around 10 thousand years (k.y.) periods of strong monsoon generally correlate with sedimentation of more altered material on the continental slope and even the shelf (Colin et al., 2010; Hu et al., 2013a; Trentesaux et al., 2006). When the timescales are centennial, or just a few thousand years, then the situation may be further complicated by the time required for clay minerals to form after a change of environmental conditions (Alizai et al., 2012). In the Indus floodplains, for example, a drying climate is associated with an incision of the river systems into the floodplain and of reworking of older more weathered material (Giosan et al., 2012), so in this situation there may be an increase in the average chemical weathering of sediment reaching the Indus River mouth despite a drying of the climate. Such processes introduce a lag time into the response of the system that may be significant when dealing with the timescales relevant to human civilization. Further south in the Godavari River basin of central India, weakening monsoon has also been associated with enhanced erosion. As the climate dried the density of the vegetation decreased, allowing faster erosion of the soils, a situation exacerbated by the onset of early agriculture, which tended to enhance soil erosion (Giosan et al., 2017). Consequently, care has to be used in the interpretation of marine proxy records because by definition these reflect the erosional response to climate change rather than climate change itself.

2.8 Cave Records as Climate Proxies

Cave records do not extend very far back into the geological past, typically only a few hundred thousand years, but they have proven extremely useful for deriving high-resolution records at sub-millennial scales, especially spanning the development of human societies since the Last Glacial Maximum (around 20 ka). In recent years, significant progress has been made in understanding changing monsoon intensities through study of cave records, specifically stalagmites, more generally referred to as speleothems. These growths can act as high-resolution climate archives because they preserve the oxygen isotope characteristics of the rainfall that permeated the rocks and formed the calcite at the time of precipitation. Crucially, they tend to form finely laminated deposits with the potential of resolving short time periods. Because weak rainfall is relatively enriched in ^{18}O compared to ^{16}O, this results in an average $\delta^{18}O$ value that is relatively positive when rainfall is light. However, as rainfall strengthens $\delta^{18}O$ values fall as the light isotopes are also precipitated. If these ratios are preserved in cave deposits then isotope analysis of the growth rings in a speleothem can be used to reconstruct the relative changes in rainfall intensity. This approach is particularly useful because the stalagmites themselves can be dated using uranium series and ^{14}C methods. Furthermore, because the growth laminae in many stalagmites are annual there is potential for a very high-resolution climate record, provided there is sufficient rainfall to form these deposits in the first place. Indeed, gaps in stalagmite records can often be related to periods of intense aridity.

Although the basic premise of climate reconstructions from speleothem isotope data is straightforward, some care has to be taken in interpreting these records because the oxygen isotope character may also be affected by temperature and the bedrock through which the water percolates. Further complications that can influence the isotopic composition of the water reaching the cave include the altitude at which the cave is located, and the distance from the ocean and the topography over which clouds have had to pass on their way to the cave. All of these processes fractionate oxygen isotopes. This means that to be effective rainfall proxies oxygen isotope ratios can only be converted into estimates of precipitation by calibrating the data for each particular cave. An example of this is shown in Figure 2.8 for the Dongge Cave in southwest China (see location on Figure 2.9). Here a series of isotopic measurements made from water in the cave can be compared with precipitation data from the surrounding region to demonstrate a clear correlation

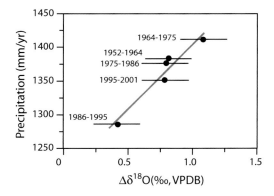

Figure 2.8 Comparison of instrumental rainfall records from the past 50 years with $\Delta\delta^{18}$O. Age ranges for each data point reflect those captured by sampling intervals in the Dongge Cave record (Wang et al., 2005). Rainfall data is the average annual rainfall at six stations between Dongge and Heshang Caves (Figure 2.4). As expected, larger differences between δ^{18}O at the two caves relate to higher annual rainfall. The observed relationship is not a simple Rayleigh Fractionation due to the effects of local terrestrial recycling of moisture (Brubaker et al., 1993; Ingraham and Taylor, 1991). Figure reproduced from Hu et al. (2008).

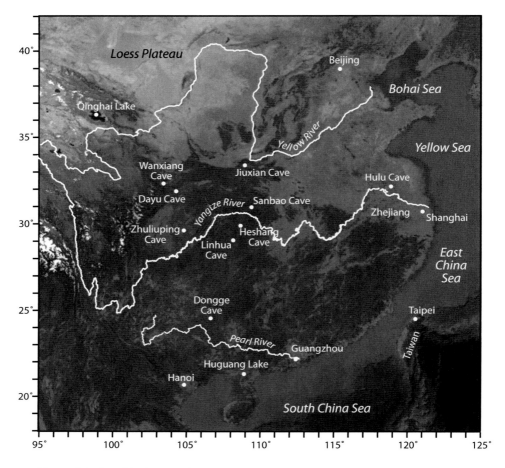

Figure 2.9 Satellite image of China and surrounding areas showing the locations of study sites mentioned in the text and the major geographic features of the area, including the courses of the major river discussed.

between recent speleothem isotope composition and average precipitation over periods of a few years. Once such a correlation has been made it is then possible to estimate past average precipitation variations from the composition of each speleothem lamina (Wang et al., 2005).

An alternative approach for estimating precipitation over a wider area is to look at the differences between the $\delta^{18}O$ values of speleothems in two different, widely separated caves at which stalagmite growth is known to have taken place at the same time. A study by Hu et al. (2008) at the Heshang Cave in the Yangtze River Valley (Figure 2.9) provided the opportunity to compare values with the existing record at Dongge Cave, which is located around 600 km to the south. In that study it was argued that the isotopic difference between the caves represented the amount of precipitation that occurred as the monsoonal rain front moved north between the two sites. As with the record for Dongge Cave alone, comparison of rainfall records and modern oxygen isotope values showed a good correlation and allowed 9,000 years of rainfall to be reconstructed across southern China. This record shows the common long-term pattern of strengthening summer rainfall between 9 ka and 7 ka, followed by a reduction after 7 ka until around 2 ka. This differencing record however produced a reconstruction with more noise than is seen in many of the case studies in East Asia. Whether this is accurate or not has yet to be fully demonstrated.

Such speleothem records have been widely reconstructed, especially in China and to a lesser extent in Arabia, where suitable carbonate bedrock exposures exist. Additional records have been derived from the eastern Himalaya where trends similar to those seen in China have been reconstructed (Sinha et al., 2005). However, there have been questions raised about the simple application of this methodology. Growth of speleothems may be seasonally biased in boreal climates so that the record does not record the summer rainfall (James et al., 2015). Other studies indicate that while $\delta^{18}O$ can appear to be a good rainfall amount proxy in some regions but in others changes in $\delta^{18}O$ are caused by changes in the source of the water (Chen et al., 2016a). The seasonal variation in temperature can also be significant (Clemens et al., 2010). Recently, a record of seawater salinity linked to monsoon dominated discharge from the Yangtze River was compared to cave records in the Yangtze catchment (Clemens et al., 2018). The cave records were seen to be dominated by orbital precession band (23 k.y.) cyclicity, while the East China Sea was dominated by cycles linked to orbital eccentricity (100 k.y.) and obliquity (41 k.y.). This implied that the cave records were not reliable rainfall proxies over certain time scales and that East Asian monsoon rainfall is controlled by greenhouse gas concentrations and ice sheet extent rather than by solar insolation.

Nonetheless, speleothems continue to be regarded by many paleoclimate specialists as being important for reconstructing monsoon variations at high resolution, not least because their long-term trends often follow variability in other paleoclimate proxies that provides confidence in their reliability.

2.9 Lake Records

Lake sediments have been used as paleo-environmental archives worldwide, and especially in Asia as indicators of monsoon strength and regional humidity. This reflects their sensitivity to environmental conditions and their ability to preserve high-resolution paleo-environmental records. Because many freshwater lakes have anoxic waters at their bottoms that are the result of stagnant circulation conditions, they hold the possibility of preserving finely laminated materials that provide detailed records of the fauna in the lake, as well as the surrounding flora. Variations in the C/N ratio can be used to constrain the changing terrestrial versus aquatic organic matter in lake sediments. Terrestrial vascular plants have much higher C/N values than aquatic algae and plankton (Meyers, 1997). When the climate is wetter then more terrestrial vegetation grows around the edge of any given lake, and this can subsequently be washed into the basin so that changing ratios of C/N can be used as a proxy for rainfall if the lake catchment remains largely stable. Furthermore, the clastic sediment in a laminated lake sequence can be used as an indicator of runoff from the surrounding countryside and thus of the strength of wet season precipitation. Contents of siliciclastic elements such as Ti and Zr in sediments can be used as proxies for eroded bedrock flux to lakes that can be compared with the biogenic sediment that is generated in the lake itself, for example, CaO (Regattieri et al., 2016; Yancheva et al., 2007). The grain size of siliciclastic sediment in the lake is also an indicator of the strength of terrestrial runoff into the lake (Sheng et al., 2015).

Oxygen isotope ratios of carbonates in lakes can further help constrain the source of the moisture in the rain, as well as the intensity of rainfall relative to the degree of evaporation (Talbot, 1990). This isotopic methodology has also been applied to siliceous biogenic material in lakes such as diatoms (Leng and Barker, 2006). Lake levels by themselves can be used to determine the balance between precipitation and evaporation. High lake levels, evidenced by stranded paleo-shorelines can be used to indicate wetter conditions, while falling levels, contracting shorelines and the precipitation of evaporite minerals, such as salts, are evidence of reduced precipitation.

2.10 Swamp Deposits

Swamps may also accumulate significant quantities of sediment that can be used to reconstruct the intensity of the monsoon in the past. Their preservation of an organic record that can be linked to precipitation strength means that swamps can be valuable in charting past rainfall and seasonality. Sedimentation history can be used to constrain paleo-environments, while pollen assemblages (which can also be derived from closed lake systems) constrain the vegetation in the surrounding areas. Both the abundance of pollen and the presence of environmentally sensitive species allow the nature of the landscape to be determined. Because some plants favor drier or more seasonal conditions, while other favor wetlands or tropical, wet environments swamp deposits can be excellent repositories of environmental information if they can be preserved (Anshari et al., 2001; Mao et al., 2006; Zong et al., 2007).

2.11 Organic Geochemistry

As well as trying to reconstruct the flora of a region based on pollen evidence, organic chemistry can also be effective in charting environmental development. The carbon isotopic composition of terrestrial plant biomass is primarily a function of the plant's specific photosynthetic pathway (C_3 or C_4, as discussed earlier) (Farquhar et al., 1989). Fortunately, these isotopic signatures manifest themselves in vascular plant epicuticular wax lipids (Tipple and Pagani, 2010) that may be transported by rivers and then preserved in sediments. Leaf wax $\delta^{13}C$ records (i.e., for C26 to C32 n-alkanoic acids) have been used extensively to reconstruct past changes in the balance of C_3 versus C_4 vegetation in drainage basins (Eglinton and Eglinton, 2008; Feakins et al., 2005) and have been applied to marine cores across monsoonal Asia (Giosan et al., 2018; Ponton et al., 2012; Yokoyama et al., 2006), as well as sediments in loess deposits (Zhang et al., 2006). The long-chain n-alkanes typically targeted by such work include C27 to C33 odd carbon number n-alkanes. These are interpreted as being derived from terrestrial higher plant leaf waxes (Eglinton and Hamilton, 1967). By contrast, C37 alkenones are derived from marine haptophyte algae (Brassell et al., 1986). Clearly resolution between these sources is critical for any meaningful reconstruction.

As for the paleosol deposits discussed earlier, the carbon isotope character of the leaf wax materials can then be used to characterize the nature of the vegetation in the source regions, especially the relative balance between C_3 and C_4 vegetation. In large catchments with diverse geography, however, care needs to be taken regarding where in the basin the leaf wax signal is derived from. Generally, this will be in the flood plains where the great mass of vegetation thrives rather than in the high mountains where much of the sediment is being derived.

These same leaf waxes can be used to independently examine the intensity of precipitation. The D/H ratio (δD) can be used as a proxy for rainfall intensity because there are clear correlations between hydrogen isotope character and rainfall in the modern day (Rozanski et al., 2013). These isotopic values may be preserved in leaf waxes but may be susceptible to fractionate as a result of evaporation in the soil between precipitation and uptake. Fractionation caused by transpiration, at least in many arid-semi-arid grasses, appears to be limited thus making this potentially a powerful rainfall proxy (Feakins and Sessions, 2010; McInerney et al., 2011). However, it has also been revealed that δD fractionation depends on photosynthetic pathways and thus on vegetation type, making simple application problematic. Bi et al. (2005) noted that the δD proxy was best used in combination with $\delta^{13}C$ data measured from the same waxes because the latter can used to be determine the photosynthetic method in each case, allowing an appropriate correction to be made. The effectiveness of the method is shown by examples from the Chinese Loess

Plateau where leaf wax δD variations correlate with magnetic susceptibility, which is linked to chemical weathering and thus climatic conditions. δD appears to be controlled by the combined effects of varying aridity, which controls the soil and leaf water evaporation, and temperature and monsoon intensity, which alter the D/H ratios of the initial precipitation (Liu and Huang, 2005).

2.12 Tree Rings

At its simplest level, the concept of using tree rings to look at monsoon intensity is relatively straightforward, with the basic premise being that in years when conditions for wood growth are favorable trees grow quickly, with thick annual bands (i.e., large volumes of new wood), whereas when conditions are harsh we would expect trees to grow very thin or no annual bands at all (Pretzsch, 2009). Care, however, must be exercised because of additional complexities. For many trees in forests, especially those growing in temperate and/or closed canopy conditions, climatic factors may not be the most important influence limiting growth. Instead, processes related to stand dynamics (especially competition for nutrients and light) may dominate the ability of individual trees to grow. Moreover, in order to generate long-term environmental records several trees normally have to be used and their records spliced together using pattern matching techniques in order to compare periods in which two or more trees were growing. Repeat analysis of trees from the same area provides us with confidence that a coherent environmental signal is being generated and is not simply the product of noise related to the growth of a single tree. While tree rings cannot normally be used to look at the distant past they are useful for reconstructing rainfall and temperature in the past one thousand years or so, often at high temporal resolution.

Environmental reconstructions based on tree ring data will be discussed in detail in subsequent chapters, but here we highlight a recent compilation of results from 327 tree ring sites across Asia (Cook et al., 2010). In this study, a gridded database was constructed based on these individual results from across the continent, although recognizing the fact that the distribution of data points is not regular because of uneven forest cover and the availability of tree species that are suitable for analysis. Figure 2.10 shows the regional patterns of drought severity during four well-known drought events during the past 500 years: the Ming Dynasty Drought (1638 to 1641 CE)(Parsons, 1970), the Strange Parallels Drought (1756 to 1768 CE)(Lieberman, 2003), the East India Drought (1790 to 1796 CE)(Grove, 2007), and the late Victorian Great Drought (1876 to 1878 CE)(Davis, 2001). Each of these events was clearly a time of widespread drought, but the intensity was unequal across Asia. For example, during the Ming Dynasty Drought of 1638 to 1641 (Figure 2.10A), the region of most intense of desiccation was in north China,

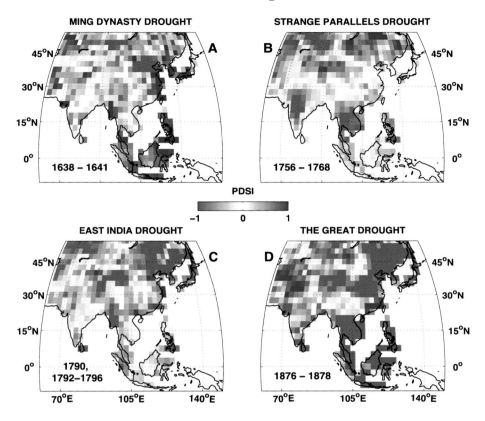

Figure 2.10 Spatial drought patterns during four historical Asian droughts. Mean Palmer Drought Severity Index (PDSI) over each of four regional droughts identified from the historical record. (A) The Ming Dynasty Drought (1638 to 1641). (B) The Strange Parallels Drought (1756 to 1768). (C) The East India Drought of the late eighteenth century (1790 and 1792–1796). In 1791, much of India appears to be slightly wet, except the region around Chennai where the drought persisted. (D) The late Victorian Great Drought (1876 to 1878). Reproduced with permission from AAAS from Cook et al. (2010).

with much of the rest of the continent staying unaffected. Indeed, even nearby eastern China experienced stronger than normal rainfall during the same time period. A more widespread drought was identified for the period 1756 to 1768, known as the "Strange Parallels Drought" (Figure 2.10B). During this event much of China was unaffected but South Asia and Southeast Asia, as well as parts of central Siberia, experienced intense desiccation that contributed to substantial societal upheaval and political reorganization. The East India Drought of 1790 to 1796 (Figure 2.10C) occurred during a great El Niño period of the late eighteenth century. Although historical accounts suggested that this affected South Asia more severely than other events, the tree ring data instead show that the strongest effects were in Myanmar, as well as across parts of the Tibetan Plateau and western China, Korea, Japan, and the Russian Far East. Some of these areas may also have been less affected because they had already developed other adaptations that made people less susceptible to drought, for instance, nomadic pastoralism on the Tibetan Plateau and throughout parts of northern China (d'Alpoim Guedes et al., 2014; Goldstein and Beall, 1991). In contrast, southern India and Sri Lanka are the most strongly affected by this particular episode.

The final event highlighted by the Cook et al. (2010) study was that in 1876–1878, known as the "Victorian Great Drought," which also coincided with one of the strongest El Niño events in the past 150 years (Davis, 2001). The tree ring data suggest that this was the strongest of the historical droughts that affected all parts of Asia, except southern China and was particularly intense in Indochina. It is believed to have resulted in the death of more than 30 million people, largely as a result of its impact on agricultural production (Davis, 2001). This was exacerbated by colonial policies, for example, export of wheat continued with a record 320,000 tons of wheat being sent to England in the peak year of the famine (1877–1878), as well as continued cultivation of cash crops, most notably cotton.

2.13 Temperature Proxies

Paleo-temperatures are often harder to reconstruct than precipitation because many of the most suitable proxies are often affected by both temperature and precipitation. Tree rings can be applied as a proxy for temperature, but only if the location of sampling can be chosen to minimize other factors. For example, if a forest grows in a region where rainfall is plentiful, then temperature may be the determining factor in controlling growth. Even in this case, ring width is not a uniform function of temperature but is biased toward the temperatures during the growing period and this must be accounted for when comparing data from different proxies. Carbon dioxide concentrations in the atmosphere may also be significant influences on ring width. In a similar fashion, studies of pollen and seeds can be used to determine the vegetation that grew in a given area in the past and the known ranges of temperatures in which the same or similar forms grow today can be used to constrain past temperatures. Adjustments may have to be made to account for changes in precipitation and seasonality. As well as simply considering the species of plant the shapes of plant leaves change with environment (Bailey and Sinnott, 1916). In particular, the proportion of woody dicotyledons with smooth (i.e., non-toothed) leaves in vegetation varies proportionately with mean annual temperature. In general, species living in warmer, wetter environments tend to have smoother leaf outlines than the smaller-

and more-toothed forms seen in temperate deciduous forests. This relationship has been quantified and schemes for each of the major continents drawn up that make extension of this process to the geological record possible. Typical errors are estimated to be around 2°C (Wing et al., 1993). More recently, this approach has been refined by the CLAMP (Climate Leaf Analysis Multivariate Program) methodology. CLAMP is a multivariate approach largely based on a data set of primarily western hemisphere vegetation that looks at a variety of leaf shape and dimensions, but especially the leaf outline (Spicer et al., 2009). This method reduces the uncertainty on the mean temperatures estimates to ±0.7°C–1.0°C and also permits estimation of the coldest month mean temperature and the warmest month mean temperature. This is important because it is the extremes that are often problematic for human societies. Regional calibration using numerous sites across China now allows improved performance in this region where the monsoon climate had caused problems for the general model (Jacques et al., 2011).

Measuring paleo-seawater temperatures is easier to achieve through a combination of oxygen isotopes and Mg/Ca data. Shells of marine organisms preserve the isotopic composition of the seawater from which they were precipitated and this in turn is a function of global ice volume and water temperatures (Kennett and Shackleton, 1976). Fortunately, Mg is preferentially incorporated in calcite as temperatures rise (Katz, 1973), which allows a correction to be made to the isotope data and the two processes separated (Lear et al., 2002). On short time scales alternative paleothermometers have been developed based on the alkenones in organic matter. The tetraether index of 86 carbon atoms (TEX86) has been used as paleothermometer (Schouten et al., 2002; Wuchter et al., 2004), although uncertainties can be significant. Moreover, there is evidence that some of the critical molecules that are assumed to be the product of marine algae production can also be synthesized onshore making simple interpretation sometimes impossible.

In any case, marine thermometers are often of only peripheral use to studies affecting human societies. Fresh-water thermometers are more powerful for such work and recently a new approach, known as "clumped isotopes," has been developed that can derive temperature from carbonate (Eiler, 2007). There is a slight thermodynamic tendency for heavy isotopes to form bonds with each other, in excess of what would be expected from a random distribution. The excess of this tendency is greatest at low temperature, with the isotopic distribution becoming more randomized at higher temperatures. This effect arises from differences in the zero point energy among "isotopologues", which are molecules that differ from one another only in the isotopes of the constituent atoms. Carbonate minerals can be analyzed to determine the relative abundances of these isotopologues. The parameter Δ47 is the measured difference in the concentration between isotopologues with a mass of 47 in a sample and a hypothetical sample with the same bulk but stochastically distributed isotopic composition. Δ47 is correlated to the inverse square of temperature (Ghosh et al., 2006). $^{13}C-^{18}O$ paleothermometry can be applied to freshwater carbonates with less ambiguity than other isotope-based methods (Huntington et al., 2010). The method is limited by the very low concentrations of isotopologues of mass 47, as well as the degree of analytical precision required, but has nonetheless made important recent contributions to paleo-environmental work.

2.14 Monsoon during the Little Ice Age

In the earlier discussion we have focused mainly on the long-term development of monsoon rainfall and temperature since the onset of the Holocene, but several proxy records now allow us to look in detail over centennial or even decadal timescales. We shall mostly review these for the critical periods dealt with in subsequent chapters, but here we show how the application of several proxies to a limited time period can be used to constrain evolving temperatures and rainfall across Asia. The G. bulloides record from Gupta et al. (2003) shows strong monsoon during the Medieval Warm Period and then a sharp decline into the Little Ice Age (Figure 2.7B), at least along the margin of Arabia. More recently, new studies from the southern tip of India also show significant changes in monsoon strength since 3 ka (Figure 2.11). Chauhan et al. (2010) measured a series of proxies from sediment in a deep-water core. $\delta^{18}O$ variations may be driven by seawater temperature or salinity, but over this time interval it is believed that salinity dominated. As a result, increases in the value of $\delta^{18}O$ have been interpreted to represent periods of enhanced aridity, resulting in less runoff and more saline oceanic waters in the southern Arabian Sea. This study also showed that dry periods were associated with reduced concentrations of Al and Ti in the sediments. Both these elements are associated with erosion of continental bedrocks and tend to increase in marine sediments when discharge in river systems was higher. Their reduction during time periods with low $\delta^{18}O$ is consistent with the idea that these were times with a weak summer monsoon. We further note that the Chemical Index Alteration (CIA), a proxy for chemical alteration of sediment (Nesbitt et al., 1980), falls during times of aridity, as might be expected given the links between weathering rates and moisture availability (West et al., 2005). This record appears to be a relatively coherent reconstruction of environments in southern India over the past 3000 years. The youngest arid phase coincides with the Little Ice Age, known from European and North American climate records (Grove, 2012;

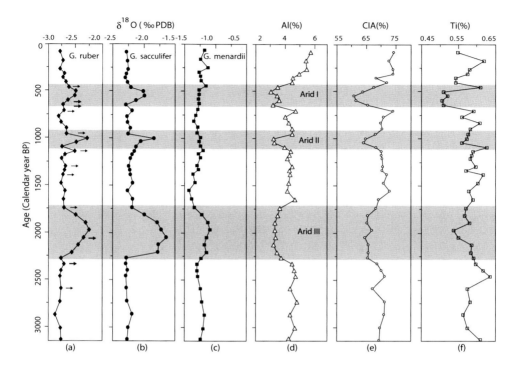

Figure 2.11 Temporal variations in G. ruber, G. sacculifer, G. menardii, CIA, Al, and Ti Heavier δ^{18}O and reduced Al and CIA during 450–650 (Arid I), ~1000 (Arid II), and 1800–2200 years BP (Arid III) are interpreted as events of reduced precipitation and fluvial flux and identified by gray shading. For better chronometric precision, ages are obtained on specific instabilities (bold arrows represent AMS ages; thin arrows represent bulk ^{14}C ages in panel (a). (Chauhan et al., 2010).

Mann et al., 2009), although there was also a humid climate and enhanced precipitation in India during the terminal stages of the Little Ice Age.

2.15 Historical Records

One advantage that can be enjoyed in looking at climatic development in East Asia is the longevity of human settlement and civilization in this region, which gives an almost unique opportunity to look at long-term changes in climate as recorded by humans, often government officials. This is particularly true in China. Zheng et al. (2006) performed a synthesis of such records, breaking China into three sections approximating a northern area around the Yellow River and Beijing, a southern area mostly around and south of the Yangtze River Valley, and a central region between the two. Since 1951 precipitation records have been kept for a number of stations

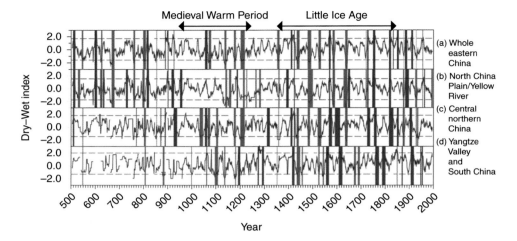

Figure 2.12 The severe persistent drought (red bar) and flood (blue bar) events for eastern China and its sub-regions during 501–2000 BCE; (a) Whole eastern China; (b) North China Plain/Yellow River Valley; (c) Central northern China/Jiang-Huai (Yangtze River) area; (d) Jiang-Nan (south of Yangtze delta) area. Gray solid line: dry–wet index series. Gray dash line: the value of the 1.645 times standard deviation higher or lower than mean of series that helps to define intervals of flooding or drought (Zheng et al., 2006).

across the country, but this reconstruction dated back to 137 BCE, providing a longer record of historical drought and floods documented in ancient records, as well as statistics for drought and flood conditions kept in local records or historical archives. Zheng et al. (2006) then derived a precipitation index that allowed the climate to be reconstructed over the past 1500 years. Figure 2.12 shows the prediction for the three areas, as well as a synthesis for the whole of eastern China, showing some coherent long-term variability. There is clearly some significant variation between regions in this reconstruction, with some areas showing opposing developments. For example, during the eleventh and thirteenth centuries there was flooding in the south, but dry conditions in the central region. Likewise, after 1550 CE there were frequent flooding events in the south, synchronous with drier conditions in north China.

Despite this spatial variability, these authors were able to identify a number of commonalities in the data. They argue that the historical records point to the existence of four dry periods since the start of the common era, from the 500s to 870s, 1000s to 1230s, 1430s to 1530s, and 1920s to 1990s. They further identified three especially wet periods, from the 880s to 990s, 1240s to 1420s, and 1540s to 1910s. They make these generalizations while recognizing that there were important variations within these longer intervals. Furthermore, they concluded that while flooding in China during the twentieth century had been as severe as that known over any of the entire documented period, but that evidence for earlier droughts of equal magnitude to those seen in the twentieth century was missing. Strong drought has largely been absent from China, except for in the north, since the latter stages of the Little Ice Age (Figure 2.12).

2.16 Synthesis and Summary

Much work remains to be done in terms of fine-tuning precipitation and temperature records, especially at high resolution. This is particularly true over the past few thousand years when interactions with humans were critical. However, significant progress has been made in reconstructing how monsoon intensity has changed in the past, especially during the Holocene, at least at the first-order level. Figure 2.7 shows a summary of some of the most reliable and detailed proxies from different parts of Asia, demonstrating some of the consistent trends detected across the continent. What appears to be beyond doubt is that the monsoon was strong during the early part of the Holocene, when global temperatures were also high and has weakened since around 8 ka. Summer rains progressively weakened and reached a minimum close to 1–2 ka, both in East and Southwest Asia, since which time there has been a modest recovery in strength. This long-term change in monsoon strength broadly parallels solar insolation at 30° north and is not strongly linked to the intensity of Northern Hemispheric Glaciation, at least not as preserved within the GISP ice core (Stuiver and Grootes, 2000). It is only the strengthening after 2 ka that appears to run counter to the solar energy trend, and that may be related to anthropogenic-induced warming (Ruddiman, 2003).

Although most of the proxies seem to show a strong summer monsoon around 8 ka, there is some dispute about when the initial intensification occurred. The Qunf speleothem record from Arabia shows the southwest summer monsoon in that region getting stronger after 10 ka, while in China the Dongge Cave record implies an earlier intensification after 12 ka, with a much longer duration of maximum intensity (Dykoski et al., 2005). If we compare the Qunf speleothem with the oceanic upwelling record, biogenic productivity offshore Oman, close to the cave location, we also see differences. The marine record implies summer monsoon winds strengthening, at least shortly after 11 ka. However, the upwelling record predicts a weakening of the monsoon between 8.6 and 9.8 ka, just as the cave record was predicting a strong period. Interestingly, this was a time when the lake record at Tso Kar in northwest India (Figure 2.13) also shows a decline from an earlier maximum (Wünnemann et al., 2010). Whether this represents a failure of one of the proxies or not is not entirely clear, but these differences may, in fact, reflect true lateral heterogeneity in the strength of the monsoon and the fact that some of

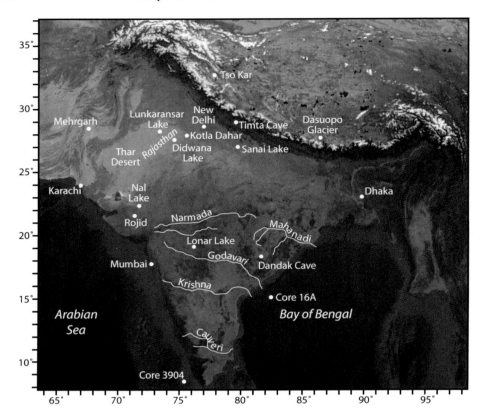

Figure 2.13 Shaded topographic map of the Indian peninsula and surrounding regions showing some of the geographic features and places mentioned in the text.

the proxies track wind strength, while others are more closely linked to precipitation. A further complexity in southwest Asia is the delivery of moisture during the winter via westerly winds that are less important further east.

Another critical transition appears to be between 5.2 and 7.2 ka when several Indian lakes were at a maximum in Rajasthan and where high humidity is also predicted at Tso Kar (Figure 2.13). The Dongge Cave record in East Asia is not at a maximum at this time, but it is nonetheless relatively strong. This period is followed by one of rapid drying across the continent. The Dongge record shows a start of this drying trend a little earlier, between 5.2 and 5.6 ka, while the Oman upwelling and Tso Kar lake records show a steep decline in monsoon strength from 5.2 to 4.6 ka. Dongge Cave shows further desiccation around 3.0 to 3.2 ka, although data density is not very high over that period. This steep decline at 3.0–3.2 ka is not shown in any other of the records, although it is a time of drying across Asia.

We conclude that monsoon rainfall strength and temperature can be reconstructed from a number of marine and terrestrial proxies that do not always agree with one another. Confidence in environmental reconstructions is best achieved through synthesis of a number of independent lines of evidence, bearing in mind their contrasting sensitivities to wind, rain, temperatures, and seasonality. The Asian monsoon behaves in inconsistent ways across the continent and there are important differences both between East and West Asia, and on smaller length scales within subregions. Areas of intense drying can be juxtaposed against regions of much wetter conditions making the reconstruction of climate relatively complicated when we are looking at millennial or smaller timescales that are particularly important when considering interactions with human society. There is clear evidence within the proxy records of periods of rapid monsoon strengthening and collapse that must have affected early human settlements across the continent.

3

Monsoon and Societies in Southwest Asia

South Asia is home to some of the earliest urban civilizations known worldwide and moreover has a relatively well-constrained climatic history spanning the Holocene, thanks to several different proxy records that largely portray a consistent evolution in environmental conditions, as described in Chapter 2. Together these factors allow us to examine how changes in climatic conditions might have impacted human farming systems and in concert how humans may have shaped and exploited their landscape and rivers to mitigate these challenges. Farming developed multiple times across South Asia during early human settlement. In this chapter, we focus our discussion on one of the key regions for which farming emerged and developed, namely the Indus River and its associated floodplains.

South Asia is largely situated south of 30°N latitude: the latitude that in East Asia compromises only the southern part of China. Despite the presence of a linked but not identical monsoon climate, this means that South Asia differs in several key regards from East Asia. While parts of high-latitude East Asia include areas where growing degree days (GDDs) are a primary limiting factor for crop growth (Figure 1.7), temperatures are largely sufficient for cultivating crops across South Asia, except for areas of higher altitude around and in the Himalaya (Figure 3.1). Water shortages, due to low rainfall on the other hand, are a crucial constraint on human farming systems in this area.

The area that was once home to the Indus Valley Civilization, which is a focus of this chapter, is remarkably arid. The area is furthermore affected by winter rains, especially in the northern areas, closest to the Himalaya as a result of moisture carried by the westerly jet (Hunt et al., 2018). These highly variable patterns of rainfall meant that humans were unable to carry out rain-fed farming for a number of important crops that were either domesticated or introduced to South Asia (Figures 3.2 and 3.3). Thus, unlike large parts of East Asia, control over water would have been a crucial component of human adaptive farming strategies in this area. This area is also home to some of the largest continuously flowing rivers

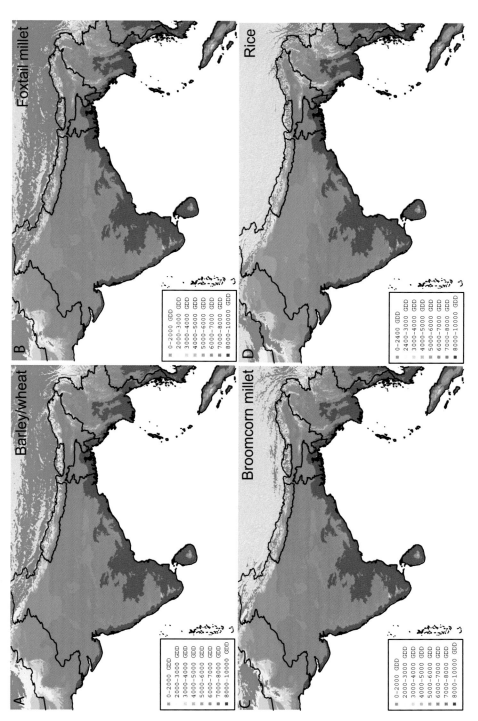

Figure 3.1 Map of available growing degree days (GDD) for a series of key crops that have been either domesticated or introduced to the South Asia. Areas in blue represent areas that are too cold for cultivation of these crops: (A) barley/wheat; (B) foxtail millet; (C) broomcorn millet; (D) rice. Data derived from Worldclim.org (Fick and Hijmans, 2017).

Figure 3.2 Annual rainfall across South Asia. Data derived from Worldclim.org (Fick and Hijmans, 2017).

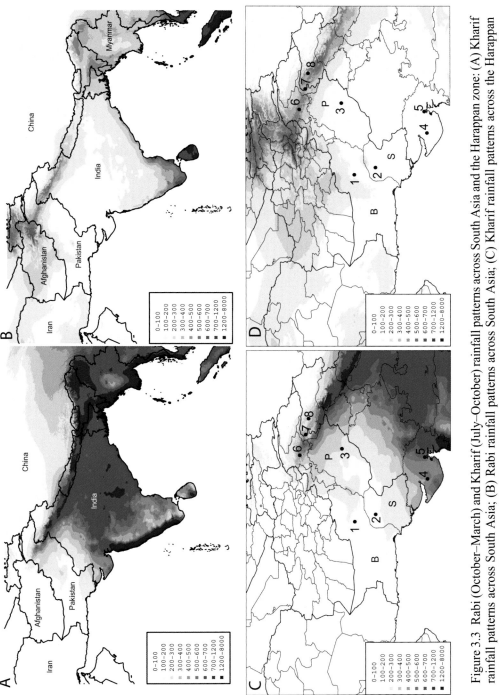

Figure 3.3 Rabi (October–March) and Kharif (July–October) rainfall patterns across South Asia and the Harappan zone: (A) Kharif rainfall patterns across South Asia; (B) Rabi rainfall patterns across South Asia; (C) Kharif rainfall patterns across the Harappan cultural zone and key sites discussed in the text: (1) Mehrgarh; (2) Mohenjo Daro; (3) Harappa; (4) Rojdi; (5) Lothal; (6) Loebanhr; (7) Burzahom; (8) Gufkral; P indicates Punjab region; (D) Rabi rainfall patterns across the Harappan cultural zone. Data derived from Worldclim.org (Fick and Hijmans, 2017).

across Asia, rivers that both provided humans with critical water for farming and also represented sources of uncertainty through ebbs in their flow and sudden changes in their course. A wide diversity of species and methods of farming have characterized South Asian farming throughout prehistory, ones that sought to meet this challenge.

We first review the major lines of evidence that constrain how summer rainfall has changed since the onset of the Holocene.

3.1 Monsoonal Weather Patterns in South Asia

Most of the rain in South Asia is delivered by the summer monsoon, which brings moisture from the Bay of Bengal progressively across the peninsula from the southeast to the northwest during the early part of the summer. In most parts of the subcontinent, the summer monsoon represents the vast majority of the precipitation (Turner and Annamalai, 2012). This is true even in the northwestern parts of the subcontinent, although here there is also moisture delivery during the winter as a result of storm systems driven by the westerly jet that lies south of the Himalayan front during that season (Hunt et al., 2018). The westerly jet brings moisture from the Mediterranean and to a lesser extent from the Caspian Sea and is the primary source of water to the glaciers of the Karakoram, which are largely shielded from the summer monsoon front by the Greater Himalaya (Karim and Veizer, 2002). Melt from these glaciers in the spring and early summer is another important source of water conveyed largely via the mainstream of the Indus River to the floodplains of northern Pakistan. Cyclones during the late summer and fall can bring heavy precipitation to eastern South Asia but are quite rare in the Arabian Sea. The summer rainfall is furthermore susceptible to disruption during monsoon breaks in the middle of the wet season that can seriously affect the volume of water delivered to populated areas. Rainfall is not uniformly distributed across South Asia but is the strongest against the topographic slope of the Himalaya, especially the frontal ranges of the Lesser Himalaya and also along the Western Ghats bordering the Arabian Sea, which also provide a topographic break, encouraging orographic rainfall (Figure 3.2)(Bookhagen and Burbank, 2006). The area covered by northwest India and Pakistan lies in a critical zone that presently receives significant amounts of both winter and summer rainfall (Petrie and Bates, 2017). The winter rain is largely focused along the southern edge of the Himalaya where the westerly jet lies during these months. Although the summer monsoon also brings rain further south, this piedmont zone, largely lying within the Punjab, thus receives rain in both seasons, making it a good area for farming. The fact that this situation has not always been like this is explored later in this chapter (Section 3.3).

3.2 Crops and Farming in South Asia

South Asia is home to a wide diversity of crops that have been cultivated in the region throughout prehistory. Different agricultural strategies have been developed throughout this region that are adapted to the phenology of the crops themselves but also patterns of rainfall across the region.

While the South Asian continent was an area of independent crop domestication, the earliest types of crops documented from this area were domesticated in the Near East, and slowly spread to northwest South Asia. These include wheat (*Triticum* sp.), barley (*Hordeum vulgare*), lentil (*Lens* sp.), and pea (*Pisum* sp.) (Weber et al., 2010c). These crops are reliant on winter rainfall and have traditionally been grown in a winter season of cultivation known as the Rabi, which is cropped from October to March and has allowed humans to take advantage of the winter westerly jets that deposit moisture on the northwestern portion of the continent.

South Asia is home to a wide variety of millets (small-seeded grasses) that were either introduced or locally domesticated and that were cultivated during the Kharif or summer season: from July to October. Local domesticates include browntop millet (*Bracharia ramosa*), raishan (*Digitaria cruciata*), hairy crabgrass (*Digitaria sanguinalis*), sawa millet (*Echinocloa colona*), little millet (*Panicum sumatrense*), kodo millet (*Paspalum scrobilatum*), yellow foxtail millet (*Setaria pumila*), and bristly foxtail millet (*Setaria verticilatta*). The exact timing and locations of the domestication of these native South Asian crops are unclear; however, research on the distribution of the wild progenitors of these crops sees a wide distribution across South Asia (Fuller, 2002). For instance, browntop millet and yellow foxtail millet are argued to have been domesticated in the southern Deccan region and little millet potentially in Gujarat (Fuller, 2002). Other species, such as kodo millet, have a wider distribution from Kerala to Rajasthan and West Bengal.

A number of other millets have also been introduced to the South Asian continent, that includes foxtail millet (*Setaria italica*) and broomcorn millet (*Panicum miliaceum*) from East Asia and a variety of African millets, such as pearl millet (*Pennisetum glaucum*), finger millet (*Eleusine coracana*), and sorghum (*Sorghum bicolor*)(Weber, 1989). Other Kharif crops include different types of different pulses, some of which may have been domesticated in South Asia, such as horse gram (*Macrotyloma uniflorum*), mung bean (*Vigna radiata*), black gram (*Vigna mungo*), sesame (*Sesamun indicum*), and rice (*Oryza* sp.) (Fuller, 2002; Zohary et al., 2012).

3.3 Paleoclimate Proxy Records in South Asia

Many of the monsoon proxies introduced in Chapter 2 were first invented and then applied to southwest Asia, and this means that there is a particularly good understanding of how rainfall and temperature have changed in this area since the onset of the Holocene. Here, we summarize what is known from these different lines of evidence over the time in which civilizations became established.

3.3.1 Oman Speleothems

Some of the best records of the South Asian monsoon are not found in the Indian Peninsular at all but are located on its periphery where the climate is very sensitive to waxing and waning of the rain. Although Oman is not a wet area, it is a place that is strongly affected by summer monsoon winds that bring rainfall to coastal areas, especially in southwest Arabia. This makes the area potentially a sensitive barometer of intensity for the southwest monsoon that affects the regions on the opposite side of the Arabian Sea. Winds that blow to the northeast from Arabia pick up moisture as they cross the ocean and deliver this to the regions of modern Pakistan and northwest India. The Hoti Cave in northern Oman shows evidence for growth between 10.5 and 6.0 ka, as well as several other interglacial periods in the deeper past (Burns et al., 2001). This pattern suggested growth during time of heavier rain and a hiatus when the climate dried. In southern Oman another cave at Qunf allowed a more detailed oxygen isotope record to be derived and showed a clear pattern of decreasing $\delta^{18}O$ from 10.4 to 7.8 ka, followed by a long-term increase after that time (Fleitmann et al., 2003)(Figure 3.4). This trend is interpreted to reflect first increasing and then decreasing precipitation in the latter part of the Holocene. There is a hiatus between 2.8 and 1.3 ka suggestive of very arid conditions at that time, followed by decreasing $\delta^{18}O$ values after 1.3 ka.

3.3.2 Marine Records

The summer winds of the southwest monsoon induce upwelling and increased organic productivity offshore Oman that resulted in clear signals in the sediment deposited at this time. Naidu et al. (1995) use data from planktonic foraminifer exacted from the Arabian Sea, which suggests that the southwest monsoon weakened from 3,500 to 1,200 years BP. Von Rad et al. (1999) used laminated muddy sediments deposited within the oxygen minimum zone on the continental margin slope of the Arabian Sea, offshore Karachi, to argue for a drop of precipitation after 4,000 to 3,500 years BP in southern Pakistan.

A regional measure of rainfall across southwest Asia was derived by Staubwasser et al. (2003). This group used a core taken from close to the Indus River mouth and examined the oxygen isotope compositions of planktonic foraminifers to constrain salinity (Figure 3.4B). Because this core had the opposite trend of many others in the Arabian Sea, it was considered unlikely that this isotopic trend

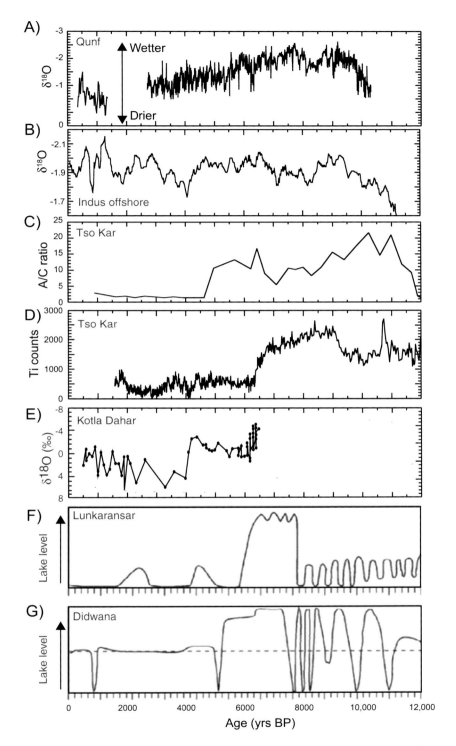

Figure 3.4 Paleo-climatic proxies for southwest Asia during the Holocene: (A) Monsoon precipitation record for the southwest monsoon as recorded at the Qunf Cave in southern Oman by Fleitmann et al. (2003), showing the long-term decline

Caption for Figure 3.4   (cont.)

in summer rainfall during the Late Harappan phase; (B) Oxygen isotopes in foraminifers from the Indus offshore region (Staubwasser et al., 2003); (C) Artemisia/Chenopodiaceae ratio (A/C) from composite core Tk 106/Tk 223. YD-B/A: End of Younger Dryas stage is marked by a dotted line (Wünnemann et al., 2010); (D) Ti XRF counts from Tso Kar (Wünnemann et al., 2010); (E) δ^{18}O VPDB (Vienna Peedee belemnite, ‰) of gastropod Melanoides tuberculate in sediment from Kotla Dahar (Dixit et al., 2014); (F) and (G) Lacustrine records from Rajasthan, India, as reinterpreted from Prasad et al. (1997) and Wasson et al. (1984). The interpretation is presented only in relative wetness terms for the different basins; the absolute water levels of lake or ground water or their transfer to values of precipitation are still problematic.

was caused by water temperature but rather by increasing salinity close to the river mouth. In turn, this implies that discharge from the Indus reduced sharply around 4,200 years BP, probably related to changes in solar insolation acting as a primary control on the summer rainfall.

In a more recent study of the winter monsoon Giosan et al. (2018) analyzed a new marine core also taken close to the Indus River mouth. They employed a planktonic foraminifer (*Globigerina falconensis*) as a proxy that is known to be sensitive to winter monsoon strength, combined with measurements of the DNA in the sediment that had accumulated since 10,000 years BP. Statistical analysis showed that much of the variability in the DNA could be attributed to sea surface temperature (SST), nutrient availability, and the sea level state. The temperature factor shows the general warming that typifies the early Holocene but then highlights a colder interval from 4500 to 3000 years BP. The chlorophyll factor tracks productivity and revealed a peak at the same time as the colder water conditions (Figure 3.5). These two factors suggest enhanced winter convective mixing between 4500 and 3000 years BP that brought colder, nutrient-rich waters to the surface (Giosan et al., 2018). This time was also one of peak *G. falconensis* abundance pointing to powerful water mixing at a time of strong winter monsoon. This study further identified two intervals of low interhemispheric thermal gradient: the early neoglacial anomaly (ENA) (4,500–3,000 years BP) and the late neoglacial anomaly (following 1,500 years BP). The strong winter monsoon during the ENA coincided with weakening summer monsoon rainfall, meaning that precipitation shifted from being strong over the Indus floodplains to be wetter closer to the Himalaya, especially toward the northeast.

Further constraints on Indian climatic evolution were derived from material recovered near the delta of the Godavari River (Figure 2.13) that allowed changing environment and flora to be reconstructed in that drainage system in central India (Ponton et al., 2012). Figure 3.6 shows a gradual increase in δ^{13}C leaf wax values

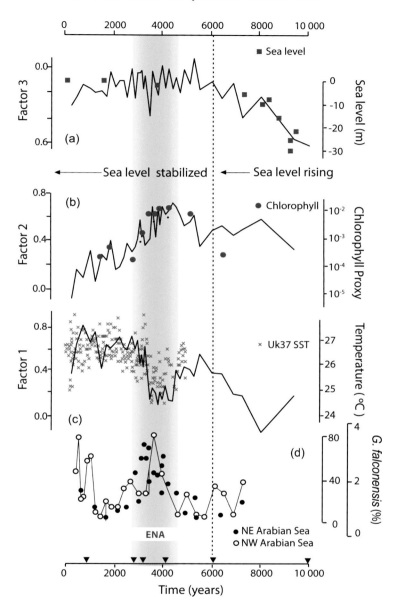

Figure 3.5 Holocene variability in plankton communities as reflected by their sedimentary DNA factor loadings (A–C) and winter mixing sensitive percentage *G. falconensis* (D) in core Indus 11C in the northeastern Arabian Sea. Relative chlorophyll biosynthesis protein abundances are also shown. Sea level reconstruction is from Camoin et al. (2004), SSTs are from Doose-Rolinski et al. (2001); and *G. falconensis* census from the northwestern Arabian Sea is from Schulz et al. (2002). Triangles show radiocarbon dates for core Indus 11C. The period corresponding to the early neoglacial anomalies (ENA) is shaded in red hues. Reproduced from Giosan et al. (2018).

since 5,000 years BP, consistent with an increasing dominance of C_4 flora and thus drying climate over that time. Increases are particularly noteworthy after around 4,000 years BP and especially after 1,700 years BP, which is an anomalously arid interval during the entire post-glacial period. Although human settlement of central India may have played a part in changing the flora, this is generally thought to be relatively modest prior to the onset of nineteenth-century deforestation (Hill, 2008). Significantly, the onset of extremely arid conditions after 1,700 years BP was paralleled by the start of significant positive excursions in $\delta^{18}O$ (Figure 3.6C). Prior to that time, these values had been relatively constant because of the high discharge from both Indian peninsular rivers and the Ganges-Brahmaputra into the Bay of Bengal. The very positive excursions that happened since that time suggested to Ponton et al. (2012) that there were phases of extreme drought that started around that time. They further correlated this recent propensity to dramatic climate variation and particularly strong drought to the onset of the construction of water tank facilities in settlements in the attempt to adapt to any regular supply of water from the summer monsoon (Gunnell et al., 2007).

3.3.3 Peat Bog Records

Phadtare (2000) collected pollen evidence from the Garhwal Higher Himalaya area and revealed that prior to 7,800 years BP the flora in this region was dominated by evergreen oaks that indicated a cool, wet climate. However, from 6000 to 4500 years BP there was a relatively warmer and humid period inferred from a bloom in the conifer population, followed by a sharp change in climate with a decrease both in temperature and precipitation during the following period (4500–3500 years BP), as shown by the relative increase in evergreen oaks again. The period of 4,500–3,000 years BP was identified as the weakest summer monsoon time since the start of the Holocene.

3.3.4 Lake Records

This climate archive has been studied in northwest India where a number of large lakes used to exist in the desert regions of Rajasthan (Figure 3.7). There is some latitudinal variability in the lake records because of changing rainfall intensity away from the coast, but this does not preclude these being used as qualitative proxies of rainfall intensity. Figure 3.4 shows records from this region that differ in detail but also show important similarities, most notably high lake levels, interpreted to reflect strong monsoon rainfall between 6,000 and 7,000 years BP (Prasad and Enzel, 2006). Lake Lunkaransar, for example, shows low and variable lake levels in the early part of the Holocene (11,000 to 7,000 years BP) and a well-

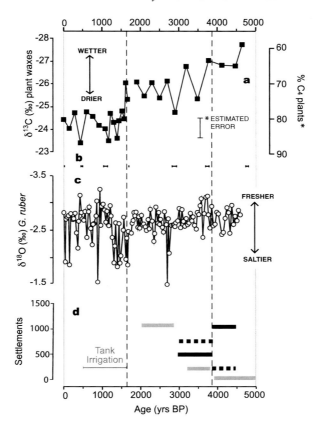

Figure 3.6 (A) $\delta^{13}C$ plant wax record from core 16A from the Godavari River mouth as the weighted average of n-alkanoic acids C26–C32. Error bar represents the maximum propagated error (1σ error) in the estimate of % C$_4$ plant cover. Vertical black dashed lines identify steps in the aridification at around 4,000 and 1,700 years BP; (B) calibrated radiocarbon ages (1σ error); (C) $\delta^{18}O$ measured on Globigerinoides ruber; values are corrected for ice volume effects in order to track seawater salinity; (D) number of settlements based on archaeological data expressed as totals over culturally defined time intervals. In solid gray, sites from the Deccan Plateau (Andhra Pradesh, Karnataka, and Maharashtra). In solid black, Indus (Harappan) sites from the dry Baluchistan, Sindh, Gujarat, Cholistan, and lower Punjab regions. In dashed black, sites from rainier upper Punjab and Haryana. The drought-prone regime in the late Holocene (after 1,700 y BP) coincides with the flourishing of water tank construction. Figure reproduced from Ponton et al. (2012).

defined maximum period around 6,300 years BP (Enzel et al., 1999) (Figure 3.4F). This peak is supported by more recent oxygen isotope work by Dixit et al. (2014) from a paleolake at Kotla Dahar (Figure 3.4E). That study showed negative $\delta^{18}O$ values at 6,300 years BP, consistent with heavy rainfall. Shortly after 4,800

Figure 3.7 Satellite of the Himalaya foreland basin showing the major tributaries of the Ganges and Indus River systems as well as geographic features mentioned in the text. Image from Google Earth.

years BP, Lake Lunkaransar experienced rapid drying, so that by 4,200 years BP it had been reduced to an ephemeral lake characterized by playa-type sedimentation. Since 4,200 years BP no additional sedimentation has occurred, consistent with the present environment that is especially arid compared to earlier in the Holocene. The decline in rainfall was better dated at Kotla Dahar than at Lake Lunkaransar, where a steep trend to drier conditions (positive $\delta^{18}O$ values) was recorded close to 4,100 years BP (Dixit et al., 2014).

More recently, Dixit et al. (2018) attempted to constrain changing monsoon rainfall intensity through the isotopic values ($\delta^{18}O$ and δD) of "gypsum hydration water" from sediments cored in paleolake Karsandi in northern Rajasthan (Figure 3.8) to infer past changes in lake hydrology. The lake water was controlled both by changing amounts of precipitation and evaporation. This record demonstrated that conditions were relatively wet in this area from around 5,100 years BP (Figure 3.8). Monsoon rainfall strengthened and the lake became fresher between 5,000 and 4,400 years BP. The environment began to dry after 4,400 years BP, so that by 3,900 years BP the area was quite arid.

A detailed environmental record has also been completed at Tso Kar, a saline lake in the Himalayan region of Ladakh, northwest India (Figure 3.7). Like the Hoti

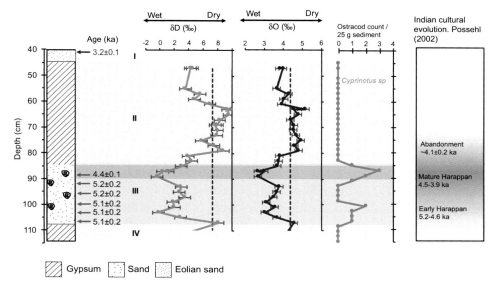

Figure 3.8 Correlation of climatic variability recorded in the lithostratigraphy, δD (orange), δ¹⁸O (blue), of paleolake Karsandi water and ostracod abundance with Indus cultural changes. The calibrated radiocarbon ages (ka) are shown with arrows pointing to their respective depths. Nearly pure gypsum deposits indicate periods of relatively lower rainfall. Roman numerals denote lithologic units. The early phase of the Indus Civilization developed during increased monsoon intensity as indicated by lower gypsum hydration water isotopes and high shell abundance after ~5.1 ± 0.2 ka. The Mature Harappan phase and peak in urbanism coincides with the lowest gypsum hydration water isotopes and highest shell abundance between ~5.0 and ~4.4 ka. Note that the subsequent decline in urbanism and disappearance of post-urban Harappan sites in this region is coincident with drying conditions, suggested by reappearance of massive gypsum with increasing gypsum hydration water isotopes and complete absence of ostracod and gastropod shells. Reproduced from Dixit et al. (2018).

Cave, this region lies on the edge of the monsoon's influence and is potentially sensitive to changes in the extent and intensity of the summer rainfalls, although it is also affected by moisture transported by the westerly jet. Surprisingly, this lake shows high values around 12,500 years BP, during the warm period of the Bølling-Allerød (Wünnemann et al., 2010). Deposition of muddy, deep-water sediments indicates that the lake level rose again about 11,800 years BP and remained high until around 8,600 years BP, a state that was interpreted to reflect the combined influence of intensified glacier melting and summer monsoon moisture supply. The lake reached its maximum level and extent between 8,500 and 7,000 years BP under warm-moist climate conditions but then shrank and became more saline in recent times (Wünnemann et al., 2010). 4,200 years BP was identified as the time when the lake was at its smallest and most saline.

Additional information about lake levels was derived from chemical data from sediments cored from within the modern basin. The relative concentration of titanium in Tso Kar lake sediments was measured by continuous X-ray fluorescence (XRF) analysis of cores. Because titanium is enriched in mafic minerals, but is not produced by biogenic activity, its relative abundance can be used as a proxy for the degree of runoff into the lake from the surrounding countryside and thus monsoon strength (see Chapter 2). The summer monsoon rains are seen to increase significantly around 9,000 years BP, and remain quite high until after 7,000 years BP when it decreased rapidly prior to 6,300 years BP (Figure 3.4D). In this study such proxies were reinforced through consideration of pollen data and in particular the relative abundance of the moist steppe flora of *Artemisia* compared to the dryer Chenopodiaceae-dominated flora (Demske et al., 2009)(Figure 3.4C).

The highest values of Artemisia/Chenopodiaceae (A/C) mark the onset of a wetter-than-present regional climate starting after around 12,000 years BP (Figure 3.4C). A major increase in the total pollen concentration was also interpreted to be indicative of a denser vegetation cover around the lake during the early Holocene climate warming. Deeper water conditions during the early Holocene are further supported by less abundant zooplankton in the sediment and increases in the freshwater algae *Pediastrum*. The interval between 10,800 and 9,200 years BP represents a time when there was a significant rise in lake level, probably linked to strong summer monsoon rains. The A/C remains high throughout much the Holocene before decreasing rather rapidly starting around 5,000 years BP. High lake levels in the mid-late Holocene were linked to melt-water from glaciers as well as the influence of local rainfall. However, it is noteworthy that the strong decline around 5,000 years BP was also seen in Lake Didwana in Rajasthan, suggesting a regional change in humidity (Figure 3.4G).

Although the Tso Kar record differs in some ways from the Rajasthani lakes, it is noteworthy that at the first-order level it also records a long-term drying from early to late Holocene and a sharp increase in dryness by 4,200 years BP. The presumed rainfall record derived from the lake levels is further supported by work on the pollen assemblages within the sediments, which testify to increased humidity after 12,000 years BP, and a maximum between 10,000 and 11,000 years BP, and then a sharp decrease after 5,000 years BP.

3.4 Rivers in South Asia: Morphology and Mythology

Our growing degree day (GDD) model (Figure 3.1) and precipitation models (Figures 3.2 and 3.3) show that, far more than temperature, water management was essential for providing sufficient resources for crops for the civilizations that arose in and around the course of the Indus River and its various tributaries. It is not surprising then that rivers play such an important role in religion and mythology.

Northern India is home to several of the world's largest rivers with a drainage divide between the southwest-flowing Indus system and the east-draining Ganges. Most of these rivers are sourced within the Himalaya, with relatively few from the south, reflecting the topography of the Himalaya and its influence in controlling precipitation (Figure 3.7). Consequently, the rivers are typically sediment-rich as they come from steep mountain valleys deeply incised because of the tectonically driven uplift of the bedrock. In southwest Asia, the Punjab region is surrounded and transected by a number of rivers with the main Indus stream to the west. The Indus, sourced in western Tibet, is the single largest stream, accounting for 58% of the modern discharge to the lower reaches, but there is also significant flux from the Himalayan-derived tributaries, with the Jhelum, Chenab, Ravi, and Sutlej, each accounting for 15, 17, 4, and 6%, respectively, of the total modern water flux (Figure 3.9) (Alizai et al., 2011a). Estimates of the preindustrial sediment discharge vary from 260 to 400 Mt/years, much less than the 530 Mt and 890 Mt for the Ganges and Brahmaputra, respectively (Karim and Veizer, 2002; Milliman and Syvitski, 1992). The lower sediment load is a simple consequence of the lower

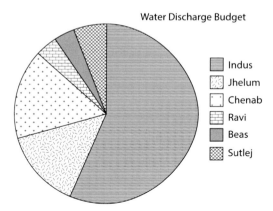

Figure 3.9 Pie diagram showing the relative pre-damming discharges from the rivers in the Indus basin. Data for the Sutlej, Ravi, Yamuna, and Beas are from the Irrigation and Power Research Institute, Punjab, India (unpublished). Data from the Indus, Jhelum, and Chenab are from personal communication with the Pakistan Commissioner for Indus Waters (Alizai et al., 2011).

monsoon rainfall in the Indus basin that results in less erosion and transport capacity, as well as being smaller in area too. The easternmost Indus tributary would have been the Ghaggar-Hakkra, which flows southwest from its origins in the Himalayan foothills, but it is now ephemeral and terminates in the sand dunes of the Thar Desert shortly after crossing the modern India-Pakistan border (Figure 3.7).

River systems hold a central role to life in South Asia providing water not only for life and farming, but also for religious practice. This makes sense given the sheer scale of these rivers in terms of their length, as well as their crucial function in bringing both the water and sediment necessary to farming in the region. Hindu devotion to the Ganges River is well known, and the Ganges, typically known as Ganga, is personified as a goddess of the same name. The Ganga, as told by the scriptural text Bhagavata Purana, was described as having sprung from the very hair of the Hindu deity Shiva (Alley, 2010). However, this only occurred after initially being sourced from the so-called Causal Ocean that surrounds the universe and was caused to leak through in the form of the Ganges after the boundary between the two was pierced by the toe of Lord Vishnu in his attempts to measure the width of the universe (Wilson, 1868). It is for this reason that the Ganges is also known as "Bhagavatpada," that is, that which emanates from the lotus feet of Bhagavan (God)(Grimes, 2004).

In being linked with a goddess, the Ganges is not unique for rivers in South Asia, and in particular, reverence is also paid to the goddess Saraswati (sometimes spelt "Sarasvati") who is the patron of knowledge, music, and arts and science (Pattanaik, 2000). Sarasvati is a particularly significant goddess in being one of the primary helpers of the central Trinity of Hinduism (i.e., Brahma, Vishnu, and Shiva) whom she aids in creating, maintaining, and destroying the universe. A river called "Saraswati" is named in the Hindu holy text the Rig Veda (written between 1700 and 1100 BCE)(Sankaran, 1999) and means literally "having many pools" in the Sanskrit (Oberlies, 1998). This river is mentioned on several occasions in all but one of the four books of the Rig Veda and usually in very positive fashion. It is described as "best mother, best river, best goddess" (Muir, 1873).

The location of the Saraswati River is discussed in these holy books as being between the Yamuna and Sutlej Rivers as they leave the Himalayan range front, close to the drainage divide (Figure 3.7), which is a location without a significant river in the modern day. In fact, this location is now occupied by the much smaller and ephemeral Ghaggar-Hakra, which flows into, but then peters out within, the dunes of the great Thar Desert (Figure 3.7). In its present condition, it is hard to relate the Ghaggar-Hakra to the stories of the great flood-giving river of the Hindu scriptures. Nonetheless, scriptural information and geographic considerations seem to place the Saraswati in what is now northwest India and Pakistan. Suggestions that the Saraswati may have flowed away from its Himalayan sources more due south toward the Rann of Kutch are clearly unrealistic because this would require crossing a significant range of hills in this direction, through which no valley cuts. The Ghaggar-Hakra would appear to be the most likely former course of the Saraswati (Sankaran, 1999).

In the related religious texts, the Yajurveda (Debroy and Debroy, 2011), the Saraswati is mentioned in a way that implies it was part of the Indus drainage system, which would be consistent with the Ghaggar-Hakra's discharge having been greater, allowing the river to cross the Thar Desert. The Yajurveda notes "Five rivers flowing on their way speed onward to Saraswati, but then become Saraswati a fivefold river in the land." According to the medieval commentator Uvata, the five tributaries of the Saraswati were the Punjab rivers: Drishadvati, Satudri (Sutlej), Chandrabhaga (Chenab), Vipasa (Beas), and the Iravati (Ravi). The Ravi, Chenab, Beas, and Sutlej are now tributaries of the Indus, which they join after flowing through the Punjab, which itself means "five rivers." Valdiya (2002) presented a reconstruction of one possible course of the Saraswati during the Vedic times (Figure 3.10). In this model, the Saraswati was clearly linked to what Valdiya calls the "Shatodru," but it is equivalent to the upper reaches of the Sutlej in the modern day. Clearly this river no longer exists, and recent fundamentalist Hindu belief is that the Saraswati River still flows underground and meets the Yamuna and Ganga at their confluence in Allahabad (Figure 3.7)(Valdiya, 2002). The question remains whether the Saraswati really existed in the recent geological past and if so, what was its importance to early farmers in the region? We shall return to this question in Section 3.5.2.

3.5 Early Farmers in the Indus Valley

3.5.1 The Development and Spread of Farming to the Indus

Sites that are related to the pre-Indus and early Indus civilizations contain some of the earliest evidence for the spread of farming technology that originated in the Fertile Crescent and was based on wheat and barley, peas, and lentils to South Asia. Early sites, known as belonging to the Aceramic Neolithic, containing these domesticates and mudbrick architecture appeared in areas such as Baluchistan following 6500 BCE (Figure 3.11).

One such site, Mehrgarh (6500–2300 BCE) (Figure 3.11) contains the earliest evidence of the spread of these Near Eastern agricultural products to northwest South Asia (Jarrige et al., 1995). The inhabitants of Mehrgarh lived in a small village consisting of mudbrick housing divided into different compartments.

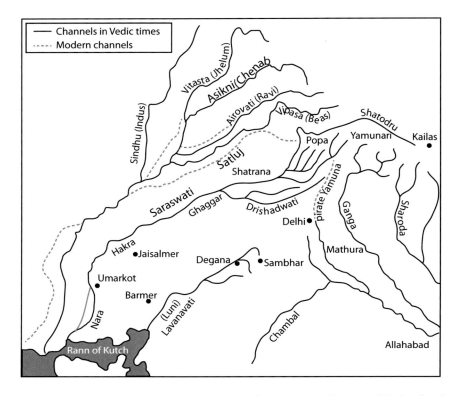

Figure 3.10 Suggested paleo-drainage map of the western Ganga and Indus flood-plains in the late Quaternary, redrawn from Valdiya (2002). Note the proposed Saraswati River running sub-parallel to the Indus as far as the Rann of Kutch.

These buildings seem to increase in size from the earliest to the latest phases of occupation of the site (Coningham and Young, 2015). Some of the rooms inside buildings were tiny and may have represented silos in which people stored grain. During the earliest phases of the occupation of these sites, woven baskets lined with bitumen were used to cook and process food. As time moved on, the inhabitants of the site developed different forms of pottery, increasing their efficiency in cooking and processing food. Despite the investment in housing, it seems likely that the inhabitants of these sites engaged in some degree of mobility. Large numbers of secondary burials in the earliest phases at Mehrgarh suggest that at least some of its inhabitants followed a relatively mobile lifestyle. They died elsewhere but their bones were still returned to their ancestral home (Wright, 2010).

Unfortunately, the only archaeobotanical analysis carried out at Mehrgarh focused on impressions of seeds derived from the mudbrick architecture and

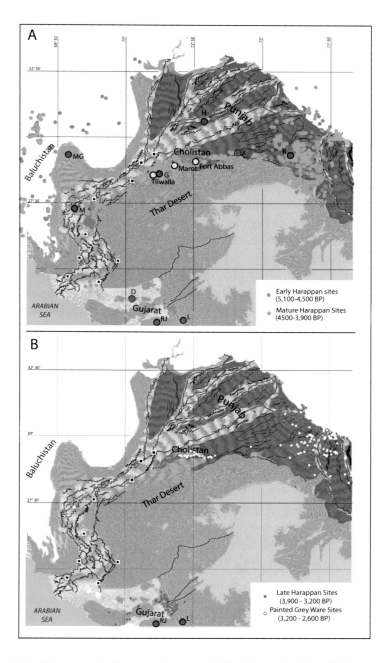

Figure 3.11 Harappan landscapes of western Indo-Gangetic plain: (A) Large-scale morphology of the region (interfluves or bar uplands in grey mask; terrace edges in dashed black lines) with river channels and chronological information. Early and Mature Harappan sites with names of some major urban centers: D – Dholavira; M – Mohenjo Daro; MG – Mehrgarh; G – Ganweriwala; H – Harappa; K – Kalibangan; R – Rakhigarhi; RJ – Rojdi; L – Lothal; (B) Late Harappan and Painted Grey Ware sites. The Painted Grey Ware Culture is the society that closely followed the Harappan but was largely displaced to the north-east relative to the earlier sites (Roy, 1984). Maps are altered from Giosan et al. (2012). Yellow dots show the locations of scientific boreholes in the study of Clift et al. (2012).

might not reflect the full range of taxa employed by the inhabitants of this key site (Constantini, 1984). Small amounts of glume wheats alongside larger quantities of barley were present at the site. Initial analysis of this material suggested that the barley may not have been fully domesticated (Constantini, 1984). However, as Fuller (2002) points out, these wild forms may simply represent weeds of barley and not an incidence of local cultivation. Other plants cultivated by the inhabitants of these sites include jujube, dates, and also potentially cotton, which may or may not have been locally domesticated (Moulherat et al., 2002).

Faunal remains from Mehrgarh have been better analyzed, and they indicate that during the earliest phases of the sites' habitation people focused on hunting wild game such as gazelle, wild goat, sheep, pig, water buffalo, onager, and deer (Meadow, 1984). By the end of the Aceramic phase however, they seem to have focused almost exclusively on sheep, goat, and cattle, suggesting that processes of domestication (or at the very least animal foddering) were underway. Meadow (1984) also documented changes in the sizes of animals: sheep and goat grew smaller and easier to manage as they became domesticated by humans. Goats, on the other hand, maintained their size across time suggesting that they may have been introduced to the region as domesticates very early on. Young goats were also often placed in Aceramic burials signifying the special importance that these animals held in the local economy from the earliest phases of the site (Meadow, 1984). The numbers and weight of cattle bone increase dramatically over time at the site, something that Meadow (1984) believes corresponds to a localized domestication of Zebu.

Little is known about how exactly the inhabitants of these early sites carried out farming; however, our review of the paleoclimate literature all points to a much wetter climate than present, up until roughly 5,000 years BP (Figures 3.4, 3.6, and 3.8). Many of these early food-producing settlements were on mountain and valley foothills: areas that have sufficient rainfall in the present day but that given the paleoclimate evidence may have benefitted from even more substantial rainfall in the past (Figures 3.3C and D). It may have thus been possible for individuals at these early sites to have carried out rain-fed forms of wheat and barley cultivation. As Wright (2010) points out, the location of these early sites likely also meant that foragers and incipient farmers could access a wide range of plant and animal species: not only those on the plains, but also those in the higher elevation mountains that surround these sites.

3.5.2 Agricultural Intensification during the Earliest Urban Societies

As time went on, sites like Mehrgarh grew in both size and complexity and eventually became centers for the production of goods like painted pottery, stones, copper, and bone products, becoming at the nexus of deeper networks of regional exchange. During the pre-urban phase (6,000–4,600 years BP) new settlements began to emerge on the alluvial plains of the Indus and its tributaries (Figure 3.11) and we see humans employing new strategies to access water in this more arid environment. For instance, there is evidence for canals along the Bolan River (Jarrige et al., 1995) potentially used for irrigation to fuel the production of wheat and barley. Wright (2010) argues that before intensive settlement occurred in the lower Indus, people constructed gabarbands (a form of dam) and humans made efforts to situate settlements near perennial springs. It is possible that the attempts to move settlements onto the alluvial plains and tap into more abundant and reliable resources from rivers may be driven by the decline in monsoonal intensity (see Section 3.2) that we see following the third millennium BCE (4,000 years BP).

Harappans moved from the simpler rain-fed systems of farming that were likely used at sites like Mehrgarh to employ a suite of intensification strategies that relied heavily on water management and that focused on exploiting the water resources available in rivers. This would have been particularly important for regions along the alluvial plain that today do not have sufficient rainfall for rain-fed farming (Fig. 3.3). This includes most of the Indus valley itself, which receives less than 300 mm of winter rainfall a year and less than 100 mm in the summer, with 5–10 mm/day in the summer (Bookhagen, 2010). Other areas, such as the Punjab, receive substantially more precipitation during the summer at present (> 500 mm) and may have been able to rely on more rain-fed farming (Figure 3.3).

Miller (2006a) outlines several agricultural water supply systems that may have been employed at urban phase Harappan sites. These include (1) a continuation of rain-fed methods of farming employed during earlier periods; (2) riverine inundation during summer monsoon floods; (3) small-scale canal irrigation to extend the area covered by river inundations; and (4) well or lift irrigation.

(A) Riverine inundation: The inhabitants of the ancient Indus and its tributaries likely cultivated areas that are close to the river and heavily inundated on a regular basis (Miller, 2006a). The Indus River and its tributaries are not deeply incised into the floodplain and regularly flood beyond the boundaries of its modern banks. In addition to providing water, riverbank overflooding also replenishes soil nitrogen and mineral content. Throughout history, humans likely took advantage of these regular flooding events. They could however be both beneficial and dangerous: floods that arrived at regular times of the year and prior to the planting of crops would have been beneficial. A strong summer monsoon would help with such regular supply. However, floods that arrived at irregular times and those of too high a magnitude, particularly those that took place after crops were planted, could destroy an entire year's harvest. Rivers can also shift to a new bed via the process of avulsion that can devastate not only an entire year's crop but permanently shift the area available for farming, sometimes to places far beyond where settlements are located.

The idea that the Harappan civilization might have been sustained by a larger early river on the edge of the Thar Desert has been proposed and often linked to the legends of the Saraswati River. A number of the largest sites linked to the Harappan civilization lie within modern-day Pakistan, and in particular a cluster of sites lie immediately north of the Thar Desert. These correspond to the early and mature phases of this culture around 5,200–4,500 and 4,500–3,900 years BP (Figure 3.11A) (Mughal, 1990; 1992). The discovery of these settlements is significant as it demonstrates that humans were able to occupy a region (the Cholistan desert) that today is only sparsely inhabited and does not have sufficient rainfall for agriculture and is primarily exploited by nomadic pastoralists. Initial indications of the presence of an ancient river that may correspond to the Saraswati were revealed by the studies of Saini et al. (2009) who used a series of boreholes spread across the plains of northwest India to identify the presence of channels in the subsurface. In particular, we note that there is evidence for a drainage system that appears to have flowed from the northeast toward the west and southwest, much like the modern rivers. Use of optically stimulated luminescence (OSL) dating allowed these authors to constrain the timing at which these channels were active to be between 20 and 30 ka, which is a time associated with relatively wet conditions prior to the Last Glacial Maximum (Herzschuh, 2006). More recently similar ages were derived from channels closer to the Himalaya and sediment provenance tracing methods were able to tie this to the Sutlej River (Singh et al., 2017), which could have formed part of the wider Saraswati River system.

The relatively old age of the channels means that they are far too old to be the rivers mentioned in the Vedic books, or to correspond to the archaeological settlement of the area that was substantially later, although Saini et al. (2009) did suggest that the Vedic Saraswati channel may be preserved within relic landforms rather than in the subsurface, much as originally proposed by Stein (1942). Because of the limited area that could be addressed in this particular study it was not possible to see whether the buried channel is connected with or related to the Ghaggar-Hakra, or how far toward the southwest the channels flowed. The more recent work by Singh et al. (2017) does however argue that the Sutlej did connect with the Ghaggar-Hakra but was diverted into its present location shortly after 8,000 years BP, implying little water discharge along the river during the time of the Mature Harappan.

Further investigation of the potential Vedic water system and its links to the Harappan settlements in Cholistan was performed by studies of these same channels further toward the southwest (i.e., downstream) in what is now Pakistan (Alizai et al., 2011b; Clift et al., 2012). A series of drill sites were designed to target the floodplains in Cholistan, close to the border between India and Pakistan. Giosan et al. (2012) showed that the channels remained active until at least around 4,500 years BP, although they were unable to determine exactly when the largest channels were active, broadly constraining them between around 49 ka and 4.5 ka. However, lack of incision of the Ghaggar-Hakra into the early Holocene floodplain argues for no major continuous flow after 8,000 years BP because the other Indus tributaries are incised after that time. This >8,000 years BP date is still well before the start of the Harappan settlements. Samples taken from the major channel sands were used to determine their source and compared with data from the modern rivers in the region to assess possible interrelationships and connections. Figure 3.12 compares the crystallization ages of zircon sand grains, measured by U-Pb methods taken from the major rivers in the Punjab, as well as the Yamuna River and Thar Desert, with those taken from the different boreholes considered in that study (Clift et al., 2012).

The U-Pb age of a zircon grain tells us when the crystal first cooled below 750°C (Hodges, 2003), and because zircons are very robust crystals they can transport a long distance while preserving evidence of their origin. Use of zircons to determine the source of sediments is a widespread method in modern sedimentary geology, which relies on the contrasting age of formation of the different bedrocks from which the rivers are deriving their load. Because each river has different rocks in their headwaters, they have relatively unique signatures that can be identified further downstream. Typically, a source or river will contain a mixture of grains with different ages, not just one or two, but it is the relative balance of these different age groups that provides the information we need to see where the sediment originally came from. Not surprisingly, there is a lot of similarity between ancient river sands from the north of the Thar Desert, as well as the modern streams, but some important differences emerge. If the drainage pattern had remained approximately stable in the recent past, then sands recovered from the Cholistan region should resemble those rivers now closest to the drill sites (i.e., the Sutlej and the Ghaggar), much as those from the Ghaggar channel in northwest India do (Singh et al., 2017).

Zircon age populations from Holocene sands from Marot, Fort Abbas, and Tilwalla (Figure 3.11) have major differences with sands carried today by both the Sutlej and the Ghaggar. Sands from Tilwalla whose sedimentation was roughly dated to between 43.6 and 4.9 ka closely resemble sands in the Beas River where 300–750 Ma zircon grains make up 61% of the load (Alizai et al., 2011a). Today the Beas River joins the Sutlej much closer to the Himalaya and would be incapable of providing such sediment to the Cholistan region, which requires a rather different

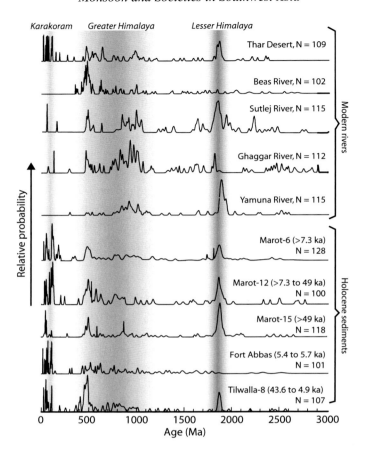

Figure 3.12 Probability density plots of U-Pb ages of zircon grains in sands from modern rivers, sand dunes, and the mid-Holocene sediments sampled from the floodplain. Note the poor match between the sands at Tiwalla, Marot, and Fort Abbas compared to the Sutlej or Ghaggar. Plotted with data from Clift et al. (2012).

drainage system in the past. Most intriguingly, the oldest sediment taken from the Marot borehole, which was dated to be older than 49 ka, showed some similarity with sediments in the modern Yamuna River. The oldest sample has consistently older zircon grains than those in the shallower part of the same location. While it contrasts with the younger sands at that location, it is statistically indistinguishable from the sands now carried by the Yamuna River and is not the product of either sedimentation from the Sutlej or reworking from the dunes of the Thar Desert.

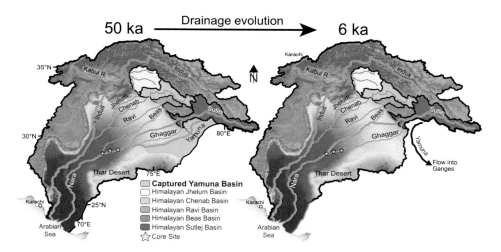

Figure 3.13 Reconstruction or drainage capture in the Indus and Ganga basins since 50,000 years ago based on the study of Clift et al. (2012). Note the loss of the Yamuna toward the east after 50,000 years ago, the capture of the Beas into the Sutlej, and the migration of the Sutlej to its modern northerly position. The diversion of the Ghaggar away from the Nara valley is poorly constrained but could be a relatively young development in the basin's evolution. Figure by Sam Van Laningham.

Thus, as well as the flow from the Sutlej highlighted by Singh (2017) there is evidence that in the deeper past a giant river existed in the Cholistan region. Until around 50,000 years BP a west-flowing Yamuna must have converged with the Sutlej and the Ghaggar in Cholistan and finally with an independent Beas River in Western Cholistan (Figure 3.13). The Beas and Sutlej Rivers began to flow together and then migrated toward the west before 5,400–5,700 years BP, likely before 8,000 years BP, leaving the regions later settled by the Early and Mature Harappan isolated and without significant water supply. Clift et al. (2012) noted that the very large river flowing in Cholistan predicted by the Saraswati legend was too old to be consistent with it being active during Harappan times. This means that the relationship between the Harappan settlements and the rivers is not closely linked to large-scale channel migration but instead to the variation in discharge in a more stable network.

A more subtle relationship with the rivers was revealed by the study of the geomorphology of river systems in this region (Giosan et al., 2012). That study noted the fact that most of the modern Punjabi rivers are now contained in wide but relatively shallow (10–15 m deep) valleys that cut into a former, now-abandoned floodplain surface constrained broadly to have been abandoned around 8–10 ka. Crucially, the Ghaggar-Hakra does not flow in such a valley, implying that its significant flow must have predated 10 ka. Incision of these valleys mostly took place prior to 5 ka because increasing aridification of the region, as shown by the lake and speleothem records discussed at the beginning of this chapter, has choked off significant flow in the rivers closer to the Thar Desert. Giosan et al. (2012) argued that agriculture flourished in this region during the Early and Mature Harappan periods because of seasonal inundation and rain-based farming while the monsoon was stronger. Thus, the Ghaggar-Hakra has never been a wide flowing river during the time of human settlement, starting 5,300 years BP with the Early Harappan, but rather was only a system of seasonally meandering channels that were fed by monsoon rain (Courty, 1995). It seems that these channels were exploited using small-scale canals to feed crops. Summer rainfall was however strong for the early part of the Harappan culture and only declined sharply after 4,200 years BP, during the Mature Harappan time, as summarized earlier.

Because the center of development north of the Thar Desert was only being supplied by the Ghaggar-Hakra, these settlements became dependent on the seasonal floods driven by monsoon rains. Although the Harappan centers developed during a period when the monsoon had been slowly weakening, it is suggested that the need to adapt to the increasing aridity contributed to the development of a complex society and to urbanization (Madella and Fuller, 2006). It was only as the climate dried even further, after 4,200 years BP, that a critical threshold was reached and the seasonal inundation farming was no longer sustainable (Dixit et al., 2014). Prior to that time the inhabitants of the Indus valley likely invested in a range of other ways of risk buffering against the risk inherent in both rain-fed and riverine inundation-based farming.

(B) Small-Scale and Well Irrigation: We know that within cities Harappans had access to highly developed water management technology. By the urban phase (4,600–3,900 years BP), highly adapted systems of water management appear to have already emerged in the large cities (20–50,000 people) that dotted the alluvial plains during this period of time (Wright et al., 2008). In addition to what may be (although their function is still highly debated) centralized structures for grain storage, impressive water management facilities were present at these sites. Ancient Harappans engineered toilets, bathing platforms, and elaborate sewage plumbing to catch the waste that was derived from these systems (Wright, 2010). Sewage systems that ran throughout the cities included contained cesspits that could be used to avoid blockage and overflooding. Harappans brought water into the cities using vertical shaft wells that were lined with brick (Wright, 2010). Water-lifting devices have existed in the region ethnographically; however, it is unclear if these were used during the Harappan period (Miller, 2006a). Evidence for sluices is also present in Harappan cities indicating that they had mastered the art of gravity control (Miller, 2006a). However, Miller (2006a) argues that wells were only infrequently used for agriculture, because other natural stands of water such as oxbow lakes were available. At Dholavira, Wright (2010) described a water management system that "included a dam constructed of blocks of stone used to retain water. The water was then channeled into terraced reservoirs that bank an outer fortification wall of the city." Finally, the inhabitants of these sites employed the extensive creation of mudbrick platforms, which Wright (2010) argues were used to protect buildings in the cities from riverine overbank flooding.

It is likely that Harappans also employed similar technology to create cultivated areas farther from rivers using natural and artificial channels and use of wells. Canals that extended the range of the inundation zone have been described in British accounts from the eighteenth century (Habib, 1999; Leshnik, 1973; Miller, 2006a). These canals required relatively little investment and used gravity to move water from the inundation zone. Miller (2006a) notes that we have no evidence for large-scale canals within the Indus basin and it is possible that the frequently shifting nature of the rivers may have made such an investment not worthwhile. For smaller-scale canals, erosion as well as modern farming practices with deeper tilling could have easily obliterated the signal of these early investments in water management and it is not surprising that little trace of them remains on the landscape today (Wright et al., 2008).

(C) Exploitation of Depressions, Oxbow Lakes, and Marshes: The margins of lakes and depressions may have been used seasonally for planting, but also as important areas for the collection of wild food and animal resources. For instance, a study by Belcher (2003) found that inhabitants of many sites in the region relied heavily on resources such as catfish caught in these environments. It is possible that aquatic foods could have been grown or foraged within these areas or that water could have been diverted from these features into smaller fields adjacent to them. The site of Harappa itself is located next to an oxbow lake (Belcher and Belcher, 2000). The initial settlement appears to have been on a point bar within a meander loop that then experienced cutoff, oxbow lake formation and the infilling of the channel, resulting in fertile alluvial plains (Schuldenrein et al., 2007).

(D) Trade: Trade played a crucial role in provisioning urban-period Harappan settlements and in buffering Harappans against risks. For instance, Belcher (2003) found that in addition to catfish that was likely locally caught in marshes and oxbow lakes, a small percentage of the fish unearthed at Harappa came from maritime environments that likely arrived at the site through the complex networks of trade that we know also brought lapis lazuli, carnelian, steatite, copper, and other stone-ware into the sites. We know that such trade also moved plant products across the Indus region. For instance, Himalayan species of tree such as elm and deodar were found in coffins at Harappa and transported long distances to the site (Madella and Fuller, 2006). Although little research has been dedicated to revealing these patterns, it is possible that a variety of the crops found in the Harappan settlements may also have been derived via interregional and long-distance trade (see later) from areas with higher rainfall such as the Punjab, buffering the Harappan ability to withstand variability within their own harvests.

(E) Diversification: Indus farmers may have reduced their agricultural risk through diversification of the crops they employed. The suite of crops initially grown at sites in Baluchistan such as Mehrgarh represent Rabi (or winter-planted) crops. These include wheat, barley, and the suite of southwest Asian legumes such as pea and lentil (Weber, 1999). The earliest types of wheat and barley that were domesticated in the Fertile Crescent were winter crops (Zohary et al., 2012), although during later periods of time, spring varieties of these crops also arose: potentially out of selective pressure to adapt to cooler climates during their spread throughout Central Asia (Jones et al., 2011a).

The amounts of water required by different varieties of winter wheat and barley are highly variable. In addition, the total amount of rainfall required by crops can be misleading as evapotranspiration can increase with warmer temperatures and higher wind. For instance, for winter wheat, the Food and Agriculture Organization of the United Nations (FAO) recommends between 450 and 650 mm for high-yielding hybrid varieties. However, for varieties from Baluchistan, Ashraf and Majeed (2006) note that minimum values could be as low as 255 mm, although their range extends to 777 mm. In the Indus basin, values of required water for wheat are listed as 271–515 mm (Kaleemullah, 2001).

Winter varieties of barley require less water than wheat does. For instance, Oram and De Haan (1995) reported that varieties cultivated in Syria can achieve maturity on as little as 200–350 mm of water. Today, both wheat and barley that are grown in the region are least partially irrigated. As Figure 3.3D shows, it would have been difficult for farmers to have relied on rainfall alone for large parts of the Harappan cultural sphere and it would have been necessary for them to develop the systems of irrigation we have described earlier in order to water their crops as they moved onto the plains of the Indus (You, 2019). While there is sufficient rainfall today to support these crops in the northern part of Baluchistan (a place where Aceramic settlements like Mehrgarh are located), it would have been crucial for the urban Harappan to master different forms of irrigation in order to be able to grow these crops.

During the Mature Harappan urban period (2600–1900 BCE), we see either the domestication or the arrival of new suites of crops that are summer planted (known as Kharif crops) including millets, rice, and other types of pulses in various parts of the Harappan cultural sphere (Weber, 1989; 1999). South Asia is home to a wide variety of millets that were either introduced or locally domesticated and that were cultivated during the Kharif or summer season. Some varieties of millet cultivated during the Kharif season can tolerate high temperatures and low water conditions, making them highly adapted to arid conditions. For instance, foxtail millet can require as little as 250–300 mm of rainfall (Saseendran et al., 2009), finger millet a minimum of 290 mm (Duke, 1983), broomcorn millet as low as 350 mm (Baltensperger, 2002), and pearl millet can be grown in low fertility soils that receive as little as 150 mm of rainfall a year (Harinarayana, 1987). There appears to be substantial regional variation in how these different crops were employed. For instance, at Rojdi in Gujarat (an area with higher rainfall), Kharif crops such as millets dominated the assemblage; however, at the more arid site of Harappa in Punjab, Rabi crops such as barley were dominant (Weber, 1989). Kharif crops thus appear to dominate areas of the Indus with higher rainfall and appear to have been primarily rain-fed.

A key question in research in the area has been whether or not there was an increase in the use of drought-tolerant Kharif crops in Harappan urban core areas such as the Punjab that corresponded to decreasing rainfall trends we have noted in Section 3.2. It is unclear for instance if the use of these crops represents an instance of multi-cropping to reduce risk at individual sites, or if these simply reflect regional differences in cropping strategies (Madella and Fuller, 2006; Petrie and Bates, 2017; Weber, 2003).

Indeed, Weber et al. (2010c) have argued that Harappan agriculture was uniquely adapted to different agro-ecological zones. They compared the ecology surrounding the sites of Lothal, Harappa, and Mohenjo Daro and noted the following patterns. Out of the three sites, Mohenjo Daro receives the lowest amounts of rainfall at only 76 mm, Harappa receives 280 mm, whereas Lothal receives 772 mm. They also have access to different sizes of river and river flow. For instance, Mohenjo Daro is located along the large Indus River valley, whereas Harappa is located close to the Ravi. Both of these rivers must have supplied substantial alluvial deposits and water to farmers. Lothal on the other hand likely had much less access to riverine irrigation and likely relied on the rain from seasonal monsoons. Mohenjo Daro and Harappa appear to have been primarily reliant on Rabi crops, which were presumably irrigated. However, at Rojdi, a site close to Lothal (Figure 3.11), Kharif crops played a more important role, although Rabi crops were still present (Weber et al., 2010c). It thus appears that Indus farmers developed highly adaptive systems that allowed them to best exploit the different types of water resources they had available to them.

It is finally worth noting that ancient farmers in the Indus catchment did not rely on grain crops alone and cultivated a range of other foods that increased their food web and resilience to changing environmental conditions. A wide range of legumes were also domesticated and cultivated in South Asia. These crops played and continue to play an important role in maintaining soil fertility as a rotational crop through their ability to fix nitrogen in their roots. Legumes domesticated in South Asia include horse gram (*Macrotyloma uniflorum*), mung bean (*Vigna radiata*), pigeon pea (*Cajanus cajan*), and the urad bean (*Vigna mungo*). Introduced domesticates include pea (*Pisum* sp.), lentil (*Lens* sp.), Hyacinth bean (*Lablab purpureus*), and grasspea (*Lathyrus sativus*). In addition to pulses, a wide variety of fruits and textile crops were also critical cultivars. These include dates (*Phoenix dactilifera*), cotton (*Gossypium* sp.), jujube (*Ziziphus* sp.), and grape (*Vitis vinifera*) (Weber, 1991). Some phytolith studies also indicate that crops such as banana (*Musa* sp.) may also have been cultivated (Madella, 2003). Other crops such as different sets of greens and even eggplants (*Solanum melongena*) may have been grown in the region. The greens of some weeds of agricultural crops, such as alfalfa (*Medicago* sp.), were consumed by both humans and animals. Likewise, we know that pastoral animals played an extremely important role at Indus civilization sites both as sources of food and also as fuel through the burning of their dung.

3.5.3 Changes in Settlement and Subsistence during the Late Harappan (or Regionalization Era)

During the Late Harappan (1900–1000 BCE), we see major changes in settlement patterns and social structure taking place in Harappan societies. Generally, we see an abandonment of (or less investment in the upkeep of) large urban settlements with evidence of central planning and the loss of a use of a centralized system of weights and scripts. We also see a decline in the very long-distance networks of trade that stretched as far as Mesopotamia and that characterized the urban period (Kenoyer, 1998). However, as Coningham and Young (2015) note, at the same time, we see the reassertion of a series of regional traditions that had begun to emerge during earlier periods of time: for these reasons, the post-urban Harappan is often known as the regionalization era.

In the Quaternary sciences, climate change, or more specifically a drying trend during the second millennium BCE (after 4,200 years BP), has often been invoked as a culprit for the "collapse" of the Indus civilizations (Giosan et al., 2012; Shinde et al., 2006; Staubwasser et al., 2003; Staubwasser and Weiss, 2006). Archaeologists see this picture differently and have argued that people did not simply abandon the area and that Harappan society did not fall apart. Rather, Harappans engaged in a series of shifts in subsistence regimes and settlement patterns that indicate people were adjusting to a new stable state (Coningham and Young 2015; Kenoyer, 1998).

Data from the large city site of Harappa show a heavy reliance on barley (Weber, 2003), particularly hulled varieties, during the urban phase with smaller amounts of wheat, pea, and lentil. Millets (*Setaria* sp.) are present in the assemblage, however only in limited contexts, suggesting that they played a secondary role. The higher emphasis on barley at the site itself may have been an attempt by the inhabitants to use a crop that increased their resilience by requiring less water input (Weber, 2003; Weber et al., 2010c). Weber (1999; 2003; 1989; 2010c), who floated over 10,000 liters of soil from the site, notes an increase in the ubiquity of Kharif crops during the Late Harappan period (3,800–3,700 years BP). He argued that this may represent an attempt to deal with decreasing precipitation during this period of time. He argues, that an increase of Kharif crops was, however, gradual and not sudden and may represent an attempt by people to widen their food webs over time. We start to see, for instance, a change in the types of diversity of crops employed at different sites throughout the region during this period of time, particularly at ones that are continuously occupied. For instance, at Rojdi, Weber (1999) found that there was a shift from finger (*Eleusine coracana*) and little millet (*Panicum sumatrense*) toward foxtail (*Setaria* sp.) types of millet. Weber (1999) argues that plants were both used more intensively by planting more variety across a range of seasons. Likewise, Madella and Fuller (2006) argue that these changes in farming strategy represent a "cultural adjustment in terms of subsistence strategies towards more drought tolerant rain-fed crops and increasingly reliance on two cropping seasons." Reddy (1997; 2003) sees similar patterns in her analysis of material from Oriyo Timbo. However, she argues that these are likely the remains of animal dung burned at the site and as a result they represent an increasing investment in pastoralism. Some of the changes in settlement location we discuss subsequently during this period of time may indeed be related to increasing investments in pastoralism. Fuller and Madella (2001) also claim to see a shift in crop-processing strategies from the Urban to the Late Harappan from larger, centralized processing to more household-based crop-processing strategies, suggesting that profound changes in social structure had also taken place.

The location of settlements also appears to shift toward areas with average higher rainfall during the Late Harappan time. This shift in settlements in the post-urban phase may simply have been because river-fed Rabi farming proved difficult in the Harappan core region and people began to move in higher densities to new ecological niches that had a history of successful Kharif methods of farming (Giosan et al., 2018; Weber et al., 2010c).

It has been suggested that the shifting courses of rivers may have played a much more important role in changing settlement patterns than lower rainfall, although as described earlier it is clear that the largest changes in the drainage systems predate the Harappan. Rather, changing discharge may have been more important (Giosan et al., 2012). Shifting river courses has figured prominently into explanations for changes in settlement patterns for a long time. Harappa itself was founded on a meander bend of the Ravi River but this now lies 7 km away to the northwest because of channel migration. Marshall (1931) first noted that changing river courses likely impacted the lives of those living in the Indus valley and these ideas were further explored by Dales (1964) who argued that changing river courses and sea level rise led to site abandonment. Rao (1973) also noted evidence of flooding at the site of Lothal that led the inhabitants to raise platforms and ultimately abandon the settlement.

Flam (1981) proposed that the Indus River, as well as its tributaries, had shifted courses several times. These shifts in river course may have resulted in several substantial shifts in settlement pattern, although the largest-scale avulsion predate urban settlement. Nonetheless, the rivers would have migrated across their floodplains due to meander loop migration, and locally the river course could jump several kilometers in just a few days when a meander loop experienced cutoff. In Cholistan, several major changes are seen in regional settlement patterns (Mughal, 1997). There are changes in the number of sites. For instance, during the urban period 174 sites were occupied, while this number drops to total of 51 during the regionalization period (Figure 3.11). Many of these sites also changed location and Mughal (1997) found that only one site shared a similar location to that of the urban period. Mughal (1997) argues that this may represent a shift in the types of subsistence strategies. Generally, these later sites move away from the floodplain and into more elevated and drier areas of land (particularly on the Ghaggar-Hakra interfluve) suggesting that people had moved away from agricultural strategies toward ones that were more focused on pastoralism (Mughal, 1997). Because these changes occurred as the Ghaggar-Hakra transformed from being continuous flow to seasonal, these migrations must represent an abandonment of the river as a source of reliable water in this region.

Other changes in regional settlement patterns also took place during this period of time. During the urban phase, large numbers of specialist sites where people focused on craft production dotted the landscape. However, these are abandoned during the regionalization era and people begin to occupy sites that served multiple functions (Mughal, 1997).

Giosan et al. (2018) argue that the ENA (4,500–3,000 years BP) "may have helped trigger the metamorphosis of the Urban Harappan civilization into a rural society through a push-pull migration from summer flood-deficient river valleys to the Himalayan piedmont plains with augmented winter rains." Flood regimes are controlled to some extent by precipitation, and in particular, the declining summer monsoon between 4,500 and 3,000 years BP may have made agriculture based on inundation less reliable (Durcan et al., 2019; Giosan et al., 2012). Giosan et al. (2018) note that following this period of time the bulk of Harappan settlements moved back toward the Himalayan foothills: areas with higher rainfall located on the upper Ghaggar-Hakra interfluve. Drying in peninsular India also explains the sharp decrease in population sites in the Gujarat region, making it a poor substitute for the core Harappan area (Figure 3.11B).

Finally, changes in the monsoon itself may have impacted other crucial aspects of Harappan civilization: trade with areas such as Mesopotamia. Mesopotamian texts refer to a place known as "Meluhha" with which they conducted maritime trade, receiving items such as lapis lazuli, gold, and a variety of precious stones including carnelian (of which they are described as being the primary exporter) (Wright, 2010). It is possible that changes in monsoonal patterns may have disrupted the wind patterns that were crucial to the trading boats that arrived at Harappan sites and as a result may have adversely affected the highly developed craft industry that flourished at many of these sites (Dixit et al., 2014; Kenoyer, 1997).

3.6 The Historic Period

The best historic climate record for the South Asian monsoon available at the present time comes from southern Oman from the location of Defore Cave (Figure 3.14) where a speleothem record is believed to record the southwest monsoon, which affects much of the Indian subcontinent (Fleitmann et al., 2003). Even though Oman is a long way from much of India this region is still influenced by the same winds and at present there is a lack of good alternative speleothems in southwest Asia.

Figure 3.15 shows the varying strength of the southwest monsoon compared to major episodes in the history of South Asia since the year 1200 CE. No straightforward correlations can be made between rainfall intensity and historic events, even though some have argued for an increase in rainfall over the centennial scale until 1700 CE (Figure 3.15) coincided with the increasing size and wealth of the Mughal Empire, which reached its cultural pinnacle with the Emperor Aurangzeb (1659–1707)(Truschke, 2017).

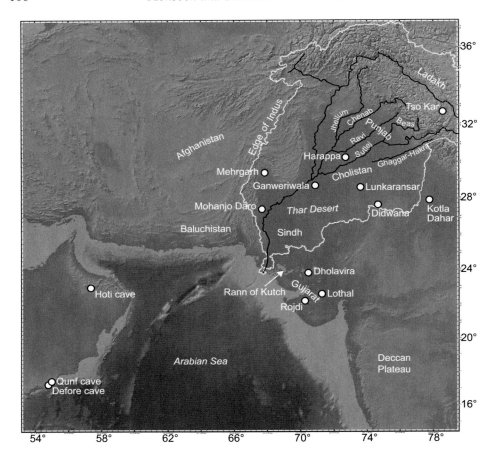

Figure 3.14 Map of the South Asia region showing the places named in this chapter.

Further information concerning the monsoon climate of South Asia has been derived from high-resolution records derived from ice cores in the Himalaya. In particular, that collected at Dasuopu (Figure 3.14) has provided a long-term history dating back to around 1440 CE (Thompson et al., 2000). The ice cores provide a relatively easily dated and detailed record of preserved oxygen isotope values. However, unlike the speleothems it appears that the oxygen isotopic composition of the ice is largely a reflection of temperature rather than rainfall intensity (Figure 3.16). The ice also collects both aerosol dust and other trace chemicals, including chloride, that can be used to track the amount of dust in the atmosphere and the amount of precipitation as reflected in chloride concentrations, which are reduced when rainfall is heavy. Thompson et al. (2000) concluded, perhaps not

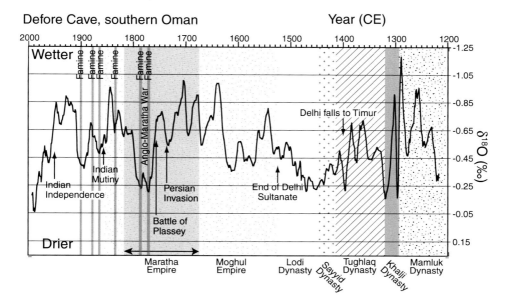

Figure 3.15 Monsoon records from Defore Cave in southern Oman, considered a proxy for the strength of rains for the southwest summer monsoon in the Arabian Sea (Fleitmann et al., 2003), smoothed over 8-year cycles. Record is compared with major historical events and periods in South Asia.

surprisingly, that times of low rainfall tended to be ones marked with increased dustiness and high concentrations of chloride. The increasing $\delta^{18}O$ values in the twentieth century indicate warmer atmospheric conditions rather than substantially weaker rainfall over that time period.

The ice core record also shows especially severe drought events, namely in 1790–1796 and in 1876–1877, confirming what is stated by historical records. The ice core records also pointed to less dramatic, but still statistically significant, reductions in monsoon rainfall during the 1640s, 1590s, 1530s, 1330s, 1280s, and 1230s. However, they were able to confirm that none of these earlier events had the severity of those seen in the late eighteenth and nineteenth centuries (Thompson et al., 2000). Some authors have tried to link these weakened monsoonal rains to a series of historic famines (Arnold, 1999; Bayly, 1988; Stokes, 1975). The first well-documented famine of the modern era occurred in 1769–1770 (Arnold, 1999). The "Bengal Famine" (1769–1773) particularly affected northeast India and the regions of Bengal that were cultivated by the British East India Company. Reference to the speleothem record shows that this was also a time of extraordinarily weak monsoons spanning several years. Another famine followed in

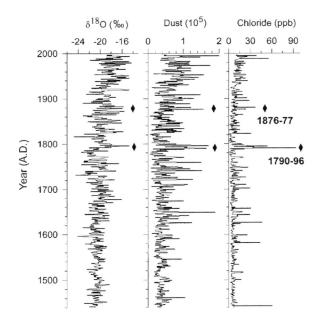

Figure 3.16 Annual averages of $\delta^{18}O$, dust, and chloride are shown from 1440 to 2000 CE in ice cores from Dasuopu, Tibet. Dust concentrations are per milliliter sample for particles with diameters >0.63 μm. Diamonds indicate the two largest historically documented monsoon failures. Figure reproduced with permission from AAAS from Thompson et al. (2000).

1783–1784, which may have been linked strong El Niño that year. That famine resulted in substantial loss of population across the area amounting to more than 30% of the population around Delhi (Stokes, 1975) and perhaps more than 60% depopulation of parts of Uttar Pradesh in northern India, and likely totaling more than 11 million deaths (Bayly, 1988).

However, a closer look at the historic record reveals that while reduced rainfall may have played a role, agricultural returns were actually high during the years of the famine (Sen, 1977). Rather, a draconian tax regime enforced by the British East India Company and speculation on grain prices caused by the new market economy in the region meant that local buyers were unable to access grain that had been produced in the region for their own consumption. For instance, Shiva (2016) notes that while in 1750, for every 1,000 units of produce, the producer paid 300, by 1830, 650 were collected, eroding farmer autonomy and ability to sell their crops for profit and or even keep enough for their own consumption. This coincided with a push by colonial governments for individuals to move away from cultivating smaller grained crops such as millet and to focus on crops

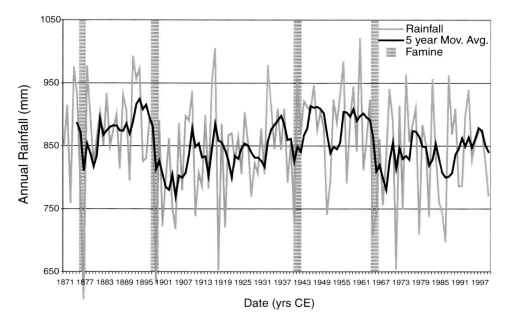

Figure 3.17 All India rainfall since 1871 showing the smoothed 5-year running average that defines cycles of higher and lower average precipitation across South Asia. Data from the Indian Institute of Tropical Meteorology (www.tropmet.res.in).

with international trade values such as wheat and barley. A focus by the colonial government on expanding poppy cultivation to produce opium to sell to China also exacerbated this crisis.

Famines continued to affect India through the nineteenth century with particularly bad ones recorded in 1837–1838, 1860–1861, 1865–1867, 1868–1870, 1876–1878, 1896–1897, and 1899–1900 (Davis, 2002)(Figure 3.17). Of these the most damaging were those in 1876–1878, also known as the Great Famine in which 5.5 million were known to have died in British-administered areas and the total death toll at somewhere between 6.1 and 10.3 million when also accounting for those living in the Princely States, outside the direct colonial rule. The Great Famine was more concentrated in the south and particularly affected regions around Chennai (Madras) and Mumbai (Bombay). The two famines right at the end of the nineteenth century accounted for the deaths of another 5 million people within the British-controlled territory in 1896–1897 and an additional 1 million in the succeeding famine (Ghose, 1982). These two events particularly affected northern and central India in the earlier episode and more toward the south in the later one. Although the famine in the Madras Presidency, as the district was called at that time, was preceded by a drought, it was made more acute by the British colonial policy of allowing continued trading and export of grain. Maintenance by the government of buffer stocks of grain made the population less resilient to climate change because they had little reserve of their own to fall back on. In two of the worst famine-afflicted areas in the Madras Presidency, grains continued to be exported throughout the famine period (Sen, 1977).

Far more than variations in the monsoon, it was the move away from traditional crops, the loss of farmer autonomy, and the presence of a destructive market economy that wrecked the greatest loss of life in recent history in South Asia.

3.7 Synthesis and Summary

Throughout the several millennia of data we review here, societies that arose in highly arid southwest Asia, most notably the Harappan, employed a wide variety of different strategies to mitigate fluctuating rainfall and river load. These strategies included increasing mobility or shifting settlement to areas with higher rainfall, diversification through use of different crops at one site, or increasing investment in pastoralism. Finally, a wide variety of small-scale irrigation systems were likely employed that were uniquely attuned to their local environment. It appears that weakening summer rainfall coupled with changes in river course led people to adapt and to move away from Urban Harappan centers, to forms of subsistence that focused on rainfed crops in areas with higher rainfall and pastoralism in more arid regions. Nonetheless, some of the strategies employed by farmers may represent useful adaptations for our future. In particular, arid adapted millets cultivated in the past may provide some of the best mitigation strategies for decreased or hard-to-predict rainfall in the future. However, the area under cultivation of these crops and genetic diversity of these crops is decreasing around the world due to the promotion of high-yielding varieties of wheat, maize, and rice developed under the Green Revolution. In South Asia, farmers have been under huge pressure to shift to these higher-yielding varieties but may, at the same time, be decreasing their long-term resilience to changing climate. Indeed today many of these once-important crops are seen as targets of elimination as they can act as weeds of other major cultivars like rice or wheat (Awan et al., 2014). We can see that in the past, humans in South Asia relied on complex systems of cultivation involving a wide variety of grain crops, which provided food security under the context of millennia of variations in the monsoon. Today, this food security is being eroded as the farmers are asked to focus on varieties that are adapted to mono-cropped and mechanized agricultural systems (see Chapter 6).

While anecdotal evidence shows that other millets are highly adapted to arid conditions very few field trials have been carried out on many of these varieties and, as a result, we are unsure of their ultimate minimum water requirements (Upadhyaya et al., 2016). Humanity may be losing out on crucial resources for food security in arid lands by not documenting these crops.

4

Origins of a Uniquely Adaptive Farming System:

Rice Farming Systems in Monsoon Asia

4.1 Introduction

Rice farming in Asia constitutes some of the most intensive farming systems on the planet (Netting, 1993). These systems have supported, and continue to support, some of the highest population densities in the world. The farmers who domesticated and grew rice throughout history developed expert knowledge that successfully allowed them to operate farming systems that are the most productive in terms of calories per hectare worldwide. Throughout both the historic and prehistoric record, Asian rice farmers have developed a dazzling array of techniques to deal with the climatic challenges that we outlined in Chapter 1: temperature, rainfall, management of river load and irrigation systems, as well as ways of addressing rising sea levels.

Aside from maize, rice is one of the most productive crops on the planet. While wheat and barley produce roughly only 500 grains per plant (or per single planted seed), rice is capable of producing over 500 seeds per panicle or over 2,000 seeds per plant (Bray and Needham, 1984)(Figure 4.1). Throughout East and Southeast Asia, wet rice (or rice grown in a paddy environment) has historically been double cropped, or placed in rotation with other crops, reducing (and most often even eliminating) fallow time. Such an approach makes this system of farming possibly the most intensive system on Earth. To those familiar with European methods of farming, it may be surprising that intensive wet rice farming systems, in fact, result in surprisingly low rates of soil nutrient depletion and erosion. Rather than deplete soil fertility, wet rice farming actually improves soil quality and fertility. Asian rice farmers have throughout history practiced a diversified, organic strategy focused on recycling both internal and external resources. The use of a wide array of nitrogen-fixing organisms allows farmers to achieve rice harvests of up to 2 tons per hectare without the application of industrially produced fertilizers (Bray and Needham, 1984). These include the use of the *Azolla* fern, which grows in paddies inhibiting weed growth, then fertilizes the paddy after its death. Farmers have also intention-ally encouraged other organisms, such as carp, ducks, crabs, and shrimp, to live

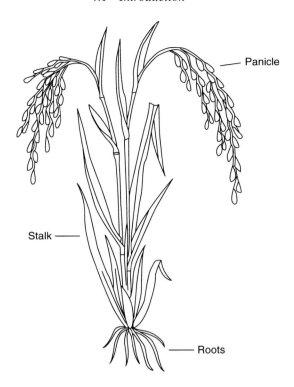

Figure 4.1 The rice plant showing structure and panicles.

within paddies that then fertilize the paddy by depositing their droppings, but who also simultaneously serve as an additional source of protein for farmers and play an important role in organic pest control, a practice that is at least 8,000 years old (Nakajima et al. 2019). Han dynasty sculptures of pigpens and toilets adjacent to fields also show that since antiquity, humans have used a combination of other fertilizers to maintain soil fertility. Netting (1993) describes how throughout Chinese history, almost all household and farm waste was expertly recycled into fertilizer, including kitchen scraps, duck, pig, silkworm, and even human excrement.

For instance, in Zhejiang Province, China (Figure 4.2), researchers found that carp fish in paddies play a variety of important roles in maintaining paddy health. As fish swim through the paddy, they bump into rice stems eating insects and pests. Overall, the presence of these fish reduced pest load by roughly 26% (Xie et al., 2011). These same fish also eat or uproot weeds, playing an important role in weed control. Xie et al. (2011) found that fish bumping into rice stalks also removed morning dew from the stalks, reducing the opportunity for fungal infections, such as rice blast disease to spread.

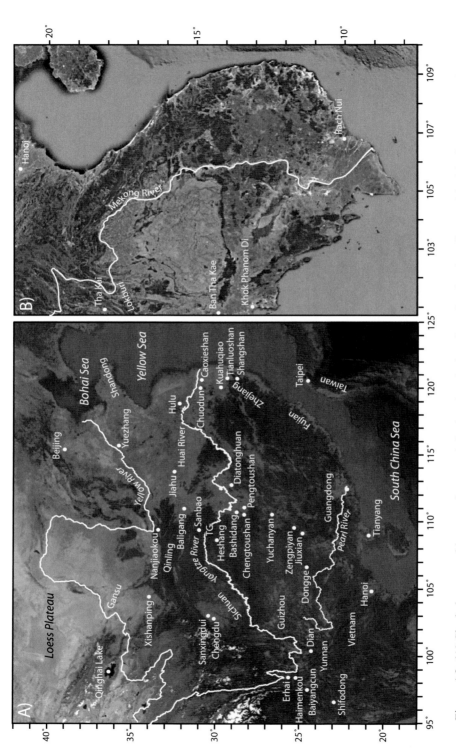

Figure 4.2 (A) Shaded topographic map of eastern Asia showing the location of major sites discussed in this chapter. TG = Three Gorges. (B) Map of Southeast Asia.

More importantly, they found that these same coupled rice–fish farming systems required 68% less pesticide than types of rice grown without this sort of ecosystem environment. Bray (1984) notes that whatever their original fertility, after several years of wet rice cultivation, fertility is increased and lasts indefinitely. Rice, of course, was only one of a few plants cultivated in rotation on these fields. In the wintertime, rice fields are planted with all sorts of vegetables and trees that bear nuts and fruits as well as mulberry trees that house silkworms for textile production (Netting, 1993).

The longevity of these systems throughout multiple episodes of changing monsoonal intensity, supporting some of the highest population densities per hectare on Earth without significant soil degradation, testifies to their resilience. In this chapter, we chart how and when rice was first domesticated and where systems of rice cultivation first emerged and spread throughout Asia. We relate this to the role that changes in monsoonal intensity may have played in the evolution of these farming systems. We examine the archaeological evidence for rice domestication, as well as the evidence for development of the wide range of different types of cultivation systems and technology that humans have employed to meet challenges represented by the changing East Asian monsoon. We first begin with a review of rice ecology and genetics.

4.2 Rice Ecology

4.2.1 Rice Genetics and Domestication

Domesticated rice (the *O. sativa* complex) is divided into two main domesticated species *O. sativa* (Asian rice) and *O. glaberrima* (African rice) (Sweeney and McCouch, 2007; Vaughan et al., 2008). Within *O. sativa*, two main species *O. japonica* and *O. indica* appear to have been separately domesticated from two different wild ancestors: the perennial *O. rufipogon* and the annual *O. nivara* whose range is confined to Southeast Asia and India (Londo et al., 2006; Sweeney and McCouch, 2007) (Figure 4.3). Fuller and Qin (2009) have argued that *Oryza sativa* subs *japonica* was likely derived from the lowland perennial ancestor *O. rufipogon* that inhabited wetlands, while *O. indica* was derived from the native rain-fed *O. nivara*. *O. rufipogon* populations are more widespread and the only wild population of rice in China, hinting that these may have been ancestral to populations of *O. japonica* (Figure 4.4).

Other differences characterize these two species/subspecies: *O. rufipogon* is not a very productive seed grower because, in wild stands, most of the plant's energy is focused into vegetative tillering. In contrast, the annual *O. nivara* is adapted to producing large numbers of seeds and grows seasonally in areas flooded by monsoons (Fuller and Qin, 2009).

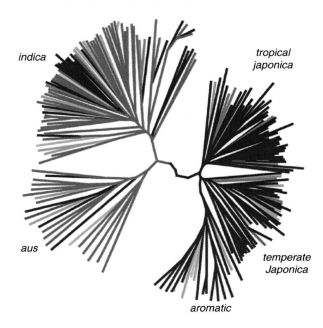

Figure 4.3 Evolutionary tree of rice, after Kovach et al. (2007).

Within *O. japonica* rices, two different varieties exist: tropical *O. japonica* (syn. *javanica*) and temperate *O. japonica* (Garris et al., 2005). Genetic research suggests that these two varieties of rice are selections from a single genetic pool that have been adapted to different ecological conditions (Sweeney and McCouch, 2007; Zhao et al., 2011). The alleles at monomorphic loci in temperate japonica are identical in size to tropical japonica, leading one to believe that temperate japonica rices may have been derived from an ancestral variety of tropical japonica (Garris et al., 2005).

Most wild rices are essentially short-day plants, meaning that they require relatively long periods of uninterrupted darkness to flower. The optimum photoperiod (or day length) of most rice varieties is 9–10 hours (Yoshida 1981). Photoperiod in the tropics varies only by a matter of minutes. However, in the most northern limits of rice cultivation, a six-hour difference in day length can occur. Tropical plants, including tropical varieties of *O. japonica*, are extremely sensitive to variations in day length. For instance, Malaysian cultivars of rice exhibit sensitivities to a day-length difference of only 14 minutes, taking 329 days to flower when planted in the winter but only 161 when planted in fall when day length was at its longest (Dore, 1959). Temperate japonica rices are relatively insensitive to photoperiod (Zhao et al., 2011), whereas tropical varieties are highly sensitive to variations in day length. In

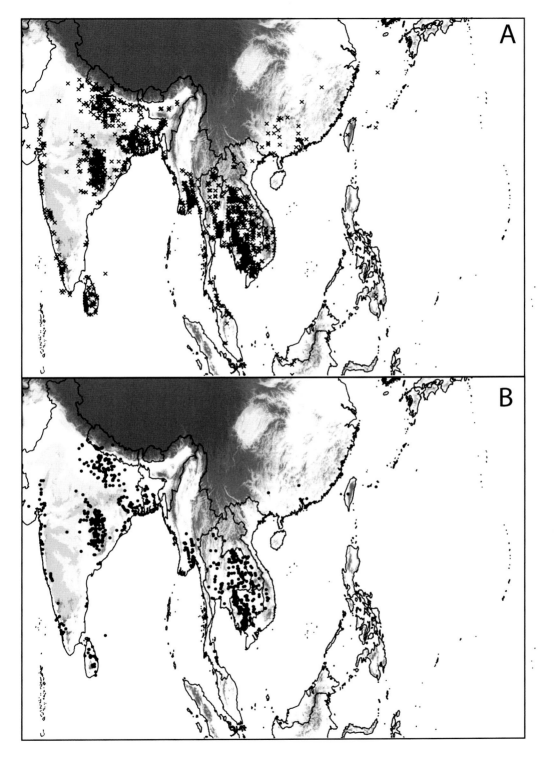

Figure 4.4 (A) Map of wild stands of *O. rufipogon* from GBIF.org (June 13, 2019) GBIF Occurrence Download https://doi.org/10.15468/dl.e6h4up (B) Wild stands of *O. nivara* from GBIF.org (June 13, 2019) GBIF Occurrence Download https://doi.org/10.15468/dl.6yfgdk.

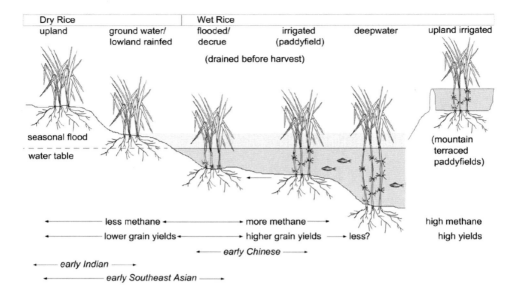

Figure 4.5 Rice ecology. Reproduced with permission from Fuller et al. (2011).

temperate varieties of rice, this is controlled by a mutation (*DTH-A4*) that is absent in wild varieties of rice, suggesting that it evolved later in time (Wu et al., 2012b). It is expected that as rice agriculture moved northward and temperate varieties of rice evolved, a high selection bias would have been placed on varieties that were less sensitive to photoperiod.

4.2.2 Rice, Temperature, and Water Requirements and Implications for a Shifting Monsoon

Rice requires substantial amounts of water to sustain its growth and as a result water stress can reduce yield or damage rice plants at any stage of growth, although the plants are particularly sensitive during heading (or flower production) (Yoshida, 1981).

Both upland (rain-fed) and lowland (irrigated or paddy) varieties of rice exist (Fuller et al., 2011) (Figure 4.5). Rain-fed varieties are grown only in areas where annual rainfall exceeds 800–1000 mm (Jacquot and Courtois, 1987; Yoshida, 1981). Upland types of rice are often grown as part of swiddening systems in parts of South, Southeast, and East Asia, where trees or shrubs are felled and burned, providing fertility for the year. In these systems, humans have engaged fallow systems of rotation to allow the soil to regain its fertility between growing seasons. Upland systems of rice cultivation are highly susceptible to the timing and intensity of rainfall and as a result may have been impacted by past fluctuations in the monsoon. As Figure 1.8A shows, the cultivation of rain-fed rice would have been limited in large parts of South Asia, with sufficient rainfall for cultivation only present west of the Ghats and in eastern India and Bengal. With the exception of parts of highland Yunnan, much of southern China receives sufficient rainfall to support systems of rain-fed cultivation.

Irrigated or paddy varieties of rice, can, however, be grown wherever humans are able to construct an artificial pond or create irrigation systems. Because they indefinitely increase soil fertility over time, these features often become permanent fixtures of the landscape.

The level and timing of application of water in irrigated types of rice must be carefully managed by farmers. Compared to the large fields we are accustomed to in dryland types of farming, such as those of wheat and barley, where fields can sometimes be hundreds of hectares in size, lowland rice farming is by necessity carried out in small paddies, where farmers can carefully monitor water levels through low bunds, dykes, or poldering. Indeed, the ideal size for most rice paddy fields is 0.1 hectare (Bray, 1994).

Rice varieties themselves present a series of adaptations to different water conditions. It is believed that photoperiod sensitivity of rice may have been an adaptation to the East Asian Monsoon (Oka, 1958). While the start of the rainy season can be somewhat unpredictable, its end is fairly fixed in time, typically around September. Photosensitive (or tropical) varieties of rice will flower even if they are planted late because flowering is triggered by a change in daytime length that takes place as the end of the rainy season draws closer (Yoshida 1981). However, non-photosensitive (or temperate) varieties require a set growing time before reaching flowering stage. They thus may be affected by water stress because they may have not finished flowering before the end of the rainy season. Most temperate varieties of rice are for this reason grown in paddy environments.

4.2.3 Rice and Temperature

Rice is a tropical plant and most varieties require temperatures of 25°C–35°C to even begin germination (Yoshida, 1981). Temperatures below 10°C will kill most varieties, making the rice plant not adapted to the cooler springs of large parts of western and northern China (Yoshida, 1981). Rice requires a substantial number of growing degree days (or accumulated heat) in order to complete its life cycle (d'Alpoim Guedes and Butler, 2014; Yoshida, 1981). Both tropical and temperate varieties exist within *O. japonica,* each of which having a different relationship with temperature. On average tropical varieties of *O. japonica* require higher numbers of accumulated temperature (or growing degree days) to complete their life cycle than temperate varieties. Based on recent studies in rice genetics, it is likely that the earliest types of rice to first be domesticated were tropical varieties of *O. japonica* (Fuller et al., 2016). As we discuss later in this chapter, temperate varieties of rice may have been an evolutionary adaptation to the fluctuating conditions of the monsoon.

4.3 The Background to Rice Domestication: Transformations during the Late Upper Paleolithic in China

In order to understand how farming societies first developed in East Asia, it is important to review the types of subsistence systems that were employed by the foragers who lived in this area during the Pleistocene/Holocene transition.

During the Last Glacial Maximum (LGM; part of Marine Isotope Stage (MIS) 2, a period defined by the oxygen isotope character of seawater), the dominantly colder and drier conditions on the steppe replaced biomes in the continental interior up to latitude 20°N, with only a few holdouts of evergreen and mixed forest along the eastern Chinese coast. For instance, pollen from a core from Tianyang lake in deep southern China (Figure 4.2) showed that during MIS3 (27,800–60,000 years BP) an evergreen temperate forest was present, however, during the LGM (27,800–14,500 years BP) an increase in grass pollen and *Artemisia* suggests that steppe-like biomes had begun to replace forest (Zheng and Lei, 1999).

At the height of the LGM (around 21,000 years BP) warm temperate forest appears to have retreated to far southern China below 24°N (Biome 6000 team, 2004; d'Alpoim Guedes, 2016), while small pockets of cool temperate forest characterized biomes across the open continental shelf that connected China, Korea, Taiwan, and Japan. Steppe-like biomes were present as far south as the northern side of Yangtze river valley, while its southern side was characterized by temperate woodland. The LGM also resulted in fundamental changes in land availability (specifically between 28,000 and 16,000 years BP), when a continental shelf that covered an area of up to 1,500,000 km^2 was exposed (d'Alpoim Guedes et al., 2016a).

The Bølling-Allerød (around 14,500–12,800 years BP) marks the end of these cooler conditions, however, a brief return to them characterized the Younger Dryas (around 12,800–11,500 years BP). It is in this climatic context that researchers initially argued that foragers may have begun to practice rice cultivation. Initially four charred rice grains and the presence of rice phytoliths at the cave site Yuchanyan (17,700–13,800 years BP) were argued to represent some of the earliest evidence of rice cultivation in China (e.g., Yasuda, 2002; Zhang and Yuan, 1998; Lu et al., 2008); however, following radiocarbon dating, these finds were found to be from much later than the time of site occupation and were intrusive (Boaretto et al., 2009). Rice phytoliths extracted from another cave site: Diatonghuan (around 12,000 years BP) were also argued to present evidence for early rice exploitation and subsequent cultivation at the site (Zhao, 1998). However, to date, there is no convincing evidence that the foragers that left remains at Bølling-Allerød and Younger Dryas sites actually cultivated rather than occasionally gathered the rice that shows up in the phytolith record.

Systematic flotation carried out at another cave site that was originally implicated in the early domestication of rice, Zengpiyan (Figure 4.2), further confirmed this picture, demonstrating that rather than relying on rice farming, the inhabitants of that site exploited a wide variety of wild foods including Chinese gooseberries, wild grapes, plums, hackberries, and hickory nuts alongside a number of unidentified charred tubers (Zhao, 2003). At Yuchanyan and Zengpiyan (Figure 4.2) zooarchaeological remains showed that the inhabitants of southern China relied on a wide variety of animals including fish, tortoises, wild boar, and wild water buffalo, although the majority of their diet included deer and various species of aquatic birds and shellfish (Prendergast et al., 2009; Zhongguo Shehui Kexueyuan Kaoguyanjiusuo, 2003).

It is possible that the pottery found at these sites could have been used to boil some of the seeds and nuts found in these same sites, increasing the efficiency of consuming these foods for foragers. Small grains are difficult to roast on a fire without either boiling or grinding into a flour (see Chapter 5 for discussion of grinding stones in northern China). While this represents an advance in the type of technology associated with food processing, these examples have shown that this is not necessarily synonymous with the beginnings of farming in the region.

Because of the intensive nature of pottery production and the difficulty in transporting it, the presence of pottery has implied a decrease in mobility to several researchers (Chen and Yu, 2017). This decrease in mobility may have been forced by demographic packing, making it difficult for foragers to operate in larger radius and resulting in a need to intensify resources around them.

Foragers with specific sets of tools appear to have mapped onto different ecological zones during the end of the Pleistocene. For instance, during MIS3 (35,000–22,000 years BP), core and flake technology was distributed as far north as 41°N in northeast China. However, during the LGM, this core and flake technology moved southward, closely mapping onto the remaining forested biomes south of the Yangtze river valley (Qu et al., 2013), suggesting that this technology was associated with exploiting resources that were housed in these warmer biomes: something that may have led foragers to intensify their resource base.

If demographic packing is what drove intensification of resources, it is possible that several climatic events may have contributed to this phenomenon. The onset of the peak LGM (around 21,000 years BP) and the southward migration of core and flake producing foragers, who once occupied areas to the North, may together have led to demographic packing of foragers in southern China (see Chapter 5). A second potential contributor to resource concentration was a rise in sea level during Meltwater Pulse 1A (14,700–14,100 years BP) that resulted in the transgression of 1,500,000 km^2 of coastal territory over the course of a few hundred years (d'Alpoim Guedes et al., 2016a) (Figure 1.10). This sea-level transgression may have forced profound cultural adaptations in foragers due to demographic packing, but also due to the loss of useful coastal habitats. Early adaptations to maritime biomes were thus likely lost under the oceans. We are thus missing a large part of the picture of late Pleistocene forager subsistence strategies in East Asia because of this inundation (d'Alpoim Guedes et al., 2016a).

4.4 Broad Spectrum Foraging, Early Exploitation of Rice and Holocene Climate

Early Holocene climate: Over the past 10 years, there has been a growing consensus that humans in south and central China only began to interact intensively with plants, such as rice, during the Holocene. Overall, the Early Holocene represents a period of time that was characterized by substantially warmer temperatures than the later Holocene. For these reasons, this period of time is sometimes referred to as the Holocene climatic optimum or Holocene Megathermal (Marcott et al., 2013). A compilation of cave isotopic records shows a very common pattern, with a maximum in presumed monsoon intensity at around 7,000 years BP followed by a general decrease until around 2,000 years BP (Cai et al., 2010) (Figure 4.6). Some of the records predict a sharp decrease in monsoon intensity between 4,000 and 5,000 years BP, for example, at Heshang, Jiuxian, and Sanbao (Figure 4.2). There is a suggestion in some of the records that monsoonal strength may have increased in the last 2,000 years, although this is not universally observed. It is interesting to note that this general pattern is seen not only in the Chinese caves but also in the Hoti Cave of Oman, arguing that the southwest monsoon and the East Asian monsoon are coupled at this timescale (Figure 4.6).

Despite the general similarity in the long-term trend of many of the cave records, we do, however, note that there are important differences, not just between East and Southwest Asia, but also within those derived only in China. It is noteworthy that while the Hoti Cave shows a steep decrease in $\delta^{18}O$ values between 11,000 and 9,000 years BP, this change is significantly younger than the trend to lower $\delta^{18}O$ values seen at Dongge Cave where this occurs at 11,000–12,000 years BP.

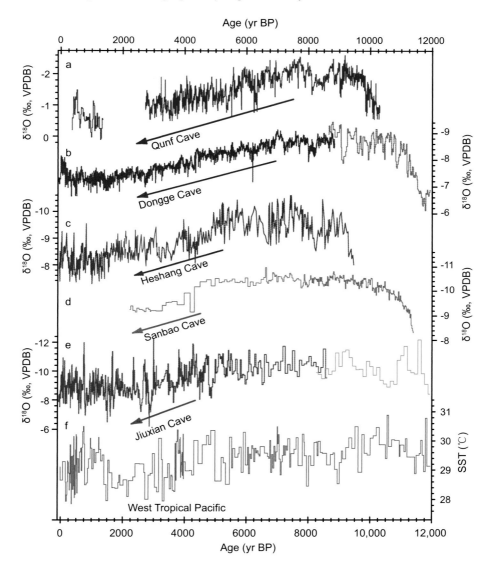

Figure 4.6 The comparison of the calcite $\delta^{18}O$ record of stalagmites from the Jiuxian (green), Hoti (dark red) (Fleitmann et al., 2003), Dongge (blue) (Wang et al., 2005); light blue (Dykoski et al., 2005), Heshang (violet; Hu et al., 2008) and (dark yellow; Shao et al., 2006; Wang et al., 2008) caves and the reconstructed SST over the West North Pacific (pink; Stott et al., 2004). All the $\delta^{18}O$ scales are reversed (increasing down). The arrows illustrate the decreasing trend of summer monsoon precipitation in each record, indicating the asynchronous changes. Figure modified from Cai et al. (2010).

Within the Chinese records it appears that there are significant differences regarding the time of the initial monsoon intensification, and indeed at one of the caves (Jiuxian) there is hardly any evidence of Early Holocene intensification at all, but rather a moderate decrease between 5,000 and 3,000 years BP (Figure 4.6e).

At the present time it is unclear whether this means that the monsoon rains only intensified locally, or whether there were some other reasons for a lag in the climate signal between the different caves. As a general rule it is hard to see whether the short-term, small-scale variations in the climate records are really representative of climate, or whether they effectively indicate noise in the data set. What is clear is that since about 6,000 years BP there has been a general, long-term drying of the climate across all of Asia, reaching a minimum around 2,000 years BP. It was further noted that this trend does not parallel the relatively stable conditions in the Greenland ice sheet (Stuiver and Grootes, 2000), implying that northern hemisphere ice volumes are not the primary control over long-term monsoon variability, despite the fact that summer monsoon tends to be weak during glacial times and strong during interglacial periods. Instead, it has been argued that the primary trigger for stronger summer monsoons is solar insolation, which peaked around 10,000 years BP and has been reducing steadily ever since (An, 2000; Burns et al., 2001; Dykoski et al., 2005). Nonetheless, it is also clear that the monsoon varies sharply over shorter time periods; variations that are not related to orbital processes, but to other factors, such as sunspots, ENSO, or the Indian Ocean Dipole.

The beginnings of rice domestication: Archaeologists have argued that human interference in the life cycle of wild rice plants may have resulted in its morphological domestication in two different areas: the lower and middle Yangtze River valley. At the middle Yangtze sites of Bashidang and Pengtoushan (9,500–8,100 years BP), the remains of houses built with wooden piles, along with pottery suggest that the inhabitants of the site were relatively sedentary (Hunan Sheng Wenwu Kaogu Yanjiusuo, 2006). Large numbers of refuse pits were also unearthed in the early phases of the site, indicating that the inhabitants spent long enough periods of time there to accumulate and dispose of considerable amounts of trash. The placement of graves at the site itself also implies that those who lived there held a strong connection to the place. Large quantities of rice were unearthed at Bashidang that were different in morphology to both known varieties of domesticated and wild rices (Hunan Sheng Wenwu Kaogu Yanjiusuo, 2006), making it possible that this rice represents a variety that may have since gone extinct (Londo et al., 2006). Based on their evaluation of the measurements published in the Bashidang report, Fuller et al. (2008) proposed that the rice unearthed at Bashidang is an extinct variety of wild rice. Unfortunately, rice spikelet (flower)

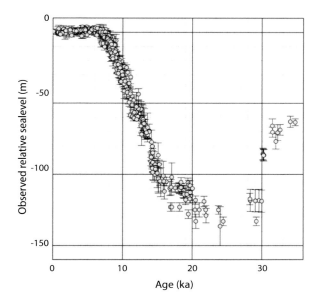

Figure 4.7 Reconstructed sea-level variations since the LGM. From Lambeck et al. (2014).

bases were not recovered from the site, making it difficult to evaluate if these specimens were undergoing domestication.

In addition to rice, substantial quantities of wild aquatic and other plants were also unearthed, including water chestnuts (*Trapa* sp.), persimmon (*Diospyros* sp.), wild plums (*Prunus* sp.), and wild soybean (*Glycine* sp.), demonstrating that the occupants of the Bashidang site still relied on a wide range of different foodstuffs (Hunan Sheng Wenwu Kaogu Yanjiusuo, 2006). Zooarchaeology has also been carried out at this site where subsistence appears to have been focused on deer hunting and fishing. Wild water buffalo (*Bubalus* sp.), chicken (*Gallus* sp.), and likely wild pig (*Sus* sp.) (represented by only three teeth) were also present in the assemblage. It is possible that humans cultivated rice at this early site; however, this likely formed part of a much broader spectrum of subsistence strategies that focused at least partially on targeting foraged and hunted aquatic resources.

A number of researchers have argued that there is evidence for early rice exploitation in the lower Yangtze River basin at sites such as Shangshan by around 10,000 years BP (Jiang and Li, 2006; Zuo et al., 2017). Unlike late Upper Paleolithic sites, and much like Pengtoushan and Bashidang, Shangshan is not located inside a cave, but rather is an open-air site, indicating that a change in settlement pattern may have taken place during the early years of the Holocene. Other features have led the

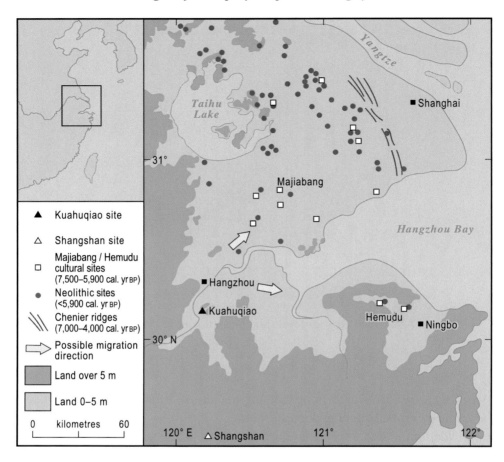

Figure 4.8 Location of the Kuahuqiao site, and others, in the lower Yangtze region. The Kuahuqiao site lies at the gateway between sites in the upland valleys, south and west of Kuahuqiao, occupied by late Mesolithic and early Neolithic hunters and foragers and sites on the coastal wetlands of Hangzhou Bay, east and north of Kuahuqiao, occupied by early Neolithic farmers such as those in Hemudu and Majiabang (white squares). Gray dots indicate locations of Songze and Liangzhu cultural sites. The yellow arrows indicate the possible migration directions of the early Neolithic communities. Ages shown are in calibrated years BP. Reproduced by permission from Zong et al. (2007).

excavators to argue that the inhabitants of this site were relatively sedentary: the presence of wooden pile dwellings and the addition of grinding stones to a repertoire that already included early pottery (Jiang and Li, 2006). Rice husk impressions were uncovered in pottery from Shangshan alongside rice phytoliths from sediment. Based on the size of these husks it was initially argued that Shangshan contained the earliest evidence of rice "domestication" (Jiang and Li, 2006).

These finds have garnered substantial criticism as others argued that rice unearthed at this site did not show morphological evidence for domestication (Fuller et al., 2008; Fuller et al., 2007). In particular, Fuller et al. (2008) have argued that size alone is not a statistically reliable means for documenting domestication, as a wide range in size exists in domesticated rice populations. He argued rather that researchers should focus on identifying increasing proportions of traits selected for by humans, such as a non-shattering rachis (Fuller et al., 2008; Fuller et al., 2009; Fuller et al., 2007). An evaluation of rice spikelet bases from Tianluoshan (Figure 4.2) shows that traits associated with domestication, such as a non-shattering spikelet base, evolved between 6,900 and 6,600 years BP (Fuller et al., 2009). However, in order for this domestication trait to show up morphologically, humans would have had to have first begun to cultivate this plant substantially earlier.

Humans in this region continued to forage important components of their diet throughout the history of early rice cultivating sites in the region. For instance, Kuahuqiao (8,000–7,400 years BP) contained evidence of canoes that may have been used for rice harvesting, along with substantial quantities of rice, however, these occur alongside significant amounts of other plants, including various types of acorn (*Quercus* spp. and *Cyclobanalopsis* sp.), peach (*Prunus persica/P. simonii*), apricot (*P. armeniaca*), and Japanese apricot (*P. mume*), as well as foxnut/gorgon (*Euryale ferox*) and water caltrop (*Trapa* sp.). Crawford (2012) points out that it is likely that all of these plants were managed by the inhabitants of sites like Kuahuqiao through niche construction. The growth of aquatic plants like foxnut, lotus, and water caltrop can be achieved by creating artificial ponds and eliminating competing species. Likewise, the fruits from nuts and peaches can be replanted to encourage growth of stands of useful trees, and undergrowth can be burned to remove competitors. Rice formed only one component of this managed landscape. The use of these wild resources was not limited in time to the past and today has continued to provide a source of resilience to Asian farmers when dealing with fluctuating water resources and climate.

Fuller and Qin (2010) argue that one reason rice exploitation may have intensified is as a result of declining oak trees (possibly as a result of climatic change) after 7,500 years BP, as noted in pollen from a sediment core at Kuahuqiao (Zong et al., 2007). However, Crawford (2012) points out that this relative oak pollen decrease is represented as a percentage of total pollen and that a relative increase in grass pollen does not necessarily signal an absolute decrease in pollen from oaks. During the Early Holocene, the early exploitation of rice was only a small part of a much wider resource base employed by foragers who targeted wetlands for their high resource density.

4.5 Development of Rice Cultivation Technology

The wild types of rice exploited by these early foragers may not have been particularly productive and this may have been a reason why they only formed a minor component of a much broader subsistence strategy. Experiments carried out by Lu (2006) in harvesting wild rice revealed that *O. rufipogon,* which reproduces primarily by tillering and not by seed, had remarkably low seed production. In Lu's (2006) experiment, panicles of wild rice contained relatively few grains, and these ripened at different times and shattered into the water causing large amounts of seed loss. In total she was only able to harvest 1.3 grains per panicle in contrast to 82.1 grains per panicle from domesticated rice. An individual would have been required to spend between 8.7 and 10.2 hours every day gathering and processing wild rice in order to avoid starvation. Compared to the 2–3 hours required for harvesting wild yams or bamboo shoots, it seems that producing ancient wild rice must have been remarkably ineffective. This has led some to argue that low yields and high risk associated with cultivated wild rices may have meant that it was first cultivated as a luxury food (Hayden, 2011). Indeed, early strategies at sites like Kuahuqiao seem to have been focused on the use of canoes for collecting what were likely perennial forms of wild rice in deeper strands of water (Zhejiang Sheng Kaogu Yanjiusuo, 2004). Hayden (2011) argues that strands of wild rice were managed through using ploughs and spades to clear seasonally dry wetland and to promote the growth of annual rather than perennial varieties of rice.

It is possible that rising sea levels in the lower Yangtze River delta after 7,700 years BP may have provided an incentive for humans to modify the habitat of this wild plant in order to prevent it from being lost to encroaching sea levels: this indeed corresponds to a period of major sea-level transgression (Lambeck et al., 2014; Peltier, 2002) (Figure 4.7). For instance, Zong et al. (2007) have argued that early systems of rice cultivation may have evolved as an adaptation to prevent encroaching sea water. Through analysis of microcharcoal and pollen, they demonstrated that farmers cleared the area surrounding the Kuahuqiao site by burning the area, either planting or encouraging the growth of rice (that was possibly fertilized with night soil) and bulrushes (*Typha* sp.: an aquatic plant whose starchy rhizomes can be consumed) at around 7,700 years BP. However, they found that after roughly one century that the area began to be regularly flooded with brackish water until 7,550 years BP when the fields were inundated by marine water (Figure 4.8). Hayden (2011) suggests that the challenges associated with encroaching marine water may have led foragers to move their rice cultivation to higher, drier, and only selectively flooded areas of the landscape (in other words, outside of the perennial marshes), which assisted in the selection of annual rather than perennial rice. These challenges may have provided an impetus for early food producers in the lower

Yangtze valley to more intensively manage the habitat of wild rice, providing the conditions that lead to its eventual domestication.

Following rice domestication in East Asia, water and the labor involved in managing rice became essential this crop's production, particularly for the paddy types of rice that have been domesticated in East Asia (Fuller and Qin, 2009). By 6,300–6,000 years BP, at sites like Caoxieshan and Chuodun in the lower Yangtze River (Figure 4.2), there is evidence that humans had already begun to employ small bunded paddies that were connected by reservoir canals (Fuller and Qin, 2009; Fuller et al., 2009). These fields required labor-intensive operations to be productive: they were small in size (roughly only 1 m wide) and appear to have been watered using buckets that lifted water from the reservoirs into the field (Fuller and Qin, 2009). Fuller and Qin (2009) argue that these fields were likely controlled by the individual households spotted throughout the fields: the houses' proximity to the field itself allowed the inhabitants of this site to monitor water levels and the crop's growth. At Chuodun, the relatively high numbers of weed flora found in the samples from these fields lead Cao et al. (2006) to conclude that systematic tillage and weed clearance had not taken place. More recently, other analyses at this site have been used to argue that the common presence of charcoal within the fields indicates that burning took place to clear the field prior to planting (Hu et al., 2013b). Large numbers of potsherds were found within the field, demonstrating that waste from habitation sites may have been placed in the fields as fertilizer (Hu et al., 2013b).

In the middle Yangtze valley (6,400–6,000 years BP), the Chengtoushan site also contains evidence for what have been identified as field systems. However, here they were longer units that measured roughly two meters wide (Fuller and Qin, 2009; Yasuda et al., 2004). In the subsequent Songze (6,000–5,300 years BP) and then the Liangzhu (5,300–4,200 years BP) periods, new agricultural innovations appeared in addition to fields, namely implements for soil tillage. Soil tillage plays an important role in weed control by deeply burying weed seeds, thus inhibiting their growth. It also aerates the soil, encouraging the growth of plants. We start to see the appearance of stone ploughs, suggesting that individuals in this area had begun to intensify rice farming, but had also begun to discover that other plant organisms that competed with the returns of ricewere thriving in their fields.

During the Songze and Liangzhu periods, sickles for harvesting rice appear in larger numbers in the archaeological record, implying that mechanisms of harvesting had also begun to become more systematized and complex (Fuller and Qin, 2009). Burning, tilling, and artificial paddy construction are agricultural adaptations that have persisted in the management of rice farming up until the present day.

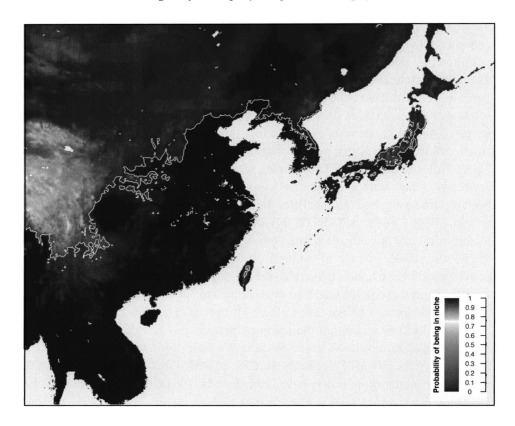

Figure 4.9 (A) Predicted extent of *O. rufipogon* during the Early Holocene or distribution of tropical japonica during the Early Holocene (around 8,200 years BP). White line indicates the range of the extent.

4.6 Climate Change and the Development of New Varieties of Rice during the Establishment and Early Spread of Rice Farming in East Asia

Warmer temperatures and stronger monsoon rains during the Early Holocene meant that the distribution of wild rice extended far beyond the boundaries occupied by tropical *O. japonica* today (Figure 4.9). Evidence from a number of sites in northern China suggests that during these warmer periods of time, wild plants were likely exploited by foragers at sites like Jiahu in the Huai river valley (9,000–8,600 years BP) (Henan Sheng Wenwu Kaogu Yanjiusuo, 1999), and at Houli sites like Yuezhuang in Shandong (Figure 4.2) by 8,000–7,000 years BP (Crawford et al., 2006). At Jiahu, systematic flotation carried out at the site by Zhao and Zhang (2009) shows that rice formed only a small percentage of the foods exploited by the site's inhabitants. Much as in the lower and middle Yangtze, other resources such as water caltrop (*Trapa* sp.), lotus root (*Nelumbo nucifera*), and various types of acorns largely dominated the assemblage, showing the continued importance of foraging. Rice showing a majority of domesticated spikelet bases, but otherwise thinner than known domesticated varieties of rice was found slightly north of the Yangtze River valley at Baligang by 8,700 years BP (Deng et al., 2015). A map of the distribution of tropical *O. japonica* shows that its range extended to cover all of these sites during the Early Holocene (Figure 4.9). Local varieties of wild tropical rice were likely managed by people in these regions and distinct trajectories to domestication may have already begun to take place.

While some argue that finds of rice at these early Holocene sites in northern China were the result of migration of early cultivators from the middle Yangtze (Zhang and Hung, 2013), Figure 4.10 demonstrates that it is equally likely that foragers began to exploit different varieties of wild rice that were able to grow locally around their habitations.

With a few exceptions, such as at Nanjiakou (around 5,900 years BP) and Xishanping in Gansu (5,000 years BP), finds of rice decrease in numbers and disappear from settlements in northern China for several thousand years following 7,000–6,000 years BP a period of time that corresponds to the shrinking of the thermal niche tropical varieties of *O. japonica* southward due to cooling temperatures at the end of the Holocene climatic optimum (Figure 4.10). Following a several thousand-year hiatus, rice reappeared in assemblages in Shandong during the Longshan period (by 4,600 years BP). However, 600 years following its introduction to the peninsula its numbers again declined in assemblages (d'Alpoim Guedes et al., 2015). By 3,500 years BP, the niche of tropical *O. japonica* had retreated from most of Shandong. Temperate *O. japonica* could still be grown in parts of Shandong. This decline in the tropical japonica niche envelops most of northern and central China, leading tropical japonica rice to be confined to deep southern China by 3,500 years BP (Figure 4.10C). This reduction in temperatures likely resulted local lineages of wild rice in northern China became extinct (d'Alpoim Guedes et al., 2015; Fuller, 2012).

Given the cooler temperatures during the later Holocene, it is possible that moving rice to northern China may have required rice to adapt to new photoperiod and temperature requirements: possibly precipitating the development of a more cold-adapted temperate variety of *O. japonica* (Gutaker et al., 2019). Indeed, Fuller and Castillo et al., (2016) have argued that rice plants at sites such as Nanjiakou are shorter and squatter and appear to correlate to a temperate morphology. Over time, shorter squatter rice grains (with a lower length–width ratio) emerge in the archaeobotanical record of China that likely correspond to temperate varieties

Tropical O. japonica   Temperate O. japonica

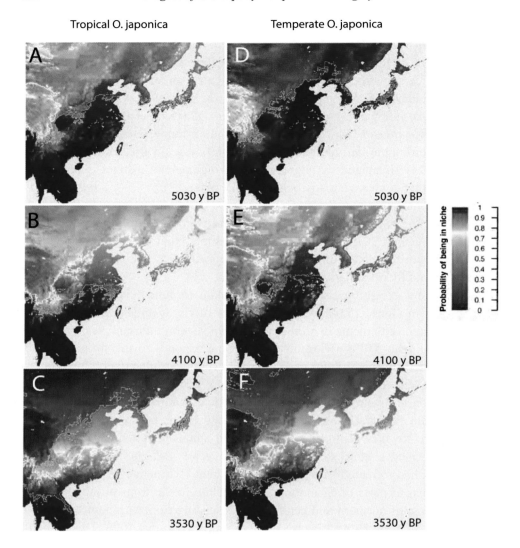

Figure 4.10   The thermal niche of tropical *O. japonica* at (A) 5,030, (B) 4,100, and (C) 3,530 years BP. Tropical *O. japonica*, temperate *O. japonica* (D) 5,030, (E) 4,100, and (F) 3,530 years BP. Figure created using methods described in d'Alpoim Guedes and Bocinsky (2018).

(Figure 4.11B). The emergence of these shorter and squatter varieties also corresponds to the movement of rice northward outside of its original center of domestication (Figure 4.11A).

Despite the heavy focus on rice in the archaeological literature, even where rice does appear in the northern Asia, it is not the only crop, nor does it appear to form

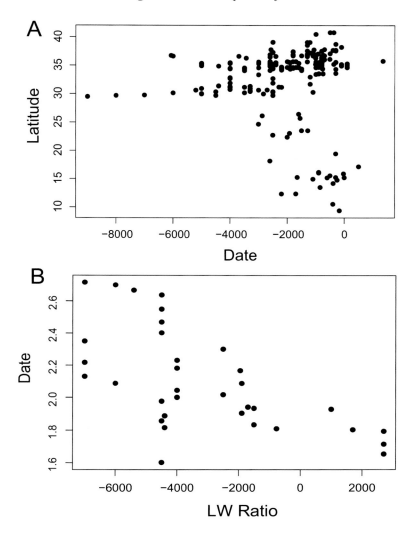

Figure 4.11 (A) Latitude of rice finds over time in East Asia showing a southward and northward spread of rice following 4,000 years BP (Fuller and Qin, 2009; Liu et al., 2007); (B) L/W plot of rice grains across China showing the emergence of a shorter squatter type of rice, likely temperate *O. japonica* throughout time.

the majority of the diet. For instance, in Korea and Shandong, rice was present in relatively small proportions compared to wheat, barley, and millets (Crawford and Lee, 2003; Crawford, 1992; d'Alpoim Guedes et al., 2015). The low probability of being in the niche, and likely lower productivity of rice in these areas may have meant that it was necessary for humans to hedge their bets, employing a diversified subsistence system.

4.7 Niche Construction, Geographic and Climatic Barriers in the Spread of Rice to Southwest and Southeast China

Spread to Southwest China: Despite its close proximity to the middle Yangtze River valley, rice farming took several millennia to spread to southwest China: into the provinces of Sichuan, Yunnan, and Guizhou. Unlike central and eastern China, southwest China is marked by its higher altitude and steep topography. Steep topography and a lack of wide basins in which river or marsh fed agriculture could be practiced may have prevented early farmers from moving their farming systems into these areas as they would likely have to first resolve several key issues surrounding water management, including moving water from valley bottoms upland. Likewise, cooler temperatures at higher altitude may have formed a barrier to the expansion of tropical forms of *O. japonica* rice (d'Alpoim Guedes, 2011; d'Alpoim Guedes et al., 2014).

These difficulties mean that it was not surprising that when rice did appear in southwest China for the first time around 4,700 years BP, it did so in the low-lying flood plain of the Chengdu Plain: an area that shares several key morphological similarities with the middle Yangtze. The region is crisscrossed by a large number of rivers that spill out from the eastern Tibetan Plateau and is low and flat, making it easy for farmers to create paddy like environments to grow rice (d'Alpoim Guedes, 2011; d'Alpoim Guedes and Butler, 2014). Clay soils, which have high water retention, and the flat environment, mean that farmers in the Chengdu Plain today grow rice in bunded paddies: one of the easiest types of paddies to construct. Bunded fields are constructed by leveling out an area designated to be the rice field and by retaining small walls (bunds) around each field.

The Chengdu Plain also differs from the rest of southwest China in terms of temperature: it is consistently warmer than the surrounding highlands of the Three Gorges, the Qinling mountains, and the Yunnan-Guizhou Plateau (Figures 4.2 and 4.10)(d'Alpoim Guedes, 2011). At roughly 4,800 years BP, a decrease in the available niche of tropical *O. japonica* took place in the middle Yangtze River valley. It is possible that the cooler temperatures created evolutionary pressure for the selection of temperate varieties of *O. japonica* to take place, and it was subsequently these varieties of rice that were moved into Sichuan.

Fuller (2012) has argued that there is a relationship between rice morphometrics and environments. In particular, shorter grains tend to be found at high latitudes or high altitudes, and are typical of temperate japonica, whereas tropical japonicas tend to have longer grains. Finds of rice from the Baodun period (4,700–3,700 years BP) sites on the Chengdu Plain fall firmly into a temperate *O. japonica* grouping making it clear that by the time rice arrived in this area, temperate varieties had already developed. A thousand years later, or by 3,700 years BP, tropical

O. japonica rice completely lost its niche across much of southwest China and much of the Yangtze River valley basin, making it likely that temperate varieties of japonica took on a new importance during this period of time (Figure 4.10B). Systematic archaeobotany throughout Sichuan indicates that the inhabitants of this area did not exclusively rely on rice: rather they also planted foxtail and broomcorn millet (d'Alpoim Guedes, 2011; d'Alpoim Guedes et al., 2013; d'Alpoim Guedes and Butler, 2014). Indeed, a look at the proportions of crops in the diet across sites of different altitudes shows that in areas of higher altitude (and presumably cooler temperatures), humans planted higher proportions of dry land crops that could tolerate lower temperatures, and be grown in areas of steeper topography where it was not possible to create rice paddies without substantial investment. A dual agricultural system, composed of both rice and foxtail millet, thus characterized agricultural subsistence patterns across southwest China. Together these crops formed a diversified food strategy, that helped buffer against risk and that could be adapted to different ecological settings, as these two different crops could be used interchangeably and in different proportions.

To the south of the Chengdu Plain, an increasing amount of work in Yunnan and Guizhou suggests that rice farming arrived in that region shortly after reaching the Chengdu Plain at around 4,600 years BP. At sites like Baiyangcun (around 4,600–4,000 years BP) and Haimenkou, rice is present alongside roughly equal quantities of both broomcorn and foxtail millet, suggesting that a diversified subsistence regime was crucial in areas of higher altitude and steeper topography (Dal Martello et al., 2018; Xue, 2010). However, following the fourth millennium BP, rice also quickly fell out of the assemblage at higher altitude sites like Haimenkou, to be replaced by colder and more arid adapted crops like wheat, barley, and millets and even wild crops like *Chenopodium* (d'Alpoim Guedes and Butler, 2014; Xue, 2010). In the Yunnan-Guizhou highlands, farmers employed diversification of their agricultural base as a means of risk reduction (Marston, 2011).

It is likely that on the Yunnan-Guizhou Plateau rice agriculture was limited in scope to selective zones, such as the flats surrounding lakes like Dian or Erhai. Here the larger flat areas around the lakes and the shores of the lake could have been turned into either floating field systems, dyked, and poldered fields or *shatian* (sand fields) on the banks of the lake. According to Bray (1994), these fields were in use on Lake Dian as late as 1990. Following the thirteenth-century, terraces expanded the areas where rice could be cultivated into higher and steeper ground.

Rice farming appears to have taken a substantial amount of time to implant itself in southern China, although here, climate itself does not appear to have been the key barrier to expansion. Instead, the hilly character of the topography is more of a challenge to the rapid establishment of paddies. Throughout coastal southern

China, we see a reliance on wild foods (such as deer, elephant, rhino, and tapir, as well as a wide range of shellfish, fish, and turtles) until at least about 5,000 years BP throughout Guangdong, Guangxi, and Fujian (Zhang and Hung 2012). In Guangxi and across northern Vietnam, a complex culture of sedentary foragers, known as "Dingshishan" occupied large sites and were dependent on reliable production of shellfish and over 84 different species of wild animals (Zhang and Hung, 2012) between 9,000 and 4,000 years BP. Zhang and Hung (2012) argue that around 6,000 years BP, a decline in temperatures may have precipitated changes in the organization of these foragers'; however, there is no indication of that these changes affected their resource distribution.

Changes in both the pottery typology and burial traditions in sites in southern China have led some authors to suggest that rice farmers from central and southwest China had begun a demographic expansion into southern China around the fourth millennium BP (Bellwood, 2005; Zhang and Hung, 2010).The timing of the spread of rice farming to southern China has only recently begun to form the object of systematic investigation. Many early finds of rice in the region were not radio-carbon dated and as a result a great amount of uncertainty exists about the timing of rice's arrival in the region. A recent systematic dating campaign at the sites of Guye and Shixia in Guangdong suggests that rice farming was introduced to the Pearl River delta between 5,000 and 4,700 years BP (Yang et al., 2016). During this period of time, the numbers of shell middens from the region reduced in number, suggesting that fundamental changes in economy were taking place. However, it is noteworthy that large-scale changes in the chemistry and mineralogy of the Pearl River, reflecting the state of the wider river basin, did not occur until around 2,700 years BP (Hu et al., 2013a). It has been argued that an increase in chemical alteration of the sediments in the river reflects widespread onset of agriculture that disrupts the natural pattern of soil erosion at the basin-wide level. Thus, rice farming at 5,000–4,700 years BP may still have been limited in the Pearl River basin (Hu et al. 2013a).

High Holocene sea-level stands may have reduced the available habitats for rice cultivation and led inhabitants of this region to focus their subsistence strategies on maritime adaptations. Rolett et al. (2011) suggest that habitats in which rice could be grown were limited for the inhabitants of the Tanshishan culture in Fujian, southeast China (5,000–4,300 years BP). Using a record of Holocene sea-level change based on sediment cores, his team demonstrated that areas that are today almost 80 km from the coastline were at the time islands where foragers in the area focused on exploiting marine biomes rather than on farming.

From southern China, it has been argued that these farming communities spread to Southeast Asia (Higham, 2002, 2005, 2013; Higham et al., 2011).

Southeast Asian foragers may not have been strangers to farming products and some of the earliest evidence for agriculture in the region comes in the form of millet from sites in the Lopburi region of Central Thailand that date to 4,300 years BP (Weber et al., 2010b). Some of the pottery from these sites, such as Tha Kae, show close similarities in style and decoration to those found at Shifodong in Yunnan (3,800–3,100 years BP) (Rispoli, 2007; Rispoli et al., 2013), further suggesting that people and their culture spread alongside farming crops. Following this, rice appears at a number of sites across Southeast Asia, including Khok Phanom Di and Ban Tha Kae on the modern Chao Phraya flood plain near Bangkok, and Rach Nui in southern Vietnam by 3,900–3,700 years BP (Castillo, 2011; Castillo and Fuller, 2010; Castillo et al., 2016; Cobo Castillo, 2018). These finds indicate that farming technology or ideas surrounding farming may have spread in several waves into Southeast Asia, one where migrating farmer pioneers interacted with local foragers (Higham et al., 2011). Unfortunately, a lack of archaeobotanical research in Vietnam, Myanmar, and Laos currently makes it difficult to truly understand how this process took place.

Zhang and Hung (2012) have argued that this southward spread of farming communities from the middle Yangtze River valley may have been climatically driven. Although they do not present a mechanism for how this change may have taken place, they argue that the end of the climatic optimum would have impacted forager populations in the region. A look at the available niche for both tropical and temperate varieties of *O. japonica* reveals that between 5,000 and 4,500 years BP in eastern China, both tropical and temperate varieties of rice could be grown with a high probability of success throughout central and southern China (Figure 4.10). Although there is material cultural evidence for rice farmers spreading from the central Yangtze valley and from eastern China into southern China, it does not appear that their initial spread was climatically driven.

Changes in climate did begin to impact the distribution of rice following 4,500 years BP (Figure 4.10). Between 4,000 and 3,500 years BP, tropical *O. japonica*, in particular, became extremely difficult to grow along the central and lower Yangtze River valley. In the west, most of Sichuan, Yunnan, and Guizhou lost their niche, and in the east large parts of Zhejiang and Jiangxi were also impacted. Temperate *O. japonica*, however, fared somewhat better (Figure 4.10). Although farming had already begun to spread to Southeast Asia by 4,000 years BP, growing population pressure in the heartland of rice domestication, combined with diminishing returns in tropical varieties of japonica rice may have contributed to greater numbers of farmers migrating south in search of more clement climates. In southern China and Southeast Asia, the spread of farming initially took place in a way that was not related to changing climate; however, climate may have played a role in intensifying a process of population expansion that had already begun.

Throughout south China and Southeast Asia, it is worth noting that rice was only a part of extremely diverse systems of subsistence that included a wide range of tubers and tree crops. For instance, recent finds at the site of Rach Nui in Vietnam indicate that despite the fact that foragers were familiar with rice and foxtail millet, these taxa made up only 2% of the total assemblage. Rather, they relied primarily on sedge tubers and nutlets (Castillo et al., 2018). Other key tree crops such as the banana (*Musa* sp.), Chinese olive (*Canarium* sp.), citrus fruits (Fuller et al., 2018), screw pine (*Pandanus* sp.), longan (*Dimocarpus longan*), kiwi (*Actidinia* sp.), litchi (*Litchi chinensis*), and sago palm (*Caryota*) also have their center of diversity in Southeast and South Asia. These fruit crops were later complimented by introductions from South Asia including the mango (*Magnifera indica*), jackfruit (*Artocarpus*), and jujubes (*Ziziphus* sp.). A range of tubers including taro (*Colocasia esculenta*) and yam (*Dioscorea* sp.) have also been critical to subsistence systems in this area. Rice only played a part in already developed systems of wild plant management that were established by foragers in the region. Unfortunately, many of these plants do not preserve well in the archaeobotanical record, however, their extensive use in the present day makes it likely that these species also played critical roles in antiquity (Weisskopf, 2018).

4.8 Great Rivers, Niche Construction, and Rice Irrigation on the Chengdu Plain

Controlling water is essential to the successful cultivation of rice. The domestication and spread of rice farming resulted in some of the most impressive anthropogenic modification of the landscape and harnessing of the power of water from China's rivers. A key example of this is human modification of rivers in Sichuan. Throughout history, the rivers that flow out of the mountains of western Sichuan have posed threats to human settlements on the Chengdu Plain. The plain's location close to the tectonically active Longmen Shan range has meant that it is vulnerable to the effects of earthquakes which have included dam bursting, landsliding, and large-scale floods (Huang and Fan, 2013). An analysis of faults along the Longmen range indicated that a large number of earthquakes have taken place in the historical past (Wen et al., 2012). According to historic records, earthquakes and resultant landslides in the Longmen Shan have created dams by blocking parts of the Min River at least ten different times between 10 BCE and 1933 CE (Wen et al., 2012). In addition to earthquakes, high rates of erosion, coupled with more intensive monsoonal rain across this range has contributed to events characterized by higher sedimentary load in rivers and sometimes their complete blockage, which can result in over-bank flooding further downstream, particularly under post-earthquake conditions (Hairong et al., 2017). In addition, the sudden decrease in altitude when the Longmen Shan reaches the Chengdu Plain creates a sediment bottleneck in the Min River, making it highly susceptible to flooding. Past periods of increased monsoonal intensity would have likewise increased both water runoff and sediment load in these rivers, increasing the probability of overbank flooding and destruction of agricultural fields and settlements.

The occupants of the Chengdu Plain have developed ingenious ways of coping with the threats posed by these rivers. Archaeological evidence suggests that early water management strategies may have begun as early as the Neolithic period of the Chengdu Plain. Flad and Chen (2013) propose that the low slope of the walls that surround early settlements on the plain indicates that their function was to provide protection from flooding. However, sometimes, it appears that rivers overwhelmed these systems and did impact settlements. A layer of river mud covering the Sanxingdui site was interpreted by archaeologists and geomorphologists as being due to the site's destruction by flooding of the banks of the Yazi River at around 3,000 years BP (Liu, 1998).

Construction of Dujiangyan Weir, just to the northwest of Chengdu, began in 285 BCE under the orders of the grandfather of the first emperor of China (Li and Xu, 2006). This weir and attendant irrigation system was designed to control the flooding of the Min River, while simultaneously irrigating a surface area of over 200,000 ha (today this network extends to over 600,000 ha) via a network of canals. This irrigation and flood management device takes advantage of the natural slope of the Chengdu Plain, which is higher in the west than in the east (Zhang et al., 2013). Dujiangyan functions by splitting the Min River into several different sections by employing a raised levee that divides the Min River into a deep and narrow inner stream and a flatter and wider outer stream. The inner stream accounts for approximately 60% of the water volume during the dry season, but this is reduced by the levee to roughly 40% during high rain events to prevent overbank flooding, while the outer stream serves to carry most of the heavier sediment load (Cao et al., 2010a). An additional weir was built using bamboo baskets filled with rocks then creates artificial whirlpools that drain additional sediment from the river. Finally, the water passes through a narrow channel that additionally prevents flooding. This network of irrigation canals has been crucial to the wet rice cultivation and is partially responsible for the Chengdu Plain being able to accommodate some of the highest population densities on Earth at over 900 people per square kilometer (Willmott, 1989).

It is possible that the origins of this system greatly predate the historic period. Recent surveys demonstrate that the inhabitants of the Baodun (4,700–3,700 years BP) and later Sanxingdui and Shi'erqiao cultures colonized the Chengdu Plain in

Figure 4.12 Rice paddies on the Chengdu Plain today.

a dispersed settlement pattern, with small hamlets dotting a landscape that was punctuated by large walled sites ranging from 7 to 245 ha in size (Flad and Chen, 2006). It is possible that this dispersed settlement pattern arose in order for farmers to be able to effectively manage dense networks of irrigation canals that fed their rice paddies (Figure 4.12). Thus, unlike the scenario painted by Fuller and Qin (2009) for the middle Yangtze region, there does not appear to have been a centralized authority coordinating the planting and harvesting of rice: rather smaller units of farmers appear to have operated together to create a highly resilient and productive farming system. Sustained by a reliable, intensifiable, and highly productive agricultural system, the Chengdu Plain became the center of important social networks, as epitomized by the spectacular bronzes from the sites of Sanxingdui and Jinsha. Following the construction of Dujiangyan, the Chengdu Plain maintained this legacy throughout the historic period by functioning as an important breadbasket for Chinese states (Sage, 1992).

4.9   The Spread of New Technology and Crops during the Historic Period

While we cannot review the entire historic period in this chapter, several key developments took place in which the changing monsoon might have played an important role in agricultural history in southern China.

By 1,690 years BP (220 CE), the entire area north of the Yangtze River valley fell out of the agricultural niche for rice (d'Alpoim Guedes and Bocinsky, 2018) (Figure 4.13) and the millets that were traditionally cultivated in this area also began to experience difficulties (see Chapter 5). Between 290 and 317 CE, a massive migration of farmers from northern to southern China took place. The Jin Capital Luoyang fell to northern invaders in 311 CE, so that much of what is now northern China was lost from the empire and the Jin Dynasty itself split in western and eastern factions. Both Dongge and Heshang speleothems (Figure 4.14) show that this took place during a particularly intense dry time interval. Most notable was a dry period starting around 280 CE and climaxing in 316 CE (Figure 4.14). An estimated one in eight farmers moved from northern to southern China during this period of time. While constant invasions into Jin territory by nomads on their margins undoubtedly played a role in this movement, the increasingly low probability of crops being in the niche reported by d'Alpoim Guedes and Bocinsky (2018) undoubtedly also played a role in this southward migration.

Following a massive defeat in 1127 CE by Jin (Jurchen), who originated from Manchuria, the Song Dynasty moved to southern China to establish the Southern Song, along with populations of refugees (Tao, 1976; Twitchett et al., 1978). Population had grown under the early Song and devastation of war in the North led to a huge southward push in population (Bray, 1994). By the time of the Song Dynasty, fear of nomadic invaders and poor agricultural returns experienced during the late Tang Dynasty meant that the greater part of China's population now lived in the south. The government was faced with the problem of feeding a greater population on a smaller amount of land. The Song government approached this challenge on several fronts: by offering farmers financial incentives to open up new land and invest in improvements and by introducing high yielding varieties of new crops.

4.9.1   Investments in Landesque Capital and Human Niche Construction in Southwest China

One of the most well-known agricultural policies (*Qing miao fa* or Green shoots policy) was implemented by Wang An-Shi and involved providing low-interest loans to farmers, setting up new agricultural colonies in what were once borderland

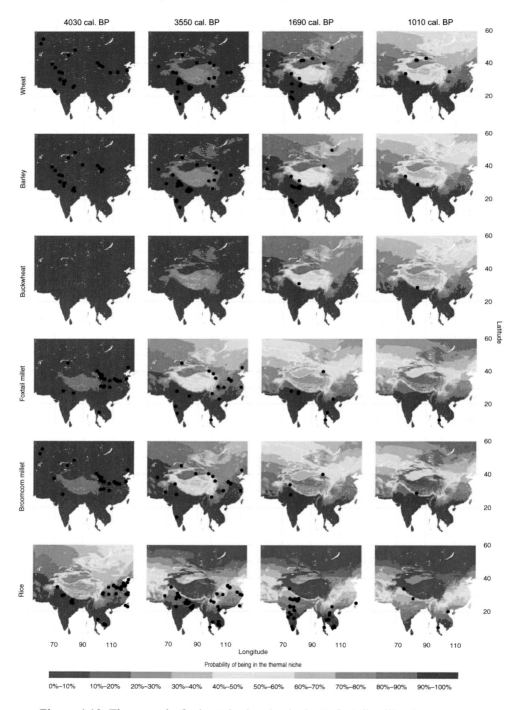

Figure 4.13 The spread of wheat, barley, buckwheat, foxtail millet, broomcorn, and rice across Asia at the end of the Holocene Climatic Optimum (4,030 years BP), 3,550 years BP, 1,690 years BP, and 1,010 years BP. Sites that have evidence of a crop at a particular year are shown as black dots. Reproduced with permission from d'Alpoim Guedes and Bocinsky (2018).

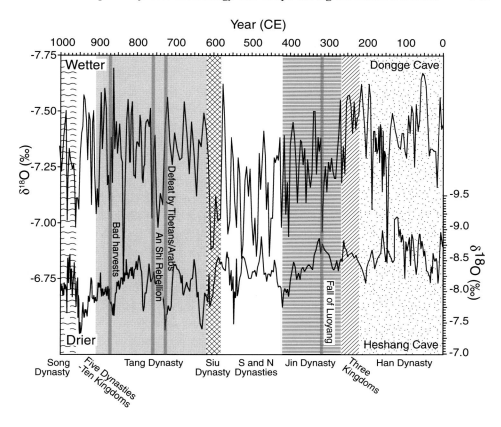

Figure 4.14 Monsoon climate records from south and central China compared with major historical events in Chinese history between 0 and 1000 CE. Shaded bands show the duration of Chinese dynasties. Data from Dongge Cave is from Dykoski et al. (2005) data from Heshang Cave is from Hu et al. (2008).

regions in the south, and investing in irrigation and terracing infrastructure (Bray, 1994). During the Song Dynasty, methods of irrigation and terracing also improved dramatically, opening up new zones of cultivation and resulting in one of the largest episodes of human niche construction for agricultural purposes known in history.

In areas such as the Chengdu Plain, the low-altitude, low-gradient landscape and ample water supply mean that rice agriculture flourished and demographic expansion in this region is historically attested as early as the Han Dynasty. On the other hand, the highlands of the Yunnan-Guizhou Plateau required both the introduction of new varieties of drought resistant rice as well as extensive terracing operations to sustain any real population expansion. Creating rice paddies in the mountainous and uneven terrain of the Yunnan-Guizhou Plateau would have required relatively large labor investments to flatten land to create paddies and terraces and to create irrigation systems capable of bringing water from rivers and lakes to areas of higher altitude, or diverting water from mountain streams. Population in this area did not truly grow until the thirteenth century, when the government began efforts to move new settlers into the region. This regime gave farmers seeds, draft animals, and helped them clear forest on the hilltops, facilitating population expansion into this region (Lee, 1982). It is largely during this period that the hilltop paddies that today characterize this region were developed.

Terraces in China are known as *san bao* (three treasures): threefold conservers that prevent erosion, conserve soil nutrients, and in many cases, result in higher-yield crops. The first allusion to irrigated terraces in southern China dates to the Tang Dynasty, where *shan tian* (mountain fields) are described by general Fan Zhou among the Man tribes of Yunnan. Similar descriptions of terraced landscapes are provided by the Tang dynasty poets Du Fu and Zhang Jiuling (Bray and Needham, 1984). Historical sources such as the *Manshu* (Book of Southern Barbarians) also suggest that the systems of terraced paddies that we associate with the landscape today do not predate the Tang dynasty (618–907 CE) (Bouchery and Lecomte-Tilouine 2010). Prior to the development of the technology of the terracing systems, it is estimated that only 6% of the total surface area of the Yunnan-Guizhou Plateau was arable, and this arable land was confined to scattered valley bottoms (Lee, 1982). Population expansion prior to 1000 CE was thus largely confined to areas such as the Dian and Erhai lakes where crops were grown along the irrigated banks (Bouchery and Lecomte-Tilouine, 2010).

4.9.2 Breeding and Modification of Rice Cultivars

Traditional varieties of rice grown prior to the Song Dynasty are reported as requiring over 150–165 days for ripening of early varieties, and between 180 and 200 days for late varieties (Zeng, 1998). According to Zeng (1998), two varieties of early ripening rice were introduced to China during the Song Dynasty. These include *Huang-lu* rice, which grows in wetland environments and has an extremely short growing season of 60–90 days, allowing easily flooded wetland/swamplands to be colonized. *Champa* (or *zhan-cheng*), an early ripening and drought-resistant rice, could be planted in the highlands and is reported as having a growing season of 60–110 days (Zeng, 1998). Champa rice was introduced to China under the reign of the emperor Zhenzong from Vietnam. This rice was more drought-resistant and could be grown in areas where earlier varieties of rice had failed, especially on hills and areas where water production was controlled with difficulty (Simoons, 1991; Zeng, 1998). In northern China, its introduction allowed rice to be grown in areas where rice productivity had been impeded because of the short growing season and in the Jiangnan area (i.e., those regions south of the Yangtze lower reaches) it allowed farmers to grow two different crops to increase productivity. By 1012 CE, Champa rice was introduced to the lower Yangtze and Huai River valleys.

4.9.3 Diversification and the Introduction of New Crops

Today in many parts of southwest China, the introduction of short season varieties of rice has allowed winter wheat to be grown as a catch crop further bolstering this region's productivity (Zhen, 2000). It is unclear when the use of wheat as a double crop began in China. According to Bray (1994), increasing mentions of double cropping system take place after introduction of Champa rice and after the fall of the Northern Song Dynasty to the Mongols in 1126 CE. However, some mentions of double cropping winter wheat or barley followed by millet are present in Fan Shengzhi's Agricultural Manual that was composed in the first-century BCE (Shi, 1956, 1974). A text found in a Western Han dynasty tomb in Yungang city in Jiangsu province contains some of the earliest historical mentions of winter wheat (Zhao and Chang, 1999). A twelfth-century work (the *Ji Le Bian*) describes that this tradition of double cropping took off shortly after the movement of refugees from the northwest to the Yangtze River delta, the Dongting lake area and the southeast coast (Zhuang, 2009). This was also partially due to taxation reasons, as farmers were only taxed on rice harvest, and by growing a second season of wheat, large untaxable profits could be made.

A series of introductions of foreign crops modified the cultivated landscape in southwest China. The patterns of adoption of these crops were complex and cultural, because climate, historical and ecological conditions played important roles in dictating how these processes took place. Short growing seasons, tolerance of poor soils, and low numbers of growing degree days meant that during the sixteenth and seventeenth centuries, crops introduced from the New World were received well in southwest China (Crosby, 2003; Jones et al., 2011b). These crops included sweet potato, potato, maize, and peanuts. These crops extended the range of cultivation in upland southwest China and in saline areas, particularly those along the Chinese coast. As early as 1959, Ho (1959) had argued that this "second agricultural revolution" made the drylands and hills of northern and western China turn from virgin land to maize and sweet potato farms, allowing a second demographic transition to take place.

Prior to the introduction of crops from the New World, it is worth briefly discussing sorghum. Sorghum was domesticated in sub-Saharan Africa sometime between 6,000 and 5,000 years BP (Dillon et al., 2007; Fuller, 2007; Weber, 1998). Following this, it is conjectured that sorghum arrived in India around 4,000 years

BP where it appears to have taken a long time to gain importance (Weber, 1998). Dillon et al. (2007) argue that sorghum was introduced to East Asia by approximately 400 CE. Known as *shu shu* in Chinese, mentions of this crop do not appear in literature until 300 CE. However, it is unclear if this mention is correct, or if the word actually referred to another crop (Bray, 1994; Bray and Needham, 1984). The Chinese name of sorghum, *shu shu* (literally millet of Sichuan), implies that it may have first been grown extensively in Sichuan. Bray suggests, however, that sorghum (also known as *gao liang*) did not become important in China until rather late (the Yuan and Ming dynasties). Modern varieties of sorghum require little heat: something that likely made it an attractive crop during the Little Ice Age.

Maize appears to have arrived in China by the sixteenth century. An account by a Portuguese priest, Alvarez Semedo, written in 1613 reports that maize was grown near Beijing for use by the emperor (Laufer, 1907). Ho (1959) argues that this crop became popular throughout China after the Ming Dynasty (1368–1644 CE) and contributed to a huge population boom during the Qing Dynasty. European explorers and botanists in the late nineteenth century found that maize was the primary crop grown throughout western Sichuan and Yunnan (Davies, 1909; Wilson, 1911) and noted that this crop was cultivated on land that was unsuitable for rice, particularly in hills and lower mountains where slopes are too steep. The fact that maize was grown on poor land appears to have led to its classification as a food associated with poverty (Anderson 1988; Gamble 1933). Despite its low prestige, this crop's function as an apparent risk-reducing or buffering crop and its ability to extend the land under cultivation led to its rapid adoption in southwest China. Maize can also grow under cooler temperatures and at higher altitudes than rice, which allowed parts of upland southwest China to benefit from a highly productive grain crop.

Chen and Kung (2011) argue that maize played an important role in population growth in Ming and Qing China. Using instrumental modeling, they proposed that maize accounted for a 21% growth in population levels. Others, however, suggest that an already growing population generated the need for the adoption of new crops and an expansion of agricultural land (Lee and Feng, 1999; Perkins, 1969). While the arrow of causality here is unclear, the fact that maize produces high yields and large grains and does not require substantial modification of the landscape likely contributed to its rapid uptake in southwest China.

The sweet potato (*Ipomoea batatas*) is another crop native to the Americas that eventually became important in southwest China. There is some discussion as to where exactly sweet potato was domesticated. The oldest remains of sweet potatoes that can be hypothesized to be domesticated come from Peru and date to roughly from 4,000 years BP (O'Brien, 1972; Zhang et al., 2000). The sweet potato is believed to have been introduced to China in 1594 CE by a Chinese businessman from Luzon in the Philippines who brought this plant to the attention of the governor of Fujian during a year of famine (Zhang et al., 2009). It thus appears to have initially been cultivated as a risk-reducing crop. Sweet potato is adapted to a high-altitude environment and poor soils. Sweet potato can also be grown on steep slopes without substantial modification of the landscape, and as a result, increased the productivity of this region dramatically by extending the range of cultivable soils (Chen and Kung, 2011). Ho (1959) argues that the adoption of sweet potato was motivated by the fact that in the eighteenth century rice had already reached its carrying capacity and farmers needed to expand into areas that could not be irrigated if they wished to increase production.

Thus, although population pressure created important motivations to find new varieties of crops that could sustain a growing population, cultural values that dictated the importance of rice meant that the emphasis was put on finding new varieties of rice or on combining rice with other crops rather than on changing crop repertoire (Bray 1994). Although climate, ecological conditions, and plant phenology influenced how humans adapted their agricultural systems to southwest China, humans themselves also actively modified and constructed the niches associated with these crops in order to ensure the continuation of foods that had important cultural meaning, such as by constructing paddies to ensure the production of rice. When cultural values demanded it, humans not only engaged in some of the most amazing feats of niche construction by investing considerable labor into terracing systems but also found ways to modify the biological properties of plants themselves, most notably rice, through breeding.

4.10 Synthesis and Summary

Rice farming traditions developed throughout prehistory in Asia have been at risk of being lost through development programs such as the "Green Revolution," that have sought to increase yields through labor-saving devices such as mechanization, field expansion, and application of commercial fertilizers (Bray, 1994; Lansing and Fox, 2011) (see Chapter 6). In this chapter, we reviewed how human niche construction and genetic modification of the rice plant itself has allowed humans throughout East Asia to successfully adapt to episodes of cooling climates, increased monsoonal intensity, and the geographic challenges posed by the new landscapes into which they moved farming. These systems have enabled parts of China, in particular, to sustain some of the highest levels of population both in prehistory and the present-day throughout multiple episodes of monsoonal variability. It is these same systems that may provide us with essential knowledge for mitigating the climatic challenges that the future might bring.

5

Dryland Farming in the Northern Monsoon Frontier

In the northern frontiers of the monsoon, farming systems based on dryland crops, such as millets and later wheat and barley, supported the rise of large states and eventually China's earliest dynasties. Unlike the ecosystem-engineered paddies we reviewed in Chapter 4 that characterize southern East Asia, the agricultural systems that typify northern Asia are largely rain-fed. As a result, these have been profoundly impacted by changes in both monsoonal rainfall and changing temperatures. Unlike the wet, rice-based systems in southern China that prevent erosion and increase soil fertility, erosion and problems with soil fertility have been a constant facet of challenges facing farmers in northern Asia.

Two crops, broomcorn (*Panicum miliaceum*) and foxtail millet (*Setaria italica*) that were domesticated in this area are known to most Europeans only as bird seed. Yet, both of these crops fueled the growth of the polities that eventually became China. Much like the other millets we reviewed in Chapter 3, these millets are remarkably tolerant of poor soils, low water, and high temperatures. Both of these crops have lost ground in some regions and time periods to two crops that were domesticated in the fertile crescent: wheat and barley.

Unlike the farming systems developed in Europe that also relied on dryland forms of farming, dryland farming systems developed in East Asia were cultivated much more intensively throughout history. In European farming systems, large portions of land were devoted to pasture or left fallow for pastoral animals. In China, on the other hand, dryland cultivation systems are often in continuous rotation of grain and vegetable production (Bray, 1986). Descriptions of agricultural systems of northern China during the Wei Dynasty in the agricultural treatise the *Qi Min Yao Shu* (ca. 535 CE), showed that the rain-fed systems that developed in East Asia were far more intensive than their western Eurasian counterparts. Farmers used elaborate systems of crop rotation and green and animal manure, allowing the cultivation of fields without fallow. The types of farming practiced in northern Asia during deeper time are unclear, although some have argued that shifting cultivation was in place until the Zhou Dynasty (690–705 CE), while others argue that intensive farming took place much earlier (Bray, 1986).

The lands that lie outside of the area where crops can be farmed are largely the territories of pastoralists: people who Chinese state builders categorized as foreigners or "barbarians." We describe the origins of subsistence regimes of people both inside and outside this cultural divide.

This chapter charts changes in human subsistence patterns over a series of major climatic periods: the end of Marine Isotope Stage (MIS) 3 (around 57,000–24,000 years BP), the Last Glacial Maximum (LGM) or MIS 2 (around 26,000–20,000 years BP), followed by the period of deglaciation that is characterized by the warmer Bølling-Allerød (around 14,500–12,800 years BP) and the colder Younger Dryas (around 12,800–11,500 years BP) interludes, and finally several transitions throughout the Holocene (11,500 years BP to present), including what appears to be a major downturn in temperature and aridification beginning at roughly 4,000 years BP (Herzschuh et al., 2006; Pei et al., 2017). Archaeological debates have centered around the impact that these changes in climate had on the development and evolution of subsistence regimes on the northern margins of the monsoon. In some cases, climate has been viewed as the primary driver in subsistence change. However, in others, archaeologists have proposed that innovations in subsistence took place under optimal climatic conditions. Throughout this chapter we chart how humans have used strategies of diversification, intensification, and niche construction, to take advantage of the characteristics of the landscape they inhabited or to adapt to climatic change.

5.1 Paleoclimate in Northern Monsoon Area from MIS3 to the LGM

Prior to the start of the Holocene, fluctuations in temperature and precipitation at the end of the Pleistocene began to substantially impact the life ways of the foragers that occupied the northern Monsoon frontier. Several studies have suggested that a warm and wet climate prevailed during the late stages of MIS 3 (around 57,000–24,000 years BP) (Shi, 2002; Yang et al., 2004; Zhao et al., 2014). A recent study combining over 40 different pollen cores implies that temperatures during the end of MIS3 were similar to those of the modern day (Zhao et al., 2014) and that northern China was characterized by conifer forests and forest meadow steppe. Monsoonal patterns may have been weaker during the late MIS3 than in the present day, making it likely that northern China was more susceptible to fluctuations in temperature than southern China. Lake records from northwest China indicate this was a period of rapid fluctuation linked to strengthened moisture flux via the westerly jet (Zhang et al., 2016a). Decreasing solar insolation during MIS3 resulted in a change from warmer, wetter conditions at 59–51.7 ka, switching to more desert-like conditions at 51.7–27.3 ka, as shown by changing vegetation and desiccation of lakes (Zhao et al., 2017b). Toward the end of MIS3, the eruption of the Aira volcano in Japan (around 28,000 years BP) may have had an important impact on the distribution and productivity of resources available for foragers (Morisaki, 2012).

During the LGM (around 26,000–17,000 years BP), temperatures in northern China dropped precipitously. For example, climate models predict that permafrost conditions extended around 10° further south than they did during preindustrial times (Liu and Jiang, 2016). Clumped isotope studies from the Chinese Loess Plateau indicate temperatures were 6°C–7°C below preindustrial levels, implying that the north China region is especially sensitive to the effects of large continental ice sheets (Eagle et al., 2013). Although competing proxies differ in their estimates of temperature changes that were derived from pollen (Ni et al., 2010; Shi, 2002; Sun et al., 2000) rather than ice core records (Shi, 2002), temperatures were likely 6°C–9° C cooler on the Tibetan Plateau, 5°C–11°C cooler in eastern China and 6°C–8°C cooler in Japan during the LGM relative to present day, comparable to the Loess Plateau data. Xerophytic shrubland, deserts, and grasslands appear to replace areas that were once forested across northern China (Ni et al., 2010; Zhao et al., 2014) (Figure 5.1).

These grassland biomes appear to have extended into areas that are now underwater (Sun et al., 2000). Until roughly 14,700 years BP, the massive continental shelf covering an area of 120–150 million km^2 was open across the length of eastern China. This area, which likely once housed essential resources for foragers, has been lost to rising sea level and became the East China Sea (d'Alpoim Guedes et al., 2016a). Because of its proximity to the ocean, warmer temperatures likely characterized this area, something that may have been attractive to early humans (Figure 5.1). Much of what we might know about early human adaptations to this area may thus be lost underwater.

5.2 Late Pleistocene Foragers in the Northern Monsoon area

Changes in climate toward the end of the Pleistocene impacted foragers who were likely highly mobile through creating changes in the distribution and abundance of the resources these foragers relied on.

During the late MIS3, there were few clear divisions in lithic technology across the territory of China: expedient core and flake technology was distributed across most of northern and southern China, up to 40° latitude in the west and almost up to 50° latitude in the warmer seaward eastern flank (Qu et al., 2013). This suggests that tools that helped humans survive in the South, were also adapted to usage in the

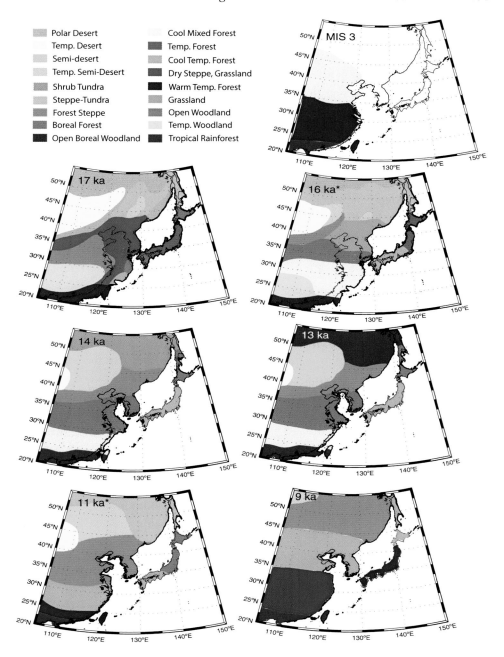

Figure 5.1 Changes in coastline and biome distribution between the MIS3 and 9,000 calibrated (cal) years BP. The panels show the predicted, gradual transgression of the shelf area as the Earth system moved out of the Last Glacial Maximum. The thin black line marks the present-day coastline for reference. The rapid coastline retreat predicted between 15,000 cal years BP and 14,000 cal years BP is due to

North: it is possible that the relatively mild temperature conditions during this period of time did not necessitate the development of new adaptive technologies. This technology maps closely onto what are predicted to have been forested biomes during the MIS3. Only in a few areas to the far west and north was a different type of technology present: one that was characterized by blades likely spread to this area from Siberia and Mongolia (Figure 5.2).

With the start of cooler temperatures during the LGM, we see a clear shift take place in lithic technology across East Asia: individuals that produced core and flake technology are pushed southward (to between 32° and 30°N latitude) where they go on to develop a number of key adaptations including early pottery (Qu et al., 2013) (Figure 5.2). Blade and microblade-based technology that had begun to arrive on the northwestern borders of China at the end of MIS3 expanded rapidly, covering most sites North of 35°N latitude, including sites in Korea by around 26,000–24,000 years BP (Bae and Kim, 2010) and in Japan by 20,000 years BP (Ikawa-Smith, 2004) (Figure 5.2).

During the LGM, the Japanese islands may have been an attractive area for foragers because the warmer oceanic climate meant that temperate forests continued to colonize these areas (Figure 5.1; see panel at 17 ka). There is currently little evidence for exploitation of marine biomes by foragers who inhabited the Japanese islands during MIS3 and the LGM, although it is highly likely that the sites at which people exploited marine resources now lie underwater. In what were once inland areas, hunting traps have been uncovered that were laid for deer and wild boar. In addition, the occasional presence of grinding stones suggests that humans exploited both nuts and tubers as early as 32,000 years BP at Tanegashima (Ikawa-Smith, 2004; Pearson, 2006).

In inland northern China, LGM conditions appear to have impacted hunter-gatherer distributions more severely than in coastal regions, with some areas appearing to be too cold to sustain even small human populations. For instance, the area north of 41°N appears to have been completely abandoned during this period of time (Barton et al., 2007) (Figure 5.3). The aridity of the climate was the principal reason that there was no major continental ice sheet

Caption for Figure 5.1   (cont.)

meltwater Pulse 1A. We use present day coastline for the MIS3. Color-coded biomes are based on Adams (1997) and BIOME 6000: The biomes for the times denoted by an asterisk (11 ka and 16 ka) are bes-guess estimates based on visual interpolation between the existing biomes (for 11 ka we additionally utilized the time slice of 10,200 cal years BP by Adams, 1997 #938}). MIS3 conditions are best-guess estimates after Zhao et al. (2014): only known biomes are interpolated. Modified after d'Alpoim Guedes et al. (2016a).

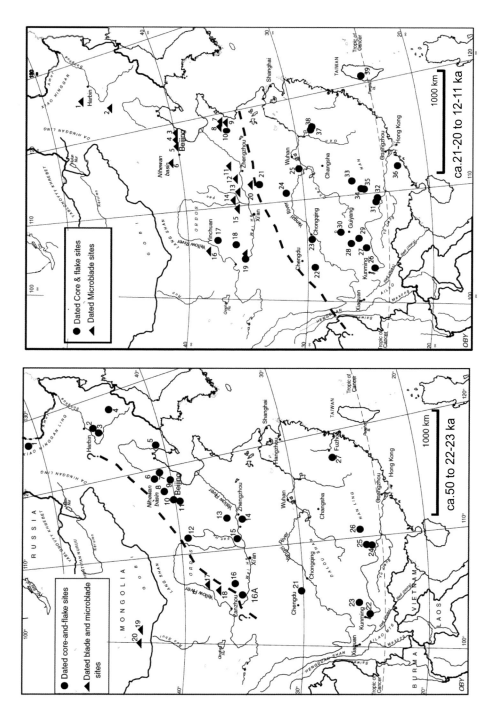

Figure 5.2 Distribution of core and flake and blade sites across northern China during LGM and MIS3. Reproduced from Qu et al. (2013).

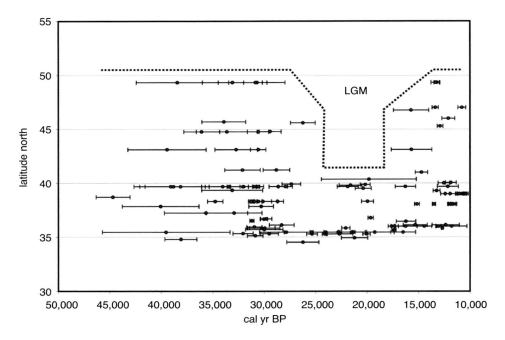

Figure 5.3 Radiocarbon data from Paleolithic North China (*n* = 116) plotted as a function of latitude. This illustrates that north China was depopulated above 41° north during the Last Glacial Maximum (24.0–18.0 ka). Diamonds represent the 2 standard deviation midpoint of each calibrated estimate, while the circles represent the 2 standard deviation range. Reproduced from Barton et al. (2007).

in northeast Asia, even at the LGM, although the presence of limited ice rafted debris in marine cores does imply a more limited ice sheet in the Kamchatka area (Bigg et al., 2008). Outside that region glaciers were limited to mountain regions.

Some have argued that the sudden appearance of microlithic technology represents a recolonization of uninhabited land following the abandonment of cold areas during the LGM (Brantingham et al., 2004; Goebel, 1999). However, others argue that the producers of this lithic technology were highly mobile hunters who followed large game as it migrated down from what was once steppe in the North during MIS3 (Barton et al., 2007; Elston et al., 2011; Yi et al., 2013). Microblade technology appears to have served a wide range of uses including the production of harpoons for fishing (such as those from a later site, Xinglongwa; Figure 5.4), production of composite hunting weapons, and possibly for the production of clothing. For instance, at Shuidonggou (Figure 5.4), finds of microlithic blades inset into a knife combined with large numbers of hare remains and bone needles

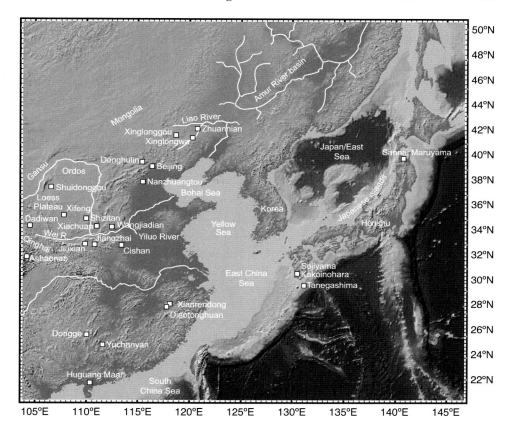

Figure 5.4 Shaded topographic map of East Asia showing the locations and geographic features discussed in this chapter.

have led some to argue that producing leather and fur clothing crucial to surviving the extreme cold was a key function of this technology (Yi et al., 2013).

Barton (2007) argues that cold temperatures during the LGM limited the radius over which these foragers were able to operate and brought once disparate groups closer together where they had to compete more intensively for resources, creating pressure to innovate through technological developments such as microliths. These cultural changes during the late Pleistocene were primarily driven by changes in group size, between group competition and mechanisms of cultural transmission (Barton 2007). Mobility as a means of dealing with climatic instability was no longer a risk-reducing option that they were able to pursue and as a result they began to utilize different ranges of strategies. Strategies employed might have included practicing storage of food, widening the diet through including new species or new adaptations at the group level, such as building extensive ties with other groups through marriage (Barton 2007). They argue that the rapid diffusion of microblades across East Asia points to the existence of groups who rapidly shared technological innovations with each other: something that was likely an adaptive strategy for dealing with their new, less than optimal environment.

In addition to cold limiting foraging radius toward the end of the LGM, changes in sea level due to Meltwater Pulse 1A (around 14,700 years BP) severely impacted the radius that these foragers were able to operate in. A coastal shelf of roughly 120–150 million km^2 was open across the East Asian seaboard for roughly 25,000 years: a period of time during which it was likely occupied by humans (d'Alpoim Guedes et al., 2016a). A recent analysis shows that the ancient coastline contained areas where the water was no deeper than 30 m and ranged in temperature from 5° to 20°C, likely housing kelp forests that contained a wide variety of marine resources for human consumption (d'Alpoim Guedes et al., 2016a). These highly productive areas likely lay at the junction of Korea, the East China Sea, Japan, and the Bohai Bay during the LGM (d'Alpoim Guedes et al., 2016a). The low-lying shelf itself may have also hosted swamps, marsh-like biomes that were productive for hunting and gathering (Figure 5.1).

Meltwater Pulse 1A, which began around 16,000 years BP and was completed at around 11,000 years BP, rapidly engulfed this shelf (d'Alpoim Guedes et al., 2016a), with sea levels rising around 20 m in just 500 years (Figure 5.1). It is possible that this rapid influx of water, derived from Antarctic melting, was partially responsible for causing the warmer conditions that characterized the Bølling-Allerød (Weaver et al., 2003). This rapid rise in sea level means that we know little about how foragers tailored their subsistence strategies to the available biomes on this continental shelf as these sites have been lost to post-LGM sea-level rise.

5.3 Intensification during the Bølling-Allerød and Younger Dryas

During the Meltwater Pulse 1A transgression, early climatic refugees likely migrated inland, creating population pressure that in addition to the fluctuating temperatures at the end of Bølling-Allerød and beginning of the Younger Dryas, may have led to a number of key adaptations. During the Bølling-Allerød, boreal forest appears to have regained ground in northern central China that would have been intermixed with a forested steppe-like environment (Figure 5.1).

Application of stable isotope measurements to leaf wax biomarkers has proven useful in the same areas of northeast Asia for tracking the evolution of environment since the LGM. It is recognized that the relative abundance of C$_3$ to C$_4$ vegetation can be affected by a number of environmental conditions that include temperature, as well as moisture and atmospheric CO$_2$. A synthetic profile was constructed by

Liu et al. (2005b) based on five core sites taken across the central Chinese Loess Plateau from which the carbon isotope composition of the dominant leaf wax compounds (C29 and C31) was determined. Consistent trends of δ^{13}C values in both leaf waxes and total organic carbon (TOC) (Figure 5.5) provide evidence that the δ^{13}Cvalues of TOC can also be used as a reliable indicator of past changes in C_3 and C_4 plants. Liu et al. (2005b) demonstrated a coherent increase in δ^{13}C values from 35,000 years BP until 8,000–10,000 years BP, at the same time that global temperatures rose after ~20,000 years BP, followed by a decrease in δ^{13}C until around 3,000 years BP (Figure 5.5). Higher isotopic ratios argue for a greater dominance in C_4 grassland flora in the Chinese Loess Plateau region and that increased C_4/C_3 ratios are positively correlated with the higher temperatures and heavier summer rainfall that characterize stronger summer monsoons. We do note that the increase in C_4, and presumably in summer rainfall initiates around 15,000 years BP, is somewhat earlier than has been inferred from other proxies discussed earlier, but at a time of warming temperatures.

Some have argued that the foragers who occupied the grassland steppes of northern central China might have taken advantage of this new abundance of plant resources and may have begun a "broad spectrum revolution."

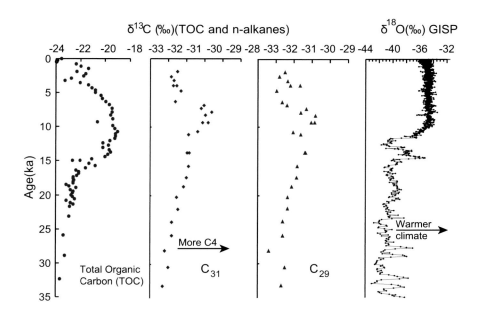

Figure 5.5 Downcore variations of δ^{13}C values of n-alkanes C29 and C31 and TOC from the Xifeng profile for the last 35 ka. The δ^{13}C values of high-molecular-weight alkanes (leaf wax) from the Xifeng profile are consistent with carbon isotopic composition of total organic carbon. Reproduced from Liu et al. (2005).

Demographic packing due to the loss of former foraging territories lost as a result of Meltwater Pulse 1A may have created an additional pressure on foragers to modify and constrain their mobility strategies to smaller radii and to begin to intensify their environment. A new set of lithic technology characterized by grinding stones appears between the end of the LGM and the Bølling-Allerød. Finds of grinding stones appear as early as 23,900–15,400 years BP at sites such as Xiachuan (Wang, 1978). At Donghulin (11,000–9,000 years BP), acorn starch on grinding stones suggests that foragers had begun to exploit nuts (Liu et al., 2010), and macrobotanical remains showed that people consumed hackberries (Hao et al., 2008a). The presence of these heavy stones suggests that people had started to become somewhat less mobile and make investments in technology that remained at one campsite. Analysis of animal bones from the same site showed that people hunted deer, pigs, and badgers, but also exploited freshwater shells, indicating a potential widening of the diet.

An analysis of charred seeds from the Shizitan site (13,800–11,600 years BP) that was occupied during the Bølling-Allerød/Younger Dryas transition showed that, in addition to nuts, these foragers had begun to intensify their production strategies by employing small seeded grasses, such as close relatives of domesticated millet, wild goosefoot (*Chenopodium* sp.), and wild barnyard grass (*Echinocloa* sp.*)* (Bestel et al., 2014). Samples of ancient starch grains from Shizitan also indicate that humans may have begun to exploit resources like acorns and possibly tubers (e.g., *Dioscorea* sp.), although the authors reference collection did not permit them to make a positive identification (Liu et al., 2011). Recent analysis of starch and of microbotanical remains from the Nanzhuangtou site (around 12,000 years BP) shows that foragers exploited wetland environments by at least 12,000 years BP: the authors found evidence of the exploitation of water caltrop (*Trapa* sp.), naids (*Najas* sp.), and pondweed (*Potagmogeton* sp.) (Yang et al., 2015c). At Zhuannian (around 10,000 years BP), macrobotanical evidence also indicated that humans ground walnuts (*Juglans* sp.) into flour on these stones (Yang et al., 2015c).

At some of these sites a new technological innovation, pottery, which allowed humans to boil foods with ease, appear. The earliest dates for pottery in East Asia are from southern China: at the sites of Xianrendong (Wu et al., 2012b)(20,000 years BP), Diaotonghuan (23,000–20,000 years BP)(MacNeish et al., 1998), and Yuchanyan (18,000–17,000 years BP)(Boaretto et al., 2009). Early finds are also present in Japan (at 16,400–11,300 years BP) and in the Amur River basin in the Russian Far East (16,500–14,100 years BP)(Kuzmin, 2006a, 2010). In northern China, the pottery appeared substantially later (12,000–9,000 years BP) at sites such as Zhuannian, Donghulin, and Nangzhuangtou (Figure 5.4).

Some have argued that this pottery represents an adaptation to the cold environment of the LGM: one that allowed hunter-gatherers to exploit a wider range of plants and animals (Bar-Yosef, 2011; Boaretto et al., 2009; Qu et al., 2013; Wu et al., 2012b; Yasuda, 2002). The earliest appearance of pottery in southern China (see Chapter 4) does correspond to the peak of the LGM; however, its distribution appears to map closely onto temperate or boreal forest biomes, whether these be in southern China, in Japan, or in early Holocene northern China (Figures 5.1 and 5.2).

The fact that pottery is widely distributed across East Asia has led scholars to suggest that cultural connections existed between the hunter-gatherers that produced it (Chi, 2002; Jordan and Zvelebil, 2009; Shelach, 2012), however, others (Keally et al., 2004; Kuzmin, 2006b) argue that this pottery was independently invented in each of these different regions. The fact that pottery is present for a long time in southern China, but begins to appear around the start of Meltwater Pulse 1A in Japan and regions affected heavily by sea-level rise, makes it possible that this technology was once employed by foragers who occupied forested and riverine biomes on the coastal shelf who retreated to formerly inland areas after sea-level rise. Analysis of lipid remains in this pottery demonstrates that foragers in different regions employed these pots to target different types of food, with some focusing on hunted game and others on fishing (Shoda et al., 2020).

We argue that it is possible that increasing population density, at least partly driven by coastal transgression, created additional competition for large game and limited available foraging radius, combined with migration of game to higher latitudes during the Younger Dryas and may have created a forcing factor that lead people to start to more intensively exploit their environment.

Following the likely demographic packing of foragers that took place after Meltwater Pulse 1A, we see particularly clear evidence for more intensive strategies in the Japanese Isles. For instance, at the Sojiyama site (13,500 years BP) in southern Kyushu, indications exist that humans had begun to process food for storage and to build permanent houses. Two houses were found at the site, as well as ventilated hearths that were likely used for smoking meat, thus allowing it to be kept throughout the year (Pearson, 2006). A large number of grinding stones and pottery were also present at the site where they are hypothesized to have been used to process nuts from deciduous and evergreen trees that Pearson (2006) argues began to recolonize the area following the end of the LGM. At the Kakoinohara site (13,500 years BP), one grinding stone weighed roughly 39 kg indicating that it was not moved in a residential mobility pattern. Rather the excavators of the site propose that it represented a summer camp, where individuals resided for several months to process nuts (Pearson, 2006). Numerous storage pits for nuts have also been unearthed in Incipient Jōmon sites throughout Japan, indicating that profound changes in forager mobility had occurred in forested biomes with nut-producing trees. Increasing evidence from a large number of sites in both Japan and Korea indicate that humans in coastal areas relied extensively on collecting shellfish and fishing (Matsui, 1996; Norton et al., 1999), much like they likely did in earlier periods of time in sites now lost to sea-level rise.

5.4 The Origins of Millet Farming in North China

Following the end of the Younger Dryas, forest mixed with steppic grasslands began to regain territory across northern China: a reconstruction of biomes at around 9,000 years BP from Adams (1997) shows that desert steppe continued to prevail around Gansu/Qinghai provinces, while cool and mixed forest occupied northern central China and Korea (Figure 5.1).

Several potential areas of domestication have been proposed for foxtail and broomcorn millet domestication. Archaeologists have debated whether or not millet domestication took place in the context of climatic instability, or whether foragers began to domesticate millets under a context of climatic stability or optimal climatic conditions. Evidence for domestication relies on documenting morphological changes in plant structure or genetics that facilitate their growth in disturbed environments and collection by human dispersers. These include an increase in grain size: a characteristic that likely developed in order to allow seeds enough reserve energy to germinate at the greater depths that tilled fields created as a habitat. Another key development is the lack of natural dispersal mechanisms, such as the development of a non-shattering rachis, which facilitates human collection once the grains ripen (Fuller, 2007; Purugganan and Fuller, 2009). Evidence for early cultivation (i.e., planting and harvesting plants that may be morphologically wild) requires other lines of evidence such as preserved fields and tools for planting like spades.

Unlike wheat, barley, and rice, detecting when exactly millets began to show morphological traits that are indicative of human management is challenging. This is primarily because the spikelet bases that would exhibit a non-shattering rachis in millet are too small to be recovered from archaeological sites. In addition, it is unclear what if any size difference there might have been between wild and domesticated millets. Based on finds of millet grains several different centers for millet domestication have been proposed that we examine here.

5.4.1 Northwestern China

A first center of domestication is argued to have taken place at the boundaries between desert steppe and forest in northwest China among groups who were limited in their mobility for game hunting by other populations (Bettinger et al., 2010; Bettinger et al., 2007). Bettinger et al. (2010; 2007) argue that cultivating and exploiting small-seeded grasses was likely a costly endeavor for these foragers, however, population pressure and climatic instability forced them to innovate and begin to exploit these grasses. However, they were not able to provide a clear timeline along which these transformations took place.

While we have already seen that foragers began to utilize small-seeded grasses as early as 13,800 years BP in central China, some of the earliest concrete evidence for humans exploiting millet in northwest China comes from Dadiwan in Gansu Province. The evidence from this site is relatively scant and late in date. Only eight carbonized grains of broomcorn millet that were dated to 8,500–7,900 years BP have been unearthed (Gansu Sheng Wenwu Kaogu Yanjiusuo, 2006): a date substantially later than the supposed beginning of the exploitation of these plants. The small quantity of these finds and the lack of systematic archaeobotany at early sites in northwest China make it impossible to determine what their importance in the diet was. It is only at the tail end of this process (by 7,900 years BP) that we have stable isotopic evidence indicating that both humans and their domestic animals consumed these crops (Barton et al., 2009). However, it is worth noting that they did so in low proportions in initial phases of the site (Barton et al., 2009).

Bettinger et al. (2007) argue that domestication took place in this area sometime between 13,000 and 7,000 years BP, a period of time during which major changes in temperature took place and when the distribution of heat-loving plants like millet likely varied widely (d'Alpoim Guedes, 2016). By 10,800 years BP millets regained their thermal niche across the area that now corresponds to northwest China. Depending on what future research in this area shows about the timing of millet domestication, this may have taken place under optimal climatic conditions if this process began after 10,800 years BP. However, if humans began cultivating these small-seeded grasses earlier (or under colder climatic conditions), then foragers were likely attempting to ensure the production of a resource that was losing ground (Figure 5.6). More research on the timing of millets domestication in northwest China is necessary to help resolve this question.

5.4.2   The Central Plains

In central China, the sites of the Peiligang Culture (around 8,000 years BP) have been posited as another center for millet domestication. The Peiligang culture site, Cishan, was excavated in 1976 and 1978, prior to the implementation of flotation technology in China. At the time shells of walnut (*Juglans regia),* hackberry (*Celtis bungeana*), and hazelnut (*Corylus* sp.) were handpicked from large storage pits

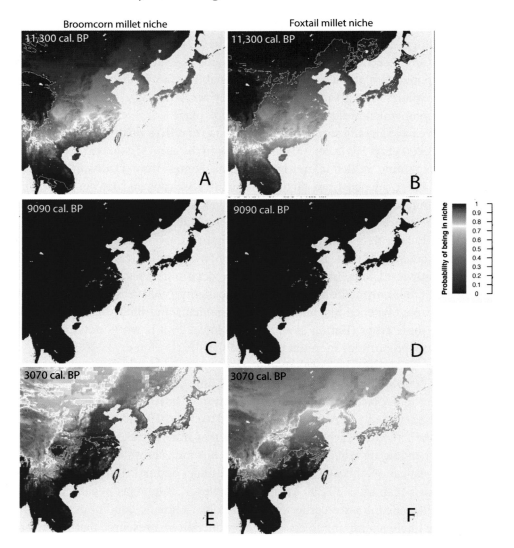

Figure 5.6 The thermal niche of broomcorn millet niche across Northern Asia at (A) 11,300 years BP, (C) 9,090 years BP, and (E) 3,070 years BP. The thermal niche of foxtail millet niche across Northern Asia at (B) 11,300 years BP, (D) 9,090 years BP, and (F). Figure created using methods described in d'Alpoim Guedes and Bocinsky (2018).

indicating that the inhabitants of this site continued to rely on foraging nuts much like their Late Pleistocene/Early Holocene predecessors (Tong, 1984). Areas of gray sediment that contained voids in the shape of foxtail millet were also found in these pits (Hubei Sheng Wenwu Guanlisuo, 1981). Using phytolith and chemical biomarkers, a 2009 study attempted to push the evidence for millet domestication at Cishan back to around 10,000 years BP (Lu et al., 2009). The authors of this paper extracted samples from a pit on the margins of the site and carried out radiocarbon dates on wood charcoal. The phytolith and chemical biomarker evidence led the authors to claim that the evidence at this site represented the earliest "domestication" of millet. However, currently there is no standard for distinguishing between morphologically wild and domesticated millet phytoliths. Likewise, it is unclear if the biomarker, *miliacin*, employed by the authors is exclusive to domesticated millets and not their wild progenitors. The reference collection used by the authors did not include wild species, so it is difficult to assess these claims. Regardless, the large numbers of millets in pits imply that humans had begun to collect more extensively (Bettinger et al., 2010) or even cultivate these crops by this time. A few handpicked remains of reported foxtail millet have been unearthed at other sites dating to roughly 8,500–8,000 years BP (Henan Working Team no. 1 of IA CASS, 1983, 1984), implying that at the very least humans collected and stored these plants by this point in time. However, analyses of starch grains on grinding stones show that millets only formed a small portion of a much wider range of plants that was exploited (Liu et al., 2013; Liu et al., 2014). While millets were present, starch grains from acorns were recovered in the largest numbers indicating that gathered foods continued to play an important role in the diet.

Although the dates on these are unclear, 474 storage pits were unearthed at Cishan that were used for the storage of food. The investment in permanent storage structures further suggests that profound adjustments in mobility took place during this period of time. The evidence for intensive exploitation from Cishan corresponds to a period of time where the probability of millets being in the thermal niche was high across central China (Figure 5.6).

5.4.3 Xinglongwa and Northeastern China

Recent data from sites of the Xiaohexi Culture have shown that small amounts of wild broomcorn millet (*Panicum miliaceum* subsp. *ruderale*) were present in the assemblage as early as 8,000 years BP alongside wild fruits and nuts. The presence of millet occurs alongside increasing evidence for sedentism in the form of permanent house structures and grinding stones. Paleoclimate records from the region suggest that this transition took place during a warm and a wet period of time (around 8,100–7,900 years BP): thus rather than the picture of resource scarcity that is argued to have driven millet domestication in northwest China, Shelach et al. (2019) argue that millet domestication in northeast China took place during a period of clement climatic conditions.

Xinglongwa Culture sites (8,200–7,300 years BP) in the Liao River valley of northeast Inner Mongolia consist of carefully planned villages composed of square houses that were organized on a grid. The excavators of these houses argue that the amount of effort put into them means that they were likely occupied on a year-round basis, indicating that a clear shift had taken place in the types of mobility strategies humans employed. It is estimated that one of the type sites, Xinlonggou was capable of supporting population levels of up to 100 individuals (Guo, 1995). Pits that were placed both inside and outside the houses indicate that humans likely relied on millets planted in the summer. The density of population supported at these sites suggests that this millet-based diet was reliable. Both broomcorn and foxtail millet have been derived from a systematic campaign of flotation carried out at Xinglonggou (8,000–7,500 years BP)(Zhao, 2005): although both millets were present, broomcorn millet appears to have formed the majority of the assemblage at the site. Some of the specimens presented in a preliminary report of the data are longer and thinner than domesticated varieties of *Panicum miliaceum*, suggesting that this assemblage may have been undergoing a process of domestication (Zhao, 2005). More complete publication of these specimens is, however, necessary to confirm this hypothesis. The preliminary report from the site states that "weeds" dominated the bulk of the assemblage at Xinglonggou. It is unclear, however, whether the weeds described were actually edible plants that may have been consumed by the inhabitants of the site. Full publication of this data and further systematic analyses at other Xinglongwa period sites should help clarify these questions.

Regardless of whether or not millets were cultivated, gathered or domesticated at Xinglongwa sites, human bones from Xinglongwa have a high C_4 isotopic signature indicating that these millets formed an important component of the diet (Zhang and Wang, 2003). Between 8,500 and 7,500 years BP (and indeed the entire Early Holocene) humans who cultivated millet in this region did so under optimal thermal conditions for these plants to grow (Figure 5.7). Temperature and likely precipitation did not limit how and when humans were able to grow these crops and indicates that their domestication took place under conditions where they could be exploited with a very high probability of success (Figure 5.6) (Shelach-Lavi et al., 2019).

It is possible that humans first began exploiting millets during the cooler and less optimal conditions of the terminal Pleistocene. However, once warmer temperatures began in the Holocene, the recolonization of the area with these new plants could have turned what was once a minor component of the diet into a highly successful endeavor. In the case of the Xinglongwa sites, it is possible that foragers who occupied the steppe tundra grassland that characterized this area during the Younger Dryas, may have begun to exploit small-seeded grasses to complement

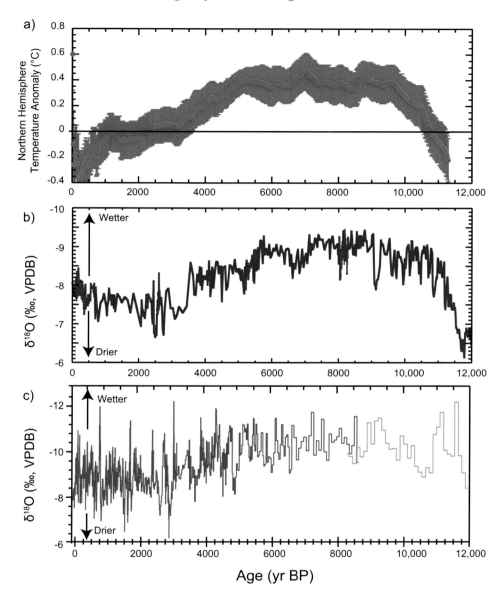

Figure 5.7 (A) Northern Hemisphere temperature record, from Marcott et al. (2013); (B) Rainfall intensity records from Dongge Cave oxygen isotopes, from Dykoski et al. (2005); and (C) Jiuxian Cave, from Cai et al. (2010).

their game hunting: partially because of the lack of nut trees and other resources in the area. The increasing productivity of small grasses following the end of the Younger Dryas may have created a pull for humans to invest in and rely more heavily on these plants.

5.5 Consolidation and Expansion during the Climatic Optimum

The climate in northwest China adjacent to the Loess Plateau has been constrained by a sediment-swamp section at Dadiwan (An et al., 2004). This location has the benefit of being positioned close to archaeological sites within the Yellow River valley, which permits the possible impact of climate change on human society to be assessed (Figure 5.8). Carbon-14 dating allowed the sedimentation history and local flora, as inferred from pollen assemblages to be reconstructed since the LGM. Pre-Holocene sedimentation appears to have been fluvial, but around 10,000 years BP swamp-like conditions were established, albeit with a flora dominated by herbaceous plants rather than trees or shrubs that favor wetter conditions than present today in that area. A strong change is noted after 8,300 years BP when the abundance of *Pinus* tree pollen increased sharply at the same time that sedimentation indicates establishment of a wetland. Taken together, these data indicate the onset of a wetter climate than that existing before 10,000 years BP. Subsequent increasing broadleaf pollen abundance, despite continued wetland conditions, is suggestive of a warmer, wetter climate between 8,300 and 7,400 years BP, continuing at least through to 6,700 years BP with a mixture of coniferous and broadleaf trees. Climatic conditions in the Yellow River valley moved back toward drier conditions between 6,700 and 6,300 years BP as pollen abundance decreased and herbaceous plants became more abundant, which was interpreted to reflect a change from a forest-steppe environment to one with more grassland steppe. The final drying seen in northwest China culminated in the accumulation of windblown loess after 4,000 years BP (Zhou et al., 2002), following the end of wetland conditions after 6,300 years BP.

5.5.1 The Yangshao Population Expansion of Mid-Holocene Optimum

During the Yangshao phase (6,500–5,900 years BP), the number of archaeological sites along the Yellow River valley and its tributaries substantially increased. Population densities increased at these sites during the Yangshao phase, reaching estimated communities totaling 400–600 people at sites like Jiangzhai (Peterson and Shelach, 2012).

Increasing numbers of archaeobotanical and stable isotopic research indicates that by this period of time millets formed a large component of the diet. For instance, by the Yangshao phase occupations at Dadiwan stable isotopic analysis indicates that millets had increased from being a relatively small percentage of the diet to then accounting for approximately 80 percent of the total consumption (Barton et al., 2009). Increasing population density may have led farmers to explore

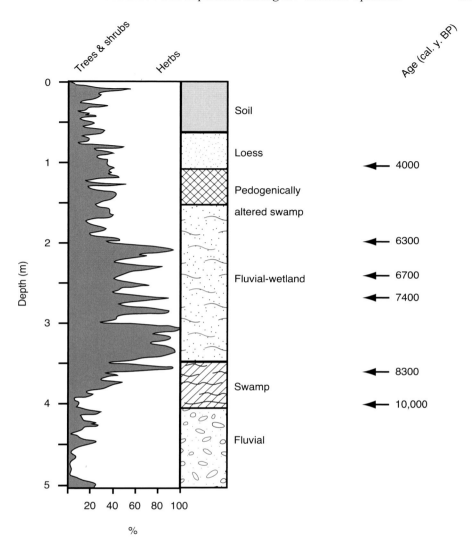

Figure 5.8 Sedimentary log of the Holocene section at Dadiwan, Loess Plateau, China (see Figure 5.4 for location) showing the transition from swamp and wetland conditions in the Early-Mid Holocene to a rapid drying after 6,300 years BP, as demonstrated by the spread of grasses and reduction in tree and shrub pollen. Redrawn with permission from An et al. (2004).

and exploit new territories. During the Miaodigou phase (5,900–5,450 years BP) Yangshao settlements spread further west into Eastern Qinghai and western Gansu, where they are sometimes known as the Shilingxia Culture (Yan, 1984). During the subsequent Majiayao period (5,400–4,650 years BP), millet farmers began a wave of expansion that lead them to upland western Gansu, Qinghai and eventually the lower altitude margins of the southeastern Tibetan Plateau (d'Alpoim Guedes, 2015; d'Alpoim Guedes et al., 2016b; Evan et al., 2011).

Majiayao farmers who appear to have spread into lower lying areas of the north and southeastern Tibetan Plateau came into direct or indirect contact with local groups of foragers who themselves began to exploit these crops, or obtained them through trade with farmers at lower elevations (d'Alpoim Guedes, 2015; d'Alpoim Guedes, 2016; d'Alpoim Guedes et al., 2016b). During this period of time (5,400–4,650 years BP), millet farming spread and intensified. An analysis of these crops thermal niche indicates that they could have been grown with a high probability of success across the surface of Northern Asia, making them a low risk option for farmers (Figure 5.6).

In the central Yellow River valley, recent geomorphological work by Rosen et al. (2017) suggests that farming activity had already begun to have an impact on the landscape during this period of time. While that study found that early Peiligang populations had little impact on the surrounding landscape and engaged in minimal deforestation, between 6,000 and 5,000 years BP, Yangshao farmers destabilized their landscape through deforestation, leading to increased erosion. This development is also supported by an analysis of environmentally derived charcoal that shows evidence for localized burning events that are not consistent with wild fires, potentially indicating that humans practiced slash and burn (or swidden) forms of farming to clear areas for new fields (Tan et al., 2011). Rosen et al. (2017) argue that this led to the alluviation of what were once deeply cut river valleys. This, in turn, allowed humans to farm on the newly created terraces in these river valleys that became seasonally flooded and continue to increase their agricultural returns (Rosen, 2008; Rosen et al., 2017).

5.5.2 Northeastern China and Korea

In northeastern China, a heavy reliance on millet in the diet appears to have continued into the Zhaobaogou (7,200–6,450 years BP) and Hongshan periods (6,450–4,950 years BP). Zhaobaogou sites are characterized by large houses with an average size of 31.78 m². Zhaobaogou itself likely housed between 200 and 380 people (Shelach, 2006). However, in this area and in contrast to the Central Plains where storage and cooking were carried out in common places, individual households appear to have carried out these activities in the northeast. By 5,300 years BP (or the Chulmun period), both foxtail and broomcorn millets had arrived in the Korean Peninsula where they seem to have been cultivated alongside adzuki bean (*Vigna angularis*) (Figure 5.6) (Crawford and Lee, 2003; Lee, 2011) and potentially domesticated soybean (*Glycine soja*) by 3,000 years BP (Early Mumun)(Lee et al., 2011).

5.5.3 *The Middle Jōmon Florescence, Chestnut Management, and Early Garden Crops*

During the Mid-Holocene, the Jōmon of Japan exploited a wide range of different plant and animal foods. An example of just a few of these includes acorns, wild adzuki beans (*Vigna angularis*), sweet chestnuts (*Castanea sativa*), horse chestnuts (*Aesculus hippocastanum*), barnyard millet (*Echinocloa crus-galli*), mulberries, (*Morus* sp.), Shiso (*Perilla* sp.), dock (*Rumex sp.)*, water chestnuts (*Trapa natans*), and walnuts (*Juglans* sp.). They also exploited trees for lacquer production.

While many of these resources were wild, there is increasing evidence that, by carrying out selection for traits that were useful to them, humans actively modified the habitat in which many of these species grew in order to increase their productivity: a form of niche construction. (Crawford, 2011). For instance, a recent study of chestnut genetics shows a bottle neck that may be indicative of human selection (Sato et al., 2003). Humans may have even introduced chestnut to environments outside of its natural range (Crawford, 2011). There is also increasing evidence that the Jōmon domesticated barnyard millet (*Echinocloa crus-galli*), or at the very least replanted seeds deep enough that they underwent an increase in size (Crawford, 2011) and had domesticated soybean by 5,000 years BP (Lee et al., 2011). Crawford (2011) also points out that a mutualistic relationship may have developed between humans and wild pigs. Pigs may have agglomerated around Jōmon settlements where food waste was plentiful, in turn, making them easy to harvest for humans at those sites.

5.6   The 4 ka Event, Collapses of Millet Agriculture on the Margins of the Tibetan Plateau and Northeast China, and Changes in Jōmon Japan

5.6.1 *Paleoclimate at 4,000 years BP*

The mid-late Holocene transition is marked by a documented decline in Northern hemisphere temperatures that have been noted in proxies from around the world (Figure 5.7) (Marcott et al., 2013). This corresponds to a decline in monsoonal intensity noted in proxies such as the Dongge speleothem from southwest China (Dykoski et al., 2005). Although most speleothem proxy records come from southern China, one proxy from Jiuxian Cave in the Qinling mountains may reflect how this change impacted northern China (Figure 5.7C) (Cai et al., 2010). In general, the decline in $\delta^{18}O$ values seems to start later at Jiuxian Cave than at more southerly locations, such as Hoti and Dongge. At Hoti Cave the climate seems to dry after 7.5 ka, while this trend started at 7.0 ka at Dongge Cave but as late as 4.7 ka at Sanbao Cave and after 4.5 ka at Jiuxian Cave. A synthesis of different paleoclimate records from across northern China, including pollen, lake level, and foraminifer records indicates that there was a decline in monsoonal intensity following 4,000 years BP (Wu and Liu, 2004) (Figure 5.9). The differences in timing across East Asia likely reflect the different response of the more tropical South Asian monsoon and the subtropical East Asian monsoon to the changing insolation forcing.

While several authors, including Wu and Liu (2004), have attempted to tie this 4,000 years BP event to changes in the archaeological record in Asia, the exact manner in which these changes in temperature and precipitation impacted human's ability to farm has only recently begun to be clarified.

5.6.2 Effects in High Latitude and Altitude Asia

The documented decline in temperatures appears to have had a fundamental impact on human's ability to farm millet in areas of high latitude and altitude Asia (Figure 5.6). Large areas where millets could previously be cultivated lost their thermal niche, including regions stretching from the Tibetan Plateau, Mongolia, northwest China, Manchuria, and northern Korea.

While millet's C_4 pathway and short growing season mean that they are able to grow in very arid conditions, both foxtail and broomcorn millets require substantial units of heat to complete their lifecycle (Cardenas, 1983; Cheng and Dong, 2010; Kamkar et al., 2006; Mann, 1946). Both foxtail and broomcorn millets have little to no-frost tolerance. Once the warm conditions that characterized the Holocene climatic optimum came to an end after 3,700 years BP, the probability of millets being in the thermal niche fell dramatically on the Tibetan Plateau (Figure 5.6). This corresponds to a fall in the numbers of radiocarbon dates across the region (d'Alpoim Guedes, 2015; d'Alpoim Guedes et al., 2015; d'Alpoim Guedes et al., 2016b) (Figure 5.10). On the southeast Tibetan Plateau, sites appear to have been abandoned and there is a clear 500-year hiatus in the radiocarbon record suggesting that millet farmers may have migrated to warmer areas or that they switched to a more ephemeral way of occupying the landscape. On the northeast Tibetan Plateau, a slightly different pattern has been discovered, where the earlier introduction of wheat and barley allowed for a continuation in occupation. This same cooling also affected the Ordos, eastern Mongolia, as well as parts of northeast China, although the potential effect that it had in these regions has yet to be fully documented in the archaeobotanical record.

5.6.3 Changes in the Late Jōmon

In Japan, the effects of this cooling event also appear to have been felt: the central mountains, which supported large populations during the Middle Jōmon period,

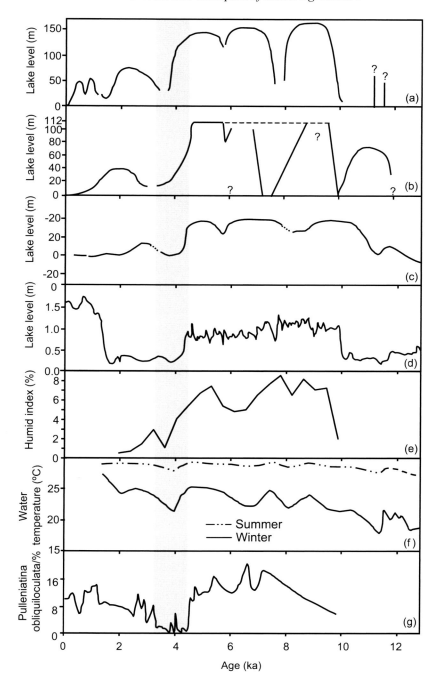

Figure 5.9 Correlation of the 4,000 years BP climatic event. (A) African Abhe Lake (Gasse, 2000); (B) African Ziway-Shala System (Gasse, 2000); (C) Daihai Lake, Inner Mongolia (An et al., 1991); (D) Yema Lake, Minqing Basin (Chen et al., 1999); (E) Arid and semiarid areas of China (Guo, 1996); (F) water temperature in Rc26-16 core (Wei et al., 1998); (G) proportion of foraminifer *Pulleniatina Obliquiloculata* in 255 core (Jian et al., 1996). Figure modified from Wu and Liu (2004).

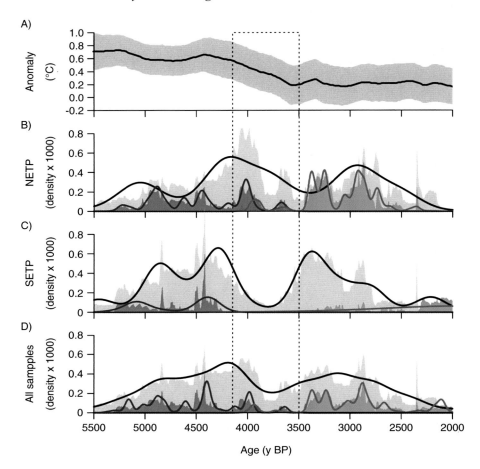

Figure 5.10 Cumulative probability of radiocarbon on the Tibetan Plateau. Radiocarbon dates on the Tibetan Plateau. (A) Marcott et al. (2013) temperature reconstruction. (B) Radiocarbon dates on the northeastern Tibetan Plateau (NETP). (C) Radiocarbon dates on the southeastern Tibetan Plateau (SETP). (D) Dates across the Tibetan Plateau. For B, C, and D, shaded areas in background are summed probability density distributions, while lines in the foreground represent a nonparametric phase model on ^{14}C ages via Gaussian mixture density estimation; lines upward sloping from left to right represent dates from millets; downward sloping lines from left to right represent dates from wheat and barley; and gray represents all available dates. The dotted box represents the transition period with both millet-derived and wheat- and barley-derived probability density distributions: 4,146–3,498 cal years BP. All radiocarbon dates have been calibrated. Modified after d'Alpoim Guedes et al. (2016b).

appear to have been depopulated along with areas of Northern Honshu (Habu, 2008; Kodama, 2003). At sites like Sannai Maruyama, this is accompanied by a decrease in the use of grinding stones, and an increase in hunting instruments, such as arrowheads, that were predominant in earlier phases of the sites. This indicated to archaeologists that humans moved away from exploiting and processing nuts that formed an important component of early Jōmon subsistence (Habu, 2008). Barnes (2015) and Kodama (2003) argued that climate likely played an important role in this shift, although Habu (2008) notes that it was impossible for archaeologists to pinpoint when this cooling trend took place.

The example of Sannai Maruyama shows that humans were behaviorally flexible and capable of using social memory to revert back to earlier patterns of subsistence in their time of need. Habu (2008) argues, in particular, that even if changing climate assisted in creating a tipping point, that it was likely their increasing overspecialization on a few resources created the tipping point that lead to major changes at the site.

5.7 Adaptations and Resilience: A New Economic Model

5.7.1 Wheat, Barley, Pastoralism and New Economic Model

Around the time that these changes in climate took place, a series of new crops and animals had begun to arrive on the eastern margins of the Northern Monsoon frontier. Wheat and barley, two crops that were originally domesticated in the Fertile Crescent, began to appear as early as around 4,200 years BP on the northeastern Tibetan Plateau (Chen et al., 2015b; d'Alpoim Guedes et al., 2016b) and have been found at other locations in China in contexts thought to date to 4,600 years BP, although these remain to be directly dated (Crawford et al., 2005). Several new domestic animals were also introduced to East Asia during this period of time: the horse, sheep, goat, and cattle (Flad et al., 2007).

Wheat and barley shared several key advantages over millets for dealing with the cooler post-climatic optimum world: both crops are frost-tolerant and unlike millets can tolerate low temperatures during several parts of their growth cycle. While the area that could be occupied by millets in high-latitude and high-altitude Asia decreased dramatically during this period of time (Figure 5.6), the thermal niche of wheat and barley meant that they were able to be cultivated across a much larger area of the Tibetan Plateau and high latitude Asia (Figure 5.11). Analysis of samples from across the southeast Tibetan Plateau suggests that once wheat and barley arrived in the area they were rapidly incorporated into agricultural systems (d'Alpoim Guedes et al., 2015; d'Alpoim Guedes et al., 2016b). For instance, at Ashaonao on the southeast Tibetan Plateau, wheat and barley formed the principal component of the diet. On the northeast Tibetan Plateau, people appear to have diversified their subsistence strategies by adopting wheat and barley, while continuing to cultivate smaller numbers of millet (d'Alpoim Guedes et al., 2015). In both areas, humans experimented with strategies of replacing the crops they

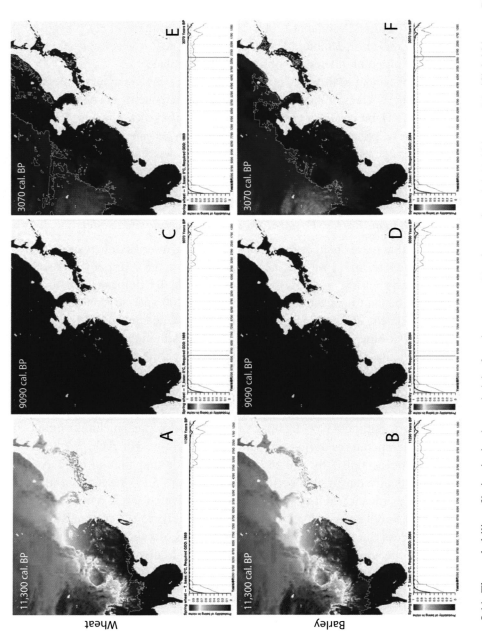

Figure 5.11 The probability of being in the thermal niche for wheat across Northern Asia at (A) 11,300 years BP, (C) 9,090 years BP, (E) 3,070 years BP. The probability of being in the thermal niche for barley across Northern Asia (B) 11,300 years BP, (D) 9,090 years BP, (F) 3,070 years BP. Figure created using methods described in d'Alpoim Guedes and Bocinsky (2018).

cultivated or diversifying their repertoire in order to cope with this climatic event. In northeast China, little systematic archaeobotanical research has been carried out, making it difficult to understand how farming systems were impacted by this change, however, increases in new domestic animals such as sheep and goat appear to have taken place (Shelach, 2009).

Although much more systematic analysis is needed, zooarchaeological work suggests that shifts in animal use took place around the margins of the Tibetan Plateau. For instance, in late Kayue Culture sites in Northwest China (around 3,500 years BP), such as Dahuazhongzhuang, we see a shift away from pig to sheep, goat, cattle, and horse (Qinghai Sheng Huangyuan Xian Bowuguan and Qinghai Sheng Wenwu Kaogudui, 1985). In the Ordos region, Shelach (2009) notes that similar patterns were present in two cemeteries where sheep, goat, and cattle overwhelmingly dominate the assemblage.

The introduction of pastoral animals also had profound impacts on the economy of the northern monsoonal frontier. For instance, the introduction of the horse from Central Asia made it easier for humans to increase their success in hunting wild animals, but also allowed people living in high-altitude and high-latitude regions of Asia to move rapidly through different altitudinal and environmental zones, increasing their ability to target and extract a wide range of different resources. Horses also allowed humans to successfully employ subsistence strategies, such as vertical transhumance and to herd the sheep, goat, and cattle that also became increasingly important components of subsistence in this area. This allowed them to access areas that while not suitable for farming, contained quality forage grasses for animals while maintaining the ability to be in trade relationships with populations in areas of lower altitude.

In Korea and Japan, archaeobotanical research has focused on identifying the earliest arrival of rice (Ahn, 2010). However, the archaeobotanical evidence shows that, in addition to rice, farmers moved a whole suite of crops including wheat, barley, soybean, and millets (Crawford and Lee, 2003; Crawford, 2011). This diverse diet likely allowed humans to hedge their bets in a cooler world and in the far northern reaches of Asia (Figures 5.6 and 5.11). Cooling climates after 4,000 years BP did not, however, substantially impact the ability to grow millets in the Central or Eastern Plains of China. Rather in this area, humans continued to rely, to a large degree, on millets and pig husbandry: it was not until much later that wheat and barley came to form substantial components of the diet.

5.7.2 Human Feedbacks and Changes in the Yellow River

The Yellow River drains some of the most arid regions of any of the large river systems of China but is nonetheless one of enormous historical importance because of its role in nurturing the centers where early states developed in China. It is also noteworthy in being one of the most heavily sediment-laden rivers in the world. Anthropogenic impacts in the Yellow River catchment date back into the Early Holocene (11,500–7,000 years BP) and increased since that time as the area was progressively more densely settled and cultivated. Historically, Chinese farmers have attempted to use a range of different tillage methods aimed at reducing erosion on the loess plains. This includes using terracing and a system of ridges and furrows along the fields that could either drain water in the case of a flood or retain in case of a drought. These systems have historic documentation as early as the third-century BCE (Bray, 1986). Measures to retain soil include employing light ploughs and harrows to create dust mulch that prevents the soil from becoming airborne. Despite agricultural patterns not changing substantially in central China after 4,000 years BP, major changes in the organization and sediment loads of the Yellow River are visible. Rosen (2008) notes that several key changes did take place in the evolution of tributaries of the Yellow River following 4,000 years BP.

Increased population density, and the need for wood to fuel the production of metal, pottery, and lime for constructing houses in the lower altitudinal ranges of the Yellow River valley following 5,000 years BP, complicates our understanding of the pollen record, which generally shows a decline in tree pollen and an increase in weedy taxa (Cao et al., 2010b) (Figure 5.12). Did changing climate impact the distribution of species or was anthropogenic cutting dominant?

In the Wei River basin (a tributary of the Yellow River), accumulations of recent loess have covered Mid-Holocene soils indicative of a drying windier climate since that time. Magnetic signatures of these soils suggest that dust accumulation had

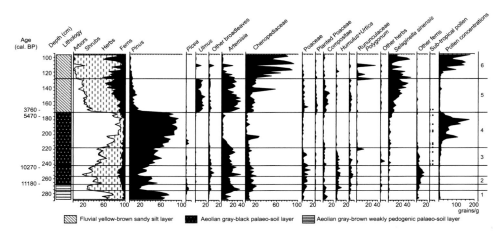

Figure 5.12 Pollen percentage diagram of profile Wangjiadian in the Yellow River basin. Reproduced from Cao et al. (2010).

increased and that weak processes of pedogenesis have been operating (Tan et al., 2011). In upland areas of the Yellow River, environmental analysis of charcoal deposits demonstrated that fires have been taking place infrequently, something that the authors interpret as being due to the increased presence of grazing animals stabilizing a steppe-like environment (Tan et al., 2011). However, in the lowlands, evidence for large-scale clearance using fire continued (Tan et al., 2011), making it possible that this change in pollen taxa may be equally the result of human disturbance of the environment, or possibly shifting cultivation or the opening of areas for new fields.

Rosen (2008) argues that in the Yiluo River valley, the alluvial valley infilling that characterized the Yangshao period in the Central Plains we described in Section 5.5 ceased between 5,000 and 4,000 years BP: there was no evidence for deposition of alluvium above deposits dating to 4,000 years BP (Rosen, 2008). She argues that this indicates stream incision was taking place during this period of time: this incision may have taken place either because rivers were stabilized in areas of higher elevation, or due to a lower base flow of the Yellow River valley because of a southward shift in the summer monsoon and decreasing precipitation. Rosen (2008) argues that this shift in the dynamics of the alluvial plains may have been a factor that resulted in a shift away from rice farming in central northern China; however, an examination of the available rice niche during this period of time indicates that temperature likely played an equally important role (Figure 4.11).

Geomorphologists have documented an increasing number of ancient flooding events. However, these are often interpreted as being the result of climatic change and not of anthropogenic modification (Huang et al., 2011; Wan et al., 2010a). One or a series of particularly large flooding events that are recorded in different sites across the Yellow River appear to have taken place at around 4,000 years BP, a time of significant climate change (Ge et al., 2010; Huang et al., 2011; Kidder and Liu, 2014; Tan et al., 2008; Wan et al., 2010a; Zhu et al., 2010). For instance, one flood on the Qishui River enveloped a Keshengzhuang settlement on a higher bank that was occupied between 4,300 and 4,000 years BP (Huang et al., 2011). Huang et al. (2011) relate the causes of this flood and sediment burial not to anthropogenic activity but rather to changes in vegetation driven by climate change, typically considered a time of rapid summer monsoon weakening both in Northeast and South Asia, but also of strong winter monsoon (Giosan et al., 2018).

Another instance of the possible influence of geological instability concerns the debate surrounding the destruction of the Qijia period Lajia site in Qinghai that took place at 3,870 years BP. Wu et al. (2016a) argue that this site was engulfed by sediments from a flood due to the breakage of a natural dam following an earthquake. The authors argue that the flood subsequently covered the archaeological settlement that had already been destroyed by tectonic activity: fissures visible in the excavations of the site were in-filled by flooding sediment. Others, however, propose that the site itself was destroyed by the earthquake and was subsequently covered by a mudslide from the gully heads behind the ruins, which was induced by heavy rainfall combined with an earthquake, not by riverine sediments (Huang et al., 2017). They also find no evidence that the burst of a paleo-dam was involved, as these workers demonstrated that this lake had dried up several thousand years prior to the destruction of the settlement (Huang et al., 2017).

It is possible that anthropogenic activities such as deforestation and destabilization of hilltop sediments during the Qijia period increased the vulnerability of the landscape to major events such as earthquakes. Coupled with a cooling and potentially drier climate, these conditions may have led humans to interact with their crops, animals, and river systems in new ways.

During later historic periods, the extent and intensity of agriculture had risen to the point that it competed with natural processes in controlling landscape evolution (Zhuang and Kidder, 2014). The Han Dynasty (221 BCE–220 CE) economy, which was centered in the Wei River valley and lower course of the Yellow River, derived most of its taxes from individual peasant households who were allocated plots of land. Larger landowners also existed who charged peasants high rents but who were also adept at evading taxes (Bray 1984). In order to reduce the size of these landed estates and increase the amount of income they were able to gain through taxes, the Han government attempted to increase agricultural production through a series of agricultural reforms. The emperor, Wu Di, oversaw a large program of canal building throughout Henan and Shanxi, as well as the Wei and Huai River valleys, considerably raising the productivity of these once somewhat arid areas. The government also provided peasants with animals, fertilizers, iron tools, and instructions on how to cultivate the land using a ridge and furrow (or *tai tian)* system, which involved deeper ploughing than earlier methods (Bray 1984). This system was particularly good at raising agricultural output in areas where irrigation was practiced, however, in regions where irrigation systems were failing or not properly employed in loess soils, erosion could quickly become an unmanageable problem.

In addition to a desire to increase agricultural output, the floodplain flood control infrastructure that was installed during the Early Dynastic period was a response to long-term increases in sediment load within the Yellow River that were largely driven by erosion of Loess Plateau sediments as that region became more heavily settled (Dong et al., 2012) and because agriculture allowed easier erosion of the top-soil in the river system. As erosion and sediment transport increased, so the flood defenses also had to be strengthened further in order to constrain the Yellow River in its original setting. This continued until a breaking point. In the floodplain records from the lower reaches of the Yellow River, Kidder and Liu (2014) demonstrated that there was a massive flooding catastrophe in the first 20 years of the first millennium CE, during the Han Dynasty.

Deforestation of the eastern Tibetan Plateau is modelled to have caused decreased transpiration and increased summer precipitation in the deforested area, as well as a wetter and warmer climate on the Tibetan Plateau in summer (Cui et al., 2007). The higher summer rainfall, in turn, would have raised the chances of seasonal catastrophic flooding. Coupled with the fact that the agricultural policy meant that peasant farmers had to invest in agricultural technology they could not afford, the landed gentry grew and large numbers of smallholder peasant farmers were dispossessed, leading to a crisis of economic refugees that had to be resettled multiple times during Han rule (Bray 1984).

5.8 Changes in the Monsoon during the Historic Period

Agriculture during the historic period also appears to be have impacted by changes in the monsoon. Both Dongge and Heshang speleothems (Figures 4.7 and 5.13) show the Jin Dynasty reached its end during a particularly intense dry stage and the Dongge record, in particular, shows decreasing monsoon strength through the Jin Dynasty (280–420 CE) with some particularly sharp, albeit short-lived dry events. Most notable was a dry period starting around 280 CE and climaxing in 316 CE (Figure 4.7). As we reviewed in Chapter 4, this corresponds to a period of time when there was a major decline in the agricultural niche in northern China, with rice, but also millets losing much of their territory (Figure 4.14). This was significant because the Jin capital, Luoyang, fell to northern invaders in 311 CE, so that much of what is now northern China was lost from the empire and the Jin Dynasty itself split in western and eastern factions. The loss of power following a time of agricultural hardship under drought and cooling temperature conditions suggests that climatic change has a role to play in guiding Chinese society at this time. Subsequently, monsoons appear to have been weak but also quite variable during the Southern and Northern Dynasties time (420–589 CE). This was a time of mass migration of Han Chinese south of the Yangtze River (Marks, 1998), and potentially spurred by intermittent droughts and poor harvests in the north that made the wetter and warmer southern regions seem more appealing for settlement.

Although there are some issues concerning their reliability as rainfall proxies and the role of temperature (Clemens et al., 2010), speleothems records have commonly

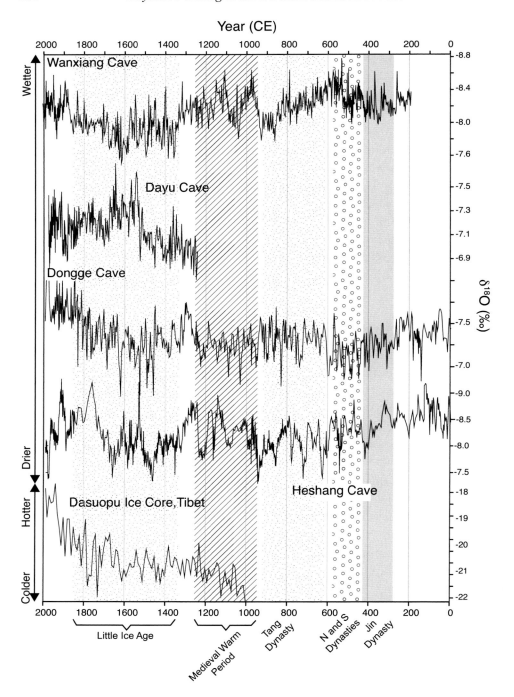

Figure 5.13 Speleothem monsoon rainfall records spanning the past 2,000 years in China compared with temperatures in the Tibetan Plateau from Thompson et al. (2000). Data from Wanxiang Cave is from Zhang et al. (2008). Data from Dayu Cave is from Tan et al. (2009). Data from Dongge Cave is from Dykoski et al. (2005). Data from Heshang Cave is from Hu et al. (2008).

been used to reconstruct both short- and long-term changes in climate. We use two such records, one from Dongge Cave in southwest China (Dykoski et al., 2005) and another from Heshang Cave (Tan et al., 2009), which is located toward the northeast. (Figure 5.13). These two records are quite different from one another over the first millennia of the common era. The oxygen isotope record from Dongge shows higher amplitude variations, as well as being of a different values compared those seen at Heshang, and this in part reflects the source of the moisture and the amount of precipitation that occurred between the marine source and the final point at which the rainfall occurs (Breitenbach et al., 2010; Dreybrodt and Scholz, 2011). Because heavy isotopes tend to fall out first the rainfall becomes progressively lighter going inland so that the isotopic composition of the rainfall at any given point reflects the integrated effects of source composition, the intensity of rainfall, and the topography over which the clouds passed.

Huguang Maar is a shallow lake that sits within a Pleistocene volcanic crater (Figure 5.4)(Sheng et al., 2017). The lake is fed by a series of small streams from a very limited catchment, and it is generally considered that this body receives very little sediment directly from the local countryside. This means that it is a good place to look at wind-blown dust that is brought into the region from central Asia, transported by the winter monsoon winds, because this is not diluted or contaminated by other clastic influxes. Yancheva et al. (2007) used high-resolution scans of the magnetic susceptibility of the sediments, coupled with X-Ray Fluorescence (XRF) scanning of the same material to identify layers that are rich in titanium (Ti) and iron (Fe). These proxies are interpreted to record the relative strength of terrestrial input into the lake from continental sources. Wind-blown dust from the Loess Plateau of northern China is thought to be particularly important. The Ti proxy is known to be high when organic carbon contents of the lake were low, but when magnetic susceptibility was high. Susceptibility is high when strong winter winds blow large amounts of Fe-rich material into the lake. Susceptibility is also enhanced when oxygen levels were high because of mixing in the water column, allowing hematite to be preserved in the sediments on the lake floor. At the same time organic carbon was destroyed before burial because of the high oxygen levels. Used together these measurements show when the winter monsoon winds used to be at their strongest.

Yancheva et al. (2007) went further in their analysis than simply looking at winter monsoon strength with their Huguang Maar environmental records. They compared their dust reconstruction with speleothem records for rainfall in southern China. This approach allowed them to see that when the winter monsoon was strong the summer monsoon was generally weak, and vice versa. Figure 5.14 shows the variability in the Ti record between around 600 CE and 950 CE. After around 710 CE (1,250 years BP) the monsoon rains appear to have weakened progressively and

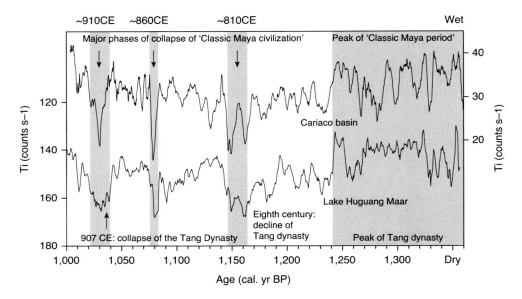

Figure 5.14 The Lake Huguang Maar paleoclimate records spanning the past 4,500 years in the context of major events in the cultural history of China. Major changes in Chinese dynasties occurred during dry phases, as indicated by the titanium (Ti) and magnetic susceptibility records from Lake Huguang Maar and applying the observed anti-correlation between the winter and summer monsoons. Modified from Yancheva et al. (2007).

reached a minimum ca. 810 CE (1,140 years BP), which was a time of decline in the economic power of the Tang Dynasty (Kurz, 2011). Final collapse of the dynasty occurred in 907 CE, again in a time of weak summer monsoon and critically following a much steeper decline than seen during the eighth-century CE.

Like the lake record at Huguang Maar, it is possible to see from the speleothem records that the Tang Dynasty declined during a time of particularly sharp reduction in monsoon rainfall (Figure 5.13), but following a time when monsoon rainfall was on average higher than at any time during the first millennia. It is possible that this may have impacted agricultural returns in northern China. During the latter part of the Tang, fear of Khitan and other nomadic invaders drove thousands more peasants to abandon their land and migrate toward southern China (see Chapter 4).

5.9 The Impact of Changing Climate on Pastoralism in the Steppe

The Dongge and Heshang cave records can also be used to look in detail at the last 800 years of Chinese history. These records again show some differences, but also have common features that give us confidence that they are a true representation of the evolving monsoonal rainfall during this time. This is not to exclude variations in temperature as also being a crucial control over agricultural practices and productivity (Lobell and Burke, 2008) not least in controlling soil moisture. Precipitation was relatively strong during the latter part of the Southern Song Dynasty (Figure 5.15), but then underwent a decrease during the thirteenth century, so that it had reached a low point by the time that the dynasty ended in 1279 CE, when it was overrun by Mongol armies under Kublai Khan (Twitchett et al., 1994). The Song had already lost control of northern China well before than time, in 1127 CE, when the region was conquered by the Jin (Jurchen) Dynasty who originated from Manchuria (Tao, 1976). This group, in turn, was displaced by the rising power of the Mongols who overran northern China completely by 1234 CE, which was also a time of weakened monsoon precipitation according to the Dongge Record (Dykoski et al., 2005)(Figure 5.15).

It is noteworthy that rainfall intensity, as tracked by the speleothem records, decreased progressively during the fourteenth century and reached a low point at the time that the Mongol Empire ended. For an empire built on the effectiveness of its mounted cavalry to defeat its enemies in battle, it is easy to understand how reduced

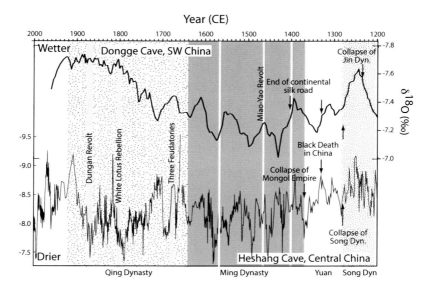

Figure 5.15 Monsoon climate records from south and central China compared with major historical events in Chinese history between 1200 and 2000 AD. Shaded bands show the duration of Chinese dynasties. Note the steep decline in monsoon strength at the end of the Ming. Data from Dongge Cave is from Dykoski et al. (2005), and from Heshang Cave is from Hu et al. (2008).

rainfall, temperature, and deteriorated pasture land, especially in the dry interior of Asia, might have serious negative effects on the ability of the government to function and especially for the army to continue its conquests across the continent. This is not to overestimate the influence of drought as the Uyghur Empire (744–840 CE), which also inhabited the Inner Asian steppe, survived a drought lasting nearly seven decades without cause societal fragmentation (Cosmo et al., 2018).

Compared to many of the following centuries, the Mongol Empire at its peak enjoyed relatively high rainfall and higher temperatures, associated with the Medieval Warm Period, especially around 1300 CE. A climate reconstruction based on tree ring records for Siberian pines from Mongolia and that reflects water supply during the warm season growth period corroborate this general picture (Pederson et al., 2014)(Figure 5.16). Moreover, recent measurements show a close association of the pine growth cycle and grass production across the steppe country. These data show a preference to abnormally wet and warm summer conditions in Mongolia during the late thirteenth century when the Mongol Empire began its rapid expansion (Pederson et al., 2014). Indeed, the record indicates that the 15 years spanning the rise of Ghengis Khan were the wettest and warmest over >1,000 years of record (Figure 5.16). Pederson et al. (2014) argued that it was this period of warm, wet conditions that allowed the Mongols to establish a strong and centralized organization and aided their initial expansion. It is, however, noteworthy that drying, which became intense after 1260 CE, coincided with the movement of the Mongol capital to China (Beijing) at that time, possibly reflecting less hospitable conditions for pasture grasses on the steppe at that time.

The reduction in moisture supply after 1260 CE would certainly have made communication across the huge distances of the Mongol Empire, stretching all the way to Europe at its zenith increasingly difficult because much of this route was through arid and semiarid regions with limited grazing at the best of times.

5.10 Synthesis and Summary

The case studies we have reviewed here demonstrate that while changes in temperature did impact the areas in which humans were able to produce crops or extract natural resources, these did not result in a "collapse" in the way that this has often been envisioned by the popular media. Humans developed new and inventive adaptive strategies as they cascaded into new basins of resilience. Some of these, particularly the barley cultivation and pastoral strategies that developed on the Tibetan Plateau, have proven to be strategies that were successful for millennia.

One key difference is worth noting between the case studies we have explored here and those we will examine in future chapters. Particularly in the beginning of

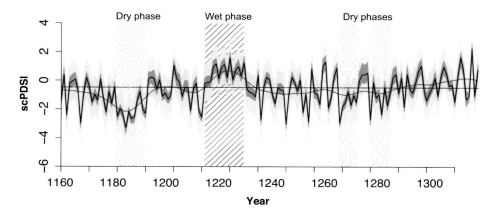

Figure 5.16 Reconstructed drought using scPDSI from 1160–1300 CE. The plot shows a 20-year spline of the annual reconstruction, the two-tailed 95 percent bootstrap confidence limits of the reconstruction and the uncertainty of the reconstruction in gray (±1 RMSE). The severe drought during the 1180s (stippled), the Mongol Pluvial (1211–1225 CE; cross hatch), and the significant drought during the movement of the empire capitol to China in 1260 CE (stippled) are shown. Modified from Pederson et al. (2014).

this chapter, we have seen that humans likely extensively employed migration as a tactic for escaping the effects of say, rising sea level or cooling and drying. They also likely used this strategy when expanding agriculture outside of its original centers of domestication. While resistance from forager groups (and other populations present) likely complicated their journeys, the development of nation-states with clearly defined borders creates new challenges for climatic and economic refugees, as does the increasingly limited available amounts of arable land for cultivation.

The two millets whose domestication we described in this chapter are increasingly being replaced by larger seeded grains such as wheat, barley, and maize. Throughout history, however, Chinese farmers have selectively bred hundreds of varieties that exhibit drought tolerance, water-logging, and wind and bird resistance. For instance, the *Qi Min Yao Shu* lists over 100 of these adapted varieties. As the biodiversity associated with traditional methods of farming in north China is lost, we are simultaneously losing potentially critical adaptations to changing climate that have served northern Chinese farmers for millennia.

6

Recent Changes in Monsoon Climate

Changes in monsoon rains over the course of the twentieth century have played an important role in exacerbating crises in agricultural production. These changes have taken place at the same time as a number of major changes in agricultural and social policies that include the "Green Revolution" and collectivized farming in China. Compared to the data we reviewed in earlier chapters, the historic weather station data collected over the course of the twentieth century allow us to better understand the nature, location, and amplitude of changes in rainfall and temperature. These changes have also taken place at the same time as large damming projects have been carried out in China, Southeast, and South Asia, which have impacted the rivers we described in previous chapters. Because of the huge amount of data available for the recent past, we only briefly discuss China and focus our discussion in this chapter on one region, monsoonal South Asia, to describe the impact that changes in the monsoon may have had on farming.

6.1 Transformations in Farming during the Twentieth Century

During the mid-twentieth century, transformations in farming technology began to be made that completely changed the face of the methods and scale of farming we reviewed in Chapters 3, 4, and 5. In Communist China, these changes took place in the form of collectivized farming. In India, these took the form of a "Green Revolution." Both of these movements shared a desire to increase agricultural productivity: one that has often come at the expense of human life, agricultural biodiversity, and resilience to changing climatic conditions.

6.1.1 Collective Farming in China

Following the establishment of the People's Republic of China in 1949, several major initiatives fundamentally changed farming systems in north and south

China. In order to more rapidly move the country from an agrarian economy to Communism, agricultural collectivization was gradually rolled out between 1949 and 1958 by establishing increasingly larger numbers of agricultural co-operatives and collective farming associations or communes (Becker, 1998; Yang, 2012). In 1958, Mao decided to accelerate efforts in the countryside by rolling out the "Great Leap Forward." One of the central ideas behind the Great Leap Forward was to allow the industrialization of China to take place alongside its agricultural development. Much like the "Green Revolution" we discuss subsequently, it aimed to mechanize and simplify traditional agricultural practices into a vision of modernity premised on achieving high productivity of a few key grain crops (Scott, 1998; Shapiro, 2001). On the communes, a number of radical and controversial agricultural innovations were promoted at Mao's behest. Many of the policies implemented by the Great Leap Forward did huge damage to the agricultural biodiversity of China that we reviewed in previous chapters.

Policies implemented included the popularization of new breeds and seeds that were adopted *en masse*, which replaced the myriad of landrace crops that Chinese farmers had developed over millennia (Chun, 1961). These new seeds focused primarily on the adoption of grain crops such as rice, wheat, and maize. Multicropping, something we saw was commonly practiced in parts of China, was also pushed to its limits, however, unlike in earlier times, less crop rotation was practiced. Some areas like the Chengdu Plain are reported to have been forced to attempt up to three crops of rice in a single year, leading to soil exhaustion and nutrient depletion. Several other policies aimed at increasing production did real damage to crop yields overall. One of these policies, close planting, pushed the idea that plants could be planted so close that children would actually be able to walk across the top of a wheat field (Hsu, 1982; Shapiro, 2001). However, in reality many of the crops planted in this manner died through lack of sun exposure and competition. Much of biodiversity housed in tree plots that surrounded agricultural farms and that provided sources of fruit, nuts, and wood for housing and shade, were also lost during this period of time, as these were cut down to fuel the backyard steel furnaces that Mao hoped would allow China to surpass the steel production of the United Kingdom. Ironically, these backyard furnaces were envisioned to allow China to manufacture tractors to plow fields.

The "kill a sparrow campaign" initiated as part of the Great Leap Forward also impacted the insect and animal biodiversity that was crucial to the functioning of the whole ecosystem models of farming that had developed over millennia in China. This campaign aimed to eliminate rats, mosquitos, sparrows, and flies. Farmers were encouraged to set up nets, use traps and pellets to kill these species, in particular, sparrows. The sudden dearth of sparrows led to a massive locust swarm in 1958, further reducing agricultural production (Shapiro, 2001).

In addition to these major changes creating disruption in farming systems, Mao had political motivations to demonstrate that China was capable of returning loans to the USSR and aimed to do so in the form of grain export. Local government officials were encouraged and sometimes arm twisted into "launching satellites" or making wildly inflated promises of how much their county or village was able to produce. When harvest time came, many villages saw their entire year's production requisitioned as a result, with nothing left for them to eat. However, between 1958 and 1960, China actually continued to be a substantial net exporter of grain, and Ashton et al. (1992; 1984) estimate that the grain export would have been sufficient to feed 16 million people. Yang Jisheng (2012) uncovered that some 22 million tons of grain were held in public granaries at the height of the famine which were sequestered for export.

On the social level, the final death knell was dealt by the policies implemented in the agricultural communes, which ended up removing risk-buffering strategies that farmers had developed. Traditional Chinese farmers relied on a wide range of techniques, such as investment in handicrafts for alternative sources of revenue or cultivating a wide range of vegetable and other "cash" and "catch" crops, as well as rearing domestic animals to provide sources of protein and eggs. Alongside stockpiling grain and even creating stores of preserved food, these practices were labelled as "rightist" or "capitalist." As the original agricultural biodiversity of their plots was destroyed by only allowing them to cultivate grain, and as all other types of food growing or animal rearing were taken outside of private hands and moved to the communal level, many families simply had no food to fall back onto once their grain was requisitioned. Movement of farmers was also restricted by the implementation of the "Hukou" system, which meant that when starvation hit, farmers were not allowed to leave their county to seek food elsewhere (Shapiro, 2001; Yang, 2012). By the end of the Great Leap Forward, an estimated at least 23 million on the low end (Peng, 1987) to 55 million people on the high end (Yu, 2005) had died from starvation (Yang, 2012), making this one of the largest famines that has ever been experienced in human history.

Government officials sought to place much of the blame for this calamity on a series of droughts and floods of the Yellow River that took place between 1959 and 1960 (Ashton et al., 1992); however, a closer look at the data shows that the area of China that suffered floods and drought during the 1950s and 1960s was substantially smaller than those that took place in the 1990s (Wang et al., 2011a; Zhang et al., 2016). While droughts and flooding in 1959 did not help the already tenuous situation, the fact that China has weathered much worse in the decades that followed suggests that the policies implemented are largely to blame for loss of life. This is a lesson that humanity should heed: humans are least able to survive episodes of climatic disruption when unjust or draconian policies are enacted.

During the Cultural Revolution (1966–1976), farmers were returned some degree of autonomy. However, the focus on increasing agricultural productivity and the area under cultivation led to an unprecedented deforestation across much of China, particularly its western margins (Shapiro, 2001), leading to an acceleration of erosional processes that continues to operate to this day (Hsu, 1982). Following Mao's death and market liberalization under Deng Xiaoping, agrarian reforms that have taken place in China since have allowed farmers to regain substantial autonomy and allowed rural economies to flourish (Hsu, 1982). However, as is the case elsewhere in the world, Chinese agricultural systems have been under pressure to become major producers of uniform varieties of grain leading to a degradation in agricultural biodiversity. The introduction of high yielding rice varieties, such as hybrid rice (or *za jiao*), has led to a substantial decrease in the biodiversity of rice landraces planted in China (Gao, 2003). Some crops that were traditionally capable of tolerating drought and high temperatures such, as the millets we reviewed in Chapter 5 are being replaced by wheat and the area under which they are cultivated has dropped substantially (Diao, 2017) For instance, Diao (2017) reports that the area under which foxtail millet was cultivated shrank from 10,000 kha in 1949 to less than 1,000 kha in 2014 (Figure 6.1).

This genetic erosion represents a great threat to humanity in dealing with the challenges faced by a changing climate as we may be forever deprived of varieties that may contain useful traits such as pest resistance and heat tolerance.

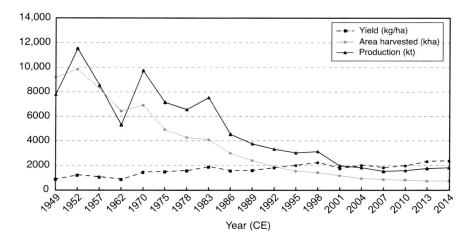

Figure 6.1 Cultivation area, yield, and production of foxtail millet in China, 1949–2014. Redrawn from Diao (2017).

6.1.2 The Green Revolution in South Asia

During the mid-twentieth century, growing population levels around the world led scholars like Paul Ehrlich to argue that large portions of the world's population would starve to death by the mid-1970s, unless population levels were brought under control and unless food production was greatly increased. Channeling earlier ideas by the British demographer Malthus, he predicted that by 1985, 1.5 billion people would die in the greatest famine known to man (Ehrlich, 1968).

During the 1950s, the hype around the potential for future famines led both government agencies and private foundations (such as the Ford and Rockefeller foundations) as well as researchers in agronomy to undertake what they said was a series of necessary transformations in the agricultural sector, a "Green Revolution" designed to increase agricultural production and to replace former farming methods. The father of the Green Revolution, dwarf-wheat breeder Norman Borlaug, won the Nobel peace prize in 1970 and was credited for saving over one billion people from starvation.

Rolled out between 1950 and the late 1960s, the Green Revolution involved several key initiatives that changed the face of farming in the areas where this technology was introduced. These included the mechanization of agriculture through an increased use of plowing, seeding, irrigation, and harvesting machines. The use of cheap nitrogenous fertilizers, pesticides, and herbicides; the development of high yielding dwarf cereals; and investments in irrigation and the development of agricultural research centers around the world (mostly funded by the US Government).

Green Revolution agriculture aimed to transform the agricultural sector by carrying out agricultural initiatives that were taught (and sold) to farmers as a package. Farmers should invest in (or be given) tractors, the right seeds to plant (height of dwarf cereals are designed to be harvested with the tractors), and purchase doses of the right fertilizers and pesticides from the government, non-governmental agencies, or larger corporations.

India, in particular, became a key focus of US efforts to roll out the Green Revolution, both at home and abroad. To see the origins of this in South Asia, we need to start somewhat earlier, in the 1950s. Begun by President Eisenhower, the "Food for Peace" or PL480 program in 1954 created a new scenario in Indian farming, which involved the United States transforming the large amount of surplus grain created on its heavily subsidized farms into "humanitarian grain" sent to India (Stone, 2002). For the United States, this made sure that there was not an oversupply of wheat on the US market that could lead to a commodity crash. In India, however, this had more devastating effects. As Stone (2002) recounts following Perkins (1997): "The huge supplies of American grain that flowed into India during the 1950s and early 1960s accomplished the function intended by Nehru's government, to keep Indian grain prices down. In fact, prices were so low that Indian domestic production stagnated. Indian farmers simply could not compete against grain sold at a loss by the American government, so they stopped trying and Indian production failed to rise fast enough to meet increasing domestic demand."

A drought in 1966 further exacerbated this situation, increasing demand for wheat imports from the United States. However, President Johnson refused to commit supplies until an agreement to roll out the Green Revolution was signed between India and the United States (Shiva, 2016).

The Intensive Agricultural Development Program (IADP) in 1960–1961, funded by the Ford Foundation in India, attempted to respond to this slack in demand (ironically created by US imports), by introducing the first Green Revolution wheat to India in 1968 and sought to replace the organic small-scale forms of agriculture that we reviewed and that had been developed in India over the past millennia into a form similar to that employed in the United States: one based on monocultures, industrial fertilizers, and pesticides. As Shiva (2016) notes, this fundamentally transformed farming from a model based on internal inputs available to farmers, to a form of farming where they were forced to participate in global market economies where they needed to acquire non-renewable resources.

While the "Great Leap Forward" and the "Green Revolution" could not be more different ideologically, they share a number of key characteristics that has served to lower the resilience of farming systems in Asia. These characteristics include: (1) a focus on increasing productivity through monocultures of crops which reduces agricultural biodiversity; (2) a lack of regard for traditional knowledge associated with farming systems, particularly practices that are critical in maintaining soil fertility and combatting erosion such as crop rotation or planting borders of trees; (3) both systems have eroded farmer autonomy and decreased their resilience by making them dependent either on government systems for food (or ability to control the food they grow) or financially on aid or on fertilizers and seeds from large multinational companies. We review further, with a focus on South Asia, how changes in precipitation, typhoon strength, and river load, intersected with these new agricultural policies to create impacts on farming systems.

6.2 Precipitation Changes in the Twentieth Century

Although there are a lot of data in geological and oceanographic proxy records to indicate changes in the intensity of the summer monsoon over the past few thousands of years, the best documented changes in climate span the more recent past when detailed historical weather observations are available. There is little doubt that the Earth's climate has been changing over the last several centuries, as indeed it has continuously through geologic time. Whether or not the more recent changes are driven by anthropogenic processes remains a controversial topic, at least in wider society, but what is less controversial is that changes have been taking place and that these have influenced those regions that lie under the influence of the Asian summer monsoon. The recent global climatic synthesis put together by IPCC Working Group 5 (Stocker et al., 2013) (Figure 6.2) considered three different compilations of changes in precipitation worldwide since the start of the twentieth century. The studies differ in their data sources and in how they extrapolate and interpret these data over the period considered. We show the GPCC model as an example. In each case the intention of the compilation was to show long-term regional changes, spanning several years, rather than anything shorter term that could be related to weather fluctuations. The duration is also long enough to have smoothed out the effects of shorter-term fluctuations, such as the El Nino-Southern Oscillation (ENSO) or the Indian Ocean Dipole.

The reconstructions show changes spanning 1901 to 2010 (Figure 6.2A), and although some regions appear to have become wetter over the twentieth century,

Precipitation change since 1901 and 1951

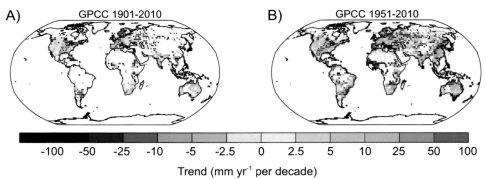

Figure 6.2 (A) Map of observed precipitation change over land from 1901 to 2010 and (B) 1951 to 2010 from the Global Precipitation Climatology Centre (GPCC) data set. Trends in annual accumulation have been calculated only for those grid boxes with greater than 70 percent complete records and more than 20 percent data availability in first and last decile of the period. White areas indicate incomplete or missing data. Black plus signs (+) indicate grid boxes where trends are significant (i.e., a trend of zero lies outside the 90 percent confidence interval). Figure modified from TFE 1, Figure 2 of Stocker et al. (2013), published by Cambridge University Press with permission of IPCC.

most notably North America and parts of Europe, it is also clear that there is a drying trend throughout much of Southeast Asia, especially in southern China. Southern India in contrast appears to have become wetter during the past century. When considering the last half of the twentieth century and the start of the twenty-first century, up until 2010, the reconstructions show some divergence. The Climate Research Unit (CRU) model shows very little change in South Asia since 1951, but instead a trend to drying in East Asia. The other two reconstructions considered by the IPCC also point to a dominant drying trend across monsoonal Asia, being locally quite intense in eastern Asia, as well as strong wettening of the climate in southern India (Figure 6.2B). Much of the change in the twentieth century has occurred since 1951. The speeding up of change as the pace of industrial development increased is consistent with the proposition that human activities may be a key force controlling monsoon intensity.

6.2.1   Climate Modelling

Reconstructing changes in monsoon intensity before the start of the twentieth century is hampered by a relative lack of meteorological measurements, with a few notable exceptions. Even those that do occur rarely go back prior to the beginning of the nineteenth century so that climate modeling has to be employed to reconstruct and understand changes in climate prior to that time. An example of such reconstructions based on climate modeling focused on the change of land use between 1700 and 1850 (Takata et al., 2009). This was a time that largely predated the Industrial Revolution and when the anthropogenic emission of CO_2 was still rather moderate. Nonetheless, it was still a time of rapid economic development, especially in the Indian subcontinent where commercial agriculture had accelerated as a result of the growing influence of the English East India Company over a diminished Moghul Empire (Fisher, 2016; Lawson, 2014). In particular, this was a time when plantations were established and expanded. Furthermore, in China there was a rapid increase in the population as a result of an expansion in the economy under the Qing Dynasty (Ramankutty and Foley, 1999). Together these two influences resulted in important changes in the extent of forest coverage, as opposed to agricultural farmland across the core region of the Asian monsoon. Critically, this was also a time when there were no particularly strong variations in ocean or solar activity, and also no major volcanic episodes (Gao et al., 2008) that might have disturbed monsoon intensity, as they have been known to do at other times (Bourassa et al., 2012). Removing the forest in order to increase cultivation of farmland resulted in an increase of surface albedo. At the same time, there was an increase in surface winds caused by the reduction of forest, and reduced precipitation across much of the area. Deforestation then generated a positive feedback with reduced soil moisture, so that local evaporation also decreased, further reducing precipitation across the region (Figure 6.3).

Climate modeling predicts that the reduction in precipitation during the second half of the twentieth century was particularly localized in southwestern China and parts of Indochina, as well as being focused over the Deccan Plateau area of western India (Bollasina et al., 2011). That this prediction has some validity has been borne out by study of ice core records in the Himalaya. At two locations, Dasuopu in the central Himalaya (Duan et al., 2004) and at Mount Everest (Kaspari et al., 2008), ice cores have been used to show a substantial increase in snow accumulation between 1500 and the mid-1800s (Figure 5.13), which is negatively correlated with summer monsoon intensity. These predict that summer monsoon precipitation in the region reduced by as much as around 20 percent during that period, which is indicative of an early form of anthropogenic climate forcing that is unrelated to industrial greenhouse gases but is nonetheless linked to changes in land use.

6.2.2 Changes in Fluvial Discharge

Changes in precipitation might be expected to cause variations in the discharge from rivers across Asia. However, the situation is more complicated than might initially be assumed, that is, that higher precipitation results in greater water discharge. In a study of global river systems, Milliman et al. (2008) compiled data from a number of major rivers worldwide and showed that there has been a common pattern of reduced discharge from rivers across much of Asia since 1951. These workers used a compilation of rainfall data between 1951 and 2003, similar to those shown by the Stocker et al. (2013) study, and, in particular, emphasized the existence of a strong drying effect across northern India, much of Southeast Asia, and northern China, with a more moderate drying being seen in southern China. Not surprisingly, this climatic trend resulted in a substantial decrease in discharge from rivers including the Indus and those in peninsular India by more than 30 percent at that time (Figure 6.4A). At the same time, strong reductions are seen in northern China in the Yellow River, while the Pearl River in southern China and the Brahmaputra show modest increases. Although Milliman et al. (2008) predicted a sharp decrease in the Yangtze, gauging data from that river suggests that it has been variable but relatively constant over decadal periods, reflecting the fact that it is located between regions with opposing temporal trends (Figure 6.4B).

Clearly, major fluctuations in discharge would have implications for agricultural production in regions that are dependent on water supply from these rivers.

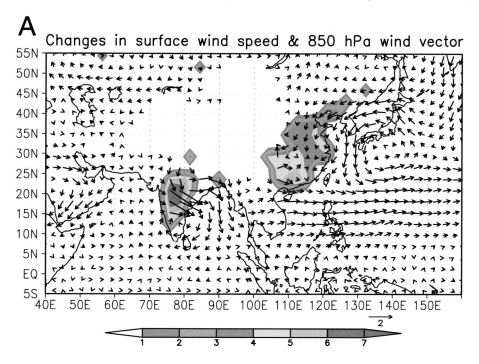

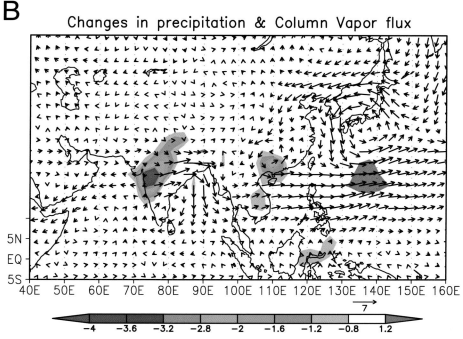

Figure 6.3 June–August mean changes between 1700 and 1850. (A) Surface wind speed (colors, in m s^{-1}) and 850 hPa wind (vectors, with unit vector 2 m s^{-1}). (B) Precipitation (colors, in mm day^{-1}) and vertically integrated vapor flux (vectors, with unit vector 7 x 10 kg m^{-1} s^{-1}). Modified from Takata et al. (2009).

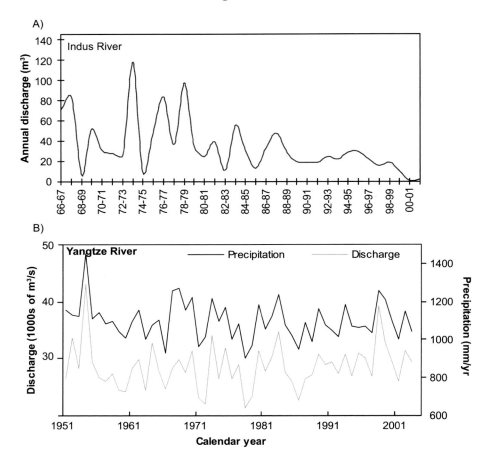

Figure 6.4 (A) History of reducing discharge in the main Indus River below the Kotri Barrage, Hyderabad (Memon, 2004). Chart shows the progressive loss of flow to the lower basin as a result of intensified agriculture in the Punjab. (B) Mean precipitation over the Yangtze catchment and discharge at Datong, near the mouth of the river since 1951 (Yang et al., 2015).

The situation is more complicated than might first be assumed because the discharge to the ocean is also influenced by the construction of dams, as well as the spread of agricultural irrigation. Damming has reduced sediment flux to the ocean in many such systems, independent of the rainfall variation (Gupta et al., 2012; Kondolf et al., 2014). Especially in southwest Asia, the damming of the major waterways is closely linked to the spread of irrigation-based farming, most notably along the Indus River, which has the largest man-made irrigation system in the world.

Gauging data from the Indus shows a steady reduction in discharge to the ocean since the mid-1960s (Figure 6.4A), but not all of this, by any means, is linked to a weakening of the summer monsoon rainfall by the same extent. This is because much of the water in the Indus is derived from snow melt, which in turn is delivered to the region via the winter westerly jet (Karim and Veizer, 2002). The Indus and to a lesser extent the Brahmaputra are also supplied by melt from rapidly shrinking glaciers whose long-term decline has serious implications for the water budget in the heavily populated downstream areas. (Immerzeel et al., 2010). By 2000, water discharge from the Indus into the Arabian Sea had reduced to the extent that it was volumetrically negligible, principally because it was being used upstream to sustain agriculture in the Punjab region (Inam et al., 2007; Kidwai et al., 2019). This has serious consequences for agriculture in the region of the delta because lack of flow allowed saltwater intrusion into that region and the salinization and loss of once highly productive farmland, as well as the demise of the mangrove swamps that used to characterize the coast region (Kidwai et al., 2019).

6.2.3 Recent Precipitation in China

Across central China the temporal trend in runoff and discharge to the ocean is less clear. Figure 6.4B shows that over the last 50 years of the twentieth-century discharge from the Yangtze was quite variable and largely followed precipitation variability within the drainage basin itself. In this case there is no long-term decline like that observed in Pakistan and, in fact, since the mid-1980s there has been a gradual modest increase into the start of the twenty-first century. Discharge from the Yangtze River to the ocean is now heavily regulated by large-scale damming projects along the course of the river (Yang et al., 2006b). Over the long term, of course, the discharge within this and every river system must be dependent on the precipitation within the catchment, which in Asia is largely driven by the summer monsoon.

Historical records can be used to reconstruct the development of monsoon intensity across China. What is clear is that northern and southern China have behaved quite differently and have often been anti-correlated with one another in terms of climatic trends. An analysis of the severity of wet and dry periods has been attempted across different parts of China (Zheng et al., 2006) and, in particular, in northern China it was possible to cross-check reconstructions of a "dry-wet index" against reconstructed precipitation for the time period 1736–1950 in the middle and lower reaches of the Yellow River. The precipitation data were derived from snow and rainfall records in the archives of the Qing Dynasty (Hao et al., 2008b; Zheng et al., 2005).

North China shows variability in monsoon precipitation with substantial variations over timescales of 20–30 years and generally good agreement between the dry–wet index and a reconstruction based on Qing Dynasty records (Figure 6.5) (Hao et al., 2008b). The first half of the twentieth century appears to have been a period of drying after a peak in moisture in 1910, following by a steep decline in precipitation until around 1930, after which a moderate increase was observed (Zheng et al., 2005). At the start of the twentieth century, precipitation appears to have been quite variable over timescales of a couple of decades. Zheng et al. (2005) further demonstrated that floods during the twentieth century were of similar intensity as those seen during earlier historical times but that twentieth-century droughts have been less severe than previously experienced.

As might be expected, the North China Plain has experienced more frequent and more intense dry periods than seen in southern or central China, with the most significant decline in precipitation observed since 1950 (Figure 6.6)(Fan et al., 2012b; Liu et al., 2005a). In comparison, southern China has seen an increase in precipitation since 1950, and no significant droughts have affected that area during the twentieth century. Because the Yangtze River lies in the center of China it is influenced by both these climatic regions and, therefore, has a more complicated discharge history than the Yellow or Pearl Rivers. In particular, it is worth noting that the years during which the great famine took place in the Great Leap Forward, 1959–1961, there was no unprecedented drought in northern or southern China. Events of similar intensity took place, both before and after this date but did not result in losses of life on the scale of that of the Great Leap Forward.

Latitudinal changes between different parts of China are best understood when we consider atmospheric circulation patterns, and, in particular, the migration history of the westerly jet stream. It has been shown that since around 1960 there has been a steady decrease in the average temperature of the troposphere over East Asia that occurs during July and August (Yu et al., 2004). As a result of this cooling

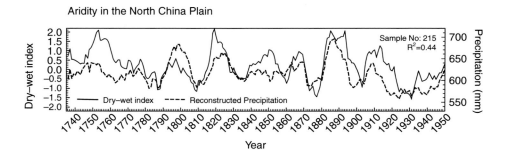

Figure 6.5 Comparison of dry–wet index and precipitation data in the North China Plain versus reconstructed precipitation series over the middle and lower reaches of the Yellow River in 1736–1950. Reproduced with permission from Zheng et al. (2006).

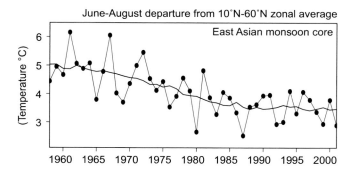

Figure 6.6 Departure of July–August (JA) mean (thin) and 10-year running mean (thick) 200–300 hPa air temperature averaged in the core East Asian monsoon region (90°E–130°E, 30°N–45°N). Summer cooling causes the upper-level westerly jet stream over East Asia to shift southward and the East Asian summer monsoon to weaken (Yu et al., 2004).

the westerly jet has been moving southward, which in turn has caused a weakening of the East Asian summer monsoon. The net effect of these changes is to increase the frequency of drought events in northern China and conversely to increase the chance of flooding in the south, especially within the Yangtze River Valley. It is this change in atmospheric circulation patterns since the middle of the twentieth century that is behind the historical climatic trends (Figures 6.5 and 6.6).

Exactly why the troposphere is cooling over Eastern Asia during the summer is not entirely apparent, but may be part of a longer-term trend that is different from the global average warming (Ramaswamy et al., 2001). This trend is particularly significant in eastern Asia because recent cooling of the lower stratosphere is only able to transfer this coolness to the troposphere in this region. Such thermal transfers are related to the presence of the Tibetan Plateau. East Asia is particularly affected because it lies downstream of convection cycles driven by the presence of the Tibetan Plateau (Ye, 1981; Yihui, 1992). Although it has been suggested that greenhouse gas emissions, particularly related to the burning of coal, might be important in driving the drying trend in northern China since 1950 (Xu, 2001), it is not clear how that process would have affected high parts of the atmosphere as identified in the Yu et al. (2004) study. What is clear is the twentieth-century trend in North China toward a weakening summer monsoon in an area with an increasingly dense population. At the same time, Japanese rainfall has experienced periods of very strong precipitation that has been associated with increasing heatwaves (Wang et al., 2019).

6.3 Evolution of Tropical Cyclones

As well as the seasonal rains of the summer monsoon, tropical cyclones can be important in delivering moisture, especially to East Asia and the Bay of Bengal, and are known to have experienced significant degrees of change in the recent historical past (Wu et al., 2005). Typically typhoons that affect these areas originate in the Philippine Sea, South China Sea, and the Western Pacific and track toward the West where they make landfall in Hainan, northern Vietnam or southwestern China (Figure 6.7)(Ho et al., 2004; Wang et al., 2007a). Alternatively, they experience bending in their trajectories and head more toward the northeast, making their way toward Japan or out into the North Pacific Ocean.

Records of typhoons that have entered the South China Sea now show that their trajectories have experienced significant changes over the past 50 years (Leung et al., 2007). Substantial periodicity in the frequency of typhoons in the South China Sea has been documented (Figure 6.8). Typhoons reached a maximum between 1988 and 1993, but this peak was followed by a significant fall in numbers. The decline has been most noteworthy in Category 4–5 typhoons, and it has been noted that the point of origin also shifted toward the west through time. Between 1961 and 1990 the stronger Category 4–5 typhoons averaged 1.9 events per year in the South China Sea, whereas since the mid-1990s this average dropped to only 0.2. Very few typhoons that originate east of 120°E now affect the South China Sea, but instead are displaced toward the north and head toward the Japanese islands (Grossman et al., 2015; Grossman et al., 2016).

The larger question is what is causing the change in the track of the typhoons, the answer to which clearly has significant societal implications, given their destructive character and ability to deliver large quantities of rainfall over short periods of time. It has been suggested that changes in the course of the typhoons in the Western Pacific are related to the cooling of the troposphere over the past two decades, as noted earlier. For example, Wu et al. (2005) hypothesized that it is this cooling trend, which results in large-scale reorganization of atmospheric circulation and, in particular, realignment of the cyclonic steering flows. Figure 6.9A shows the spatial distribution of tropical cyclones during the typhoon season, which coincides with the summer monsoon. This compilation spans the years 1970–2006 and highlights the fact that historically the northern South China Sea was the place for the highest occurrence of tropical cyclones in East Asia (Tu et al., 2009). Another zone of high occurrence stretches from the northern Philippines toward Japan, reflecting the traces shown for 2011 (Figure 6.7). There has been a long-term decline in cyclone occurrence in the northern South China Sea, especially since the late 1990s, as a result of cyclones tending to be steered toward the north after that time (cf. Figure 6.8). Those typhoons that do affect the northern

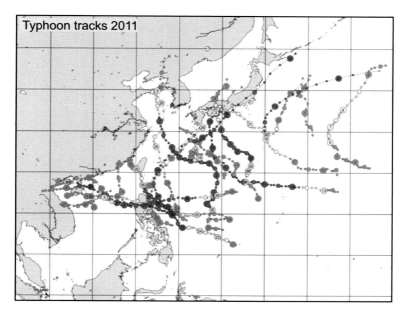

Figure 6.7 Map of tracks of typhoons affecting the western Pacific in 2011. Numbers denote the date of the month that the storm occupied this location at 0000 UTC. Blue – Tropical Depression (Class 2), mid gray – Tropical Storm (Class 3), light gray – Severe Tropical Storm (Class 4), Dark gray – Typhoon (Class 5). Downloaded from Digital Typhoon website (Kitamoto, 2011).

South China Sea have originated more within that basin in the recent past rather than coming from the Pacific Ocean. Tu et al. (2009) showed the difference in the occurrence of tropical cyclones between 1970–1999 and 2000–2006 (Figure 6.9B). This plot clearly shows the reduction in typhoon activity in the South China Sea, but a sharp increase in the East China Sea and those areas extending toward Japan. In 2019, Japan was struck by "Super-typhoon" Hagibis, which caused nine billion dollars worth of damage and killed 89 people in the Kanto region of central Japan (Cappucci and Freedman, 2019). There is a separate area of increasing high typhoon intensity in the Western Pacific, although this does not impact human settlements. If this recent trend in typhoon steering continues then this will have significant implications for the many populous areas along the eastern Chinese coast and those in southern Japan and Korea.

Rainfall in southern China has been increasing over the latter part of the twentieth century and tropical cyclones have a part to play in this. In their study of cyclone intensity and rainfall across the southern China area, Kim et al. (2012) argued that when looking at the total summer rainfall, and in particular the early part of the summer, that tropical cyclones were not

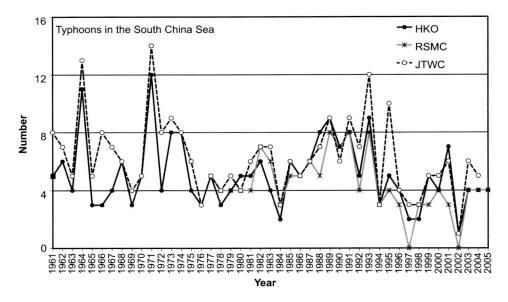

Figure 6.8. Annual number of typhoons occurring in the South China Sea based on Hong Kong Observatory (HKO), Regional Specialized Meteorological Centre (RSMC) Tokyo, and Joint Typhoon Warning Centre (JTWC) best tracks, respectively. Modified from Leung et al. (2007).

important in driving the longer-term rainfall trends. However, precipitation linked to typhoons tends to become much more important in August toward the end of the regular monsoon season. The impact of cyclones has been most dramatic along the southern coast of China. There is evidence that these have been intensifying in the more recent past. Kim et al. (2012) considered two time periods, 1981–1992 and 1993–2002. Although both times showed significant increases in monthly summer rainfall, it was apparent that only a small fraction of the precipitation was related to tropical cyclones during 1981–1992, but that since that time typhoon rainfall had become much more important, especially along the southernmost coast. Some of this extra rainfall has been related to increased influence of typhoons that formed over the South China Sea late in the summer season (Wang et al., 2007a), as opposed to in the open Pacific where the majority of East Asian typhoons have their origins (Joint Typhoon Warning Center, 2002). Although total rainfall has increased in southern China because of the typhoons originating within the South China Sea in June and July, this is balanced by the synchronous reduction in cyclone activity over that same period in southern China area, as a result of large-scale atmospheric changes across the region, as discussed earlier.

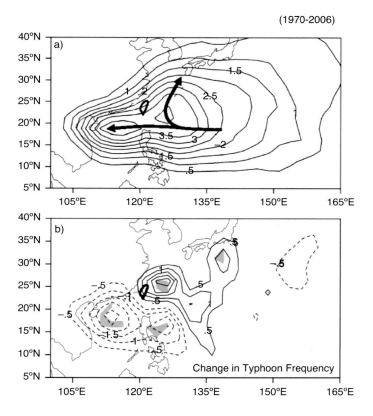

Figure 6.9. (A) June–October (JJASO) typhoon frequency climatology averaged over the period of 1970–2006. The contour interval (CI) is 0.5 per season (JJASO) per grid box (2.5° × 2.5°). The bold arrows represent the majority of typhoon paths in the western North Pacific–East Asian region. The (B) June–October typhoon frequency differences for the period of 2000–2006 minus the period of 1970–1999 from the Regional Specialized Meteorological Center, Tokyo. The contour interval is 0.5; shading denotes that the difference between the mean of the two epochs is statistically significant at the 5 percent level. Figure modified from Tu et al. (2009).

6.4 Rainfall in South Asia

A simplified image of the variability in recent rainfall in South Asia can be derived by consideration of the All India summer monsoon Rainfall Index (AIRI) (Parthasarathy et al., 1992) (Figure 6.10). This is a useful general indicator of the strength of monsoon rain across India but lacks the resolution to be able to define regional variations and, in particular, those aspects related to the dynamics of the monsoon because it focuses entirely on rainfall and not on wind speed or atmospheric pressure differences. The AIRI itself is an area average of 29 different subdivisional rainfall measurements, which themselves are an average of all the stations in a given district. This approach demonstrated that when looking at the largest scale there were stronger than normal monsoons during the 1880s, 1915–1925 and 1930–1945, while rainfall was substantially less than average at 1899–1905, in the 1960s and around the turn of the twenty-first century (Zhou et al., 2010)(Figure 6.10). Such trends are slightly out of phase compared to the reconstruction for China (Figure 6.5). Looking at the entire subcontinent it is generally argued that there has been an overall decrease in summer monsoon precipitation of the past 50 years (Dash et al., 2009; Joseph and Simon, 2005), Although whether this is true across all of South Asia is not apparent when looking only at the AIRI.

A more detailed regional breakdown of rainfall trends in the latter part of the twentieth century was provided by Lau and Kim (2010), who divided up the country into five different sectors and plotted total annual rainfall for each of these since 1960 (Figure 6.11). They further refined their analysis by dividing the early and later parts of the summer monsoon season so that changes in the onset of the monsoon could be identified. There are some common trends that apply to much of the Indian subcontinent, yet regional differences emerge. Since 1960 there appears to be a strengthening of rainfall during the May–June season, but the degree of strengthening varies from region to region and is strongest in the Central Northeast and Northwest, with almost no change seen in the Northeast, West Central, or Peninsular areas (Figure 6.11A). In contrast, there has been a decrease in intensity during the latter part of the monsoon season, although the degree of change is much more modest than that seen in the early phase (Lau and Kim, 2010). Most regions were fairly constant in their rainfall until 2006, with the clearest decrease in intensity being in the West Central and Peninsular areas. It is

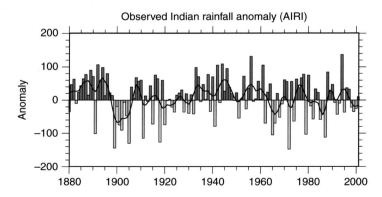

Figure 6.10 The observed June–August (JJA) mean all Indian monsoon rainfall index (AIRI) given as anomalies (in mm/day, w.r.t 1961–1990). Thick solid lines represent data smoothed with a 10-year filter. From Zhou et al. (2010).

Regional changes in Indian rainfall

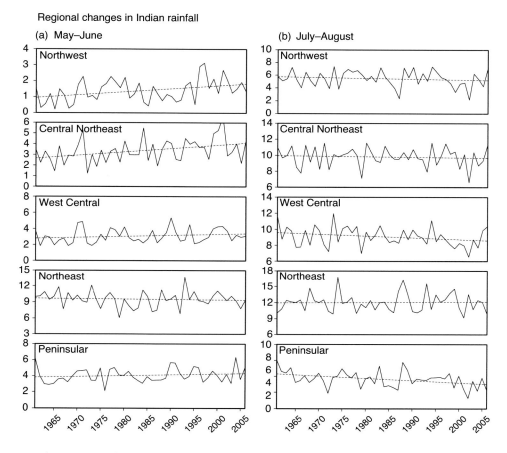

Figure 6.11 Time series and linear regression lines (dotted) of each of the five regional rainfall areas over India for (a) May–June, and for (b) July–August, during the latter half of the twentieth century. Units are in mm/day. Modified from Lau and Kim (2010).

noteworthy that changes from year to year were of much greater magnitude than these long-term trends stretching over more than 40 years, which reflects the fact that monsoon intensity has historically been tied to variations in the ENSO phenomenon (Azad and Rajeevan, 2016; Krishnaswamy et al., 2015) and to a lesser extent the Indian Ocean Dipole (Kumar et al., 2006).

Lau and Kim (2010) argue that the drying trend in South Asia during the late twentieth century was not caused by global warming, but was instead associated with an increase of aerosols in the atmosphere, especially in industrial parts of northern India. It is well understood that black carbon and dust may have strong impact on the intensity of monsoon rainfall and wind speed through increasing absorption of the heat in the troposphere (Padma Kumari et al., 2007; Ramanathan et al., 2005). High aerosol concentrations have the potential to decrease monsoon strength because they block solar radiation from reaching the ground and thus weaken the generation of the low-pressure system that draws moist air inland from the ocean.

It should, however, be recognized that there is a link between atmospheric aerosols, especially desert-derived dust and ENSO with El Niño conditions linked to lower aerosol concentrations across northern India (Kim et al., 2016). Conversely, when aerosols are abundant this is associated with more moisture transport and heavier precipitation in northern India during April–May followed by reduced rainfall later in the season across all of South Asia. Monsoon strength and aerosols have been linked by the "elevated heat pump" hypothesis, which argues that accumulation of dust and black carbon over the Indo-Gangetic Plain and the foothills of the Himalaya during the pre-monsoon season results in intensified warming in the middle and upper troposphere and subsequently leads to increased rainfall over northern India in late spring and early summer (Gautam et al., 2009; Lau and Kim, 2006). This trend is best shown in the Northwest and Central Northeast regional trends (Figure 6.11). Increased aerosol concentrations in recent times may also account for why there is a common decrease in the later part of the monsoon season because if there are more clouds in the atmosphere in the early part of the monsoon then this increases the blocking of solar radiation from reaching the ground and results in an earlier reduction in the low-pressure system and thus also in rainfall. Continued industrialization of northern India would be expected to accentuate this trend, further driving monsoon summer rainfall down across the flood plains.

Certainly there is good evidence in the recent past that aerosol emissions have been rising across northern India, as the region develops economically (Kühn et al., 2014; Li et al., 2017a). However, it should be remembered that the area has been affected by particle flux even before industrialization began. Dust has been delivered from the deserts in Arabia, as well as closer by from the Thar Desert of what is now western India and southern Pakistan during the summer monsoon when winds are blowing from the southwest toward the Himalaya (Glennie and Singhvi, 2002; Singhvi et al., 2010). Dust aerosols transported to northern India may sweep up the finer black carbon aerosols generated by industry. The black carbon can then form a coat over the larger dust particles, making them more absorbent and thus increasing their albedo and ability to cool the surface of the Earth (Ramanathan and Carmichael, 2008).

Observations made in the New Delhi area show that concentration of aerosols in the atmosphere has been increasing steadily since around 1980 (Lau and

Kim, 2010). Atmospheric visibility has decreased since that time, and this is a measurement directly linked to aerosol concentrations in the lower levels of the atmosphere. Furthermore, the Aerosol Index estimated from satellite data over the Indo-Gangetic plain has shown a steady increase over the same time period (Figure 6.12), consistent with this trend.

The impact that aerosols have had on the Indian Summer Monsoon has also been addressed through modeling using the U.S. National Oceanic and Atmospheric Administration (NOAA) Geophysical Fluid Dynamics Laboratory (GFDL) CM3 GCM (Donner et al., 2011), which includes aerosol–cloud interactions and aerosol indirect effects (Bollasina et al., 2011). This model attempted to simulate the monsoon since 1860 to 2005 using a series of runs incorporating the different proposed forcing functions, that is, solar variations, volcanoes, greenhouse gases, ozone, aerosols, and land use. Although a model run that used all the forcing functions did a good job of reproducing the observed climate trends (i.e., drying over northern India and wettening in the south), it was noteworthy that a model driven only by aerosol-related processes did an equally good job of predicting the climate change. As a result, it was concluded that it was the rise in aerosol emissions that has been the primary driver for Indian monsoon intensity changes over the last century (Bollasina et al., 2011). In contrast, the impact of rising greenhouse gas concentrations would tend to generate increased rainfall across the entire area, contrary to the observations.

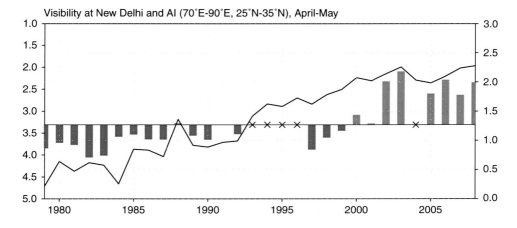

Figure 6.12 Time series of Aerosol Index from Total Ozone Mapping Spectrometer (TOMS), Earth Probe Total Ozone Mapping Spectrometer (EPTOMS), and Ozone Measuring Instrument (OMI) averaged over the Indo-Gangetic Plain (70–90°E, 25–35°N), in increasing relative units in the right ordinate and visibility (solid black line) over New Delhi in decreasing relative units in the left ordinate, from 1979 to 2007. Missing data in Aerosol Index (AI) are indicated by symbol "X." From Lau and Kim (2010).

A more detailed analysis of twentieth-century Indian summer monsoon development is made possible by using empirical orthogonal functions (EOFs). EOFs are a statistical analytical tool that is similar to principal component analysis and allows temporal and spatial patterns to be isolated in data. In the study of Mishra et al. (2012) two such EOFs were defined. EOF1 was chosen to correlate with the All India Rainfall, as well as with ENSO oscillation strength. Until recently weak summer monsoon rain were associated with periods of El Niño (Krishnamurthy and Goswami, 2000; Kumar et al., 1999; Wainer and Webster, 1996). EOF2 instead looked at the rainfall anomalies of opposite polarity between northern India, principally the Indo-Gangetic Plain, and those in southern India over the peninsula. EOF1 mirrors the overall trends in rainfall patterns, with the driest areas during the twentieth century highlighted in northwest India and the wettest close to the Ganges–Brahmaputra Delta (Figure 6.13A).

The EOF2 model shows strong spatial variability, with positive values in southern India and negative values over the Indo-Gangetic Plain (Figure 6.13B), which corresponds to the drying trend since 1950 seen across northern India (Figure 6.13D), accounting for a decrease of 4–5 percent (Ramanathan et al., 2005). EOF2 is linked to the pattern of sea surface temperature variation in the Indian Ocean, which in turn correlates with ENSO. This link is consistent with the long-term drying of the plains being linked to ENSO intensity. The principal component of EOF2 (PC2) negatively correlates with the negative rainfall anomalies across northern India, at least since 1950 (Figure 6.13C). In contrast, the positive rainfall anomalies observed over southern India, in association with a positive polarity of PC2, are consistent with convergence of low-level wind anomalies over that region. Summers after the peak of El Niño events are associated with a weakening of the westerly jet over the northern Indian Ocean, which in turn helps maintain high sea surface temperature anomalies and so reduces the land–ocean contrast that drives the monsoon, thus reducing rainfall. However, it is not obvious why the greater warmth of the northern Indian Ocean during these summers should favor reduced monsoon rainfall over northern India. Instead, it seems more likely that the linkage between the ENSO cycle and monsoonal rainfall anomalies in the subsequent summer is made through the atmosphere (Mishra et al., 2014).

6.5 River Drainage Evolution

6.5.1 The Kosi River

As well as being affected by changing rainfall, past water supply has also been controlled by reorganization of river systems, especially in the form of drainage capture and avulsion. A particularly dramatic example of this was seen in 2008 in

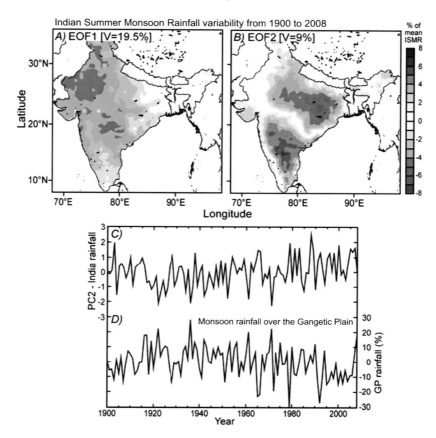

Figure 6.13 (A and B) The two leading empirical orthogonal functions (EOFs) of June–September (JJAS) Indian Summer Monsoon Rainfall variability for the 109-year period from 1900 to 2007. Shading indicates the rainfall anomalies, expressed as a percentage of the climatological mean Indian Summer Monsoon Rainfall (ISMR), observed in association with a PC amplitude of +1 SD. V indicates the fraction of the domain-integrated variance explained by the two modes. (C) The standardized PC2 (the time series of the expansion coefficient of EOF2) and (D) monsoon season rainfall averaged over the Gangetic Plain (21°N to 30°N, 75°E to 89°E) expressed as a percentage of the climatological mean rainfall. Reproduced and modified from Mishra et al. (2012).

Nepal and India when the Kosi River (Figure 6.13) experienced a major change in the direction of its flow. The Kosi had long been recognized as a drainage where the river channel was prone to significant shifts, something that it has in common with many of the major rivers draining from the Himalaya into the Ganges. Wells and Dorr (1987) reconstructed the evolving path of the Kosi River based on historical records and showed a progressive migration from the east toward the west between

1736 and 2008 (Figure 6.14). They concluded that the river's tendency to readjust the position of its channel between the point at which it leaves the Himalaya and its confluence with the Ganges was mostly driven by stochastic and autocyclic processes involving the reduction in slope along the path of the active channel as a result of sedimentation along its levees (Bowman, 2019). In doing so, this makes other drainage pathways more attractive. Although earthquakes might be expected to have a role in this readjustment very little evidence could be found to show that that had actually been the case, at least over the study period.

Flooding events, in particular, appear to be times at which a river is susceptible to readjustment into a new course. The 2008 reorganization of the Kosi River occurred during a flood that was driven by heavy monsoon rains and that resulted in a breach within man-made embankments that had been constructed in the Nepalese part of the river to keep the flow in its previous, quasi-stable configuration. However, once the embankments had been breached the river rapidly adjusted to a new course, resulting in flooding of wide areas of Nepal and northern India (Figure 6.14B). Although this was a natural phenomenon it caused a significant amount of damage and resulted in the deaths of 250 people in India alone (Coggan, 2008). Almost three million people had to be evacuated in India, while in Nepal around 53,000 people were affected (UN Office for the Coordination of Humanitarian

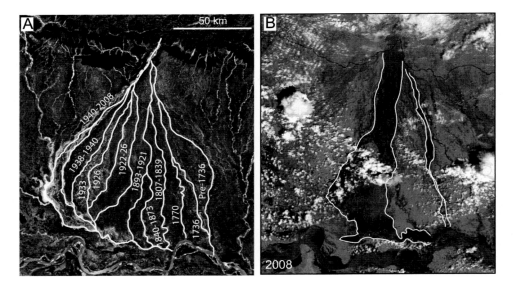

Figure 6.14 (A) Map showing the migration of the Kosi River in northern India and Nepal since the early eighteenth century (Wells and Dorr, 1987). (B) Image of the Kosi River in 2008 after its rapid avulsion to a new central location on the alluvial fan (NASA Terra MODIS image from NASA Visible Earth).

Affairs, 2009). In India more than 340,000 hectares of land were inundated, destroying large areas of crops (Coggan, 2008). This example demonstrates the potential large-scale impact of drainage reorganization on modern settlements.

6.5.2 Damming and Irrigation in the Indus Basin

Western India and what is now Pakistan have some of the driest conditions of any within monsoonal Asia. Consequently, it is in these regions that the greatest efforts have had to be made to try and manage the scarce water resources in order to sustain the societies developed there. Although some adjustment can be made in terms of the type of crops that are grown in the region, management of the water resources and especially the Indus River and its tributaries have been particularly important in allowing the development of large urban centers. In recent times this has taken the form of large-scale dam construction that was originally initiated under British colonial rule to sustain the plantation agriculture (Gilmartin, 2015; Kidwai et al., 2019). This trend then accelerated after independence in order to sustain additional growth of the agricultural economy within the floodplains, particularly under Green Revolution initiatives.

The first major barrage on the Indus was constructed near the Pakistani city of Sukkur. This barrage was built during the British Raj from 1923 to 1932 and named Lloyd Barrage after the then Governor of Bombay Sir George Lloyd (Naqvi, 2012). After independence, the barrage was renamed after the city and still functions as a critical part of the water management system in the north of Sindh province. Sukkur Barrage forms the controlling structure of what is arguably the largest irrigation system in the world. A network of canals extends over a length of 9,923 km and distributes water over more than 20,000 km², thus allowing advanced agriculture in what is otherwise a semi-desert region to the west of the dunes of the Thar Desert (Figure 3.14). A number of other major dams now controls the flux of water through the Indus system, including the Jinnah Barrage completed in 1946, the Kotri Barrage at Hyderabad completed in 1955, the Taunsa Barrage in Punjab completed in 1958, the Guddu barrage in northern Sindh completed in 1962, and the Chashma Barrage in Punjab, which was completed in 1971.

While these dams have generally been beneficial to agriculture, there have been negative side effects related to the general improvement in irrigation. Because the barrages have reduced water flux to the lower reaches of the river they have resulted in major water shortages in the southern parts of Sindh province and, in particular, in the delta region (Briscoe et al., 2006; Syvitski et al., 2013). Irrigation channels, which were cut into the delta plain and that were intended to allow exploitation of the region for agriculture, are having the reverse effect as seawater has been drawn landward through these channels causing the farmland that was previously fertile to become salinated, poisoned, and unusable (Day et al., 2019). The lack of fresh water in the delta region has also had significant negative effects on the fisheries (Hussain et al., 2010), which support three quarters of all the people living in that area, presently around 900,000. Sharp drops in the productivity of the fisheries have resulted from the fact that many of the species that thrive in the delta and adjacent offshore spend part of their life cycle in the mangrove swamps, which used to typify the coast of the delta. However, the lack of fresh water has resulted in the loss of significant areas of mangrove swamp, totaling >70,000 ha during the past 50 years (Rasul et al., 2012; Walsh, 2010).

Nonetheless, the mangroves of the Indus Delta still represent the largest desert climate mangrove forest in the world. Inam et al. (2007) noted that before 1955 when the dam was constructed there were no days in which there had been no water flow at the Kotri Barrage. Between 1960 and 1967, the number of zero-flow days had increased to as much as 100, while after the construction of the Mangla dam on the Jhelum River tributary the number of zero flow days had increased to 250 by 1975. This has impacted a natural resource that is both culturally and economically important to thousands of people. Mangrove swamps are also critical for allowing humanity to adapt to some of the effects of changing climate. The roots of mangrove forests help stabilize coastlines by trapping sediment and slowing the movement of water: factors that are critical in reducing the impacts of storm waves, flooding and sea level rise. These forests also trap carbon dioxide and globally these biomes avoid the release of approximately 13 million metric tonnes of carbon (Miteva et al., 2015).

6.5.3 *Indus Floods of 2010*

The most dramatic negative consequences of damming were seen in 2010 as a result of the major flooding in the Indus caused by the unusually strong monsoon of that summer. The rains in the summer of 2010 were particularly intense because of the coincidence of a strong La Niña that year in the Pacific Ocean, which caused stronger rains in South Asia (Khalid et al., 2018). This was also coupled to an unusual configuration of the jet stream that year, which resulted in it not migrating in the typical fashion. Rather the jet stream remained in a fixed location driving continuous rains in South Asia and especially northern Pakistan, while at the same time delivering heat waves in central Russia (Marshall, 2010). Exactly why the jet stream became stuck in this spectacular fashion is a point of continuing research, although initial indicators suggest that this may be related to weakened solar activity during the winter and spring of that year (Lockwood et al., 2010). The discharge at Sukkur Barrage increased much more sharply in August 2010 compared to a typical year, such as 2013 (Figure 6.15). The very sharp increase in discharge after August 6 and that culminated at a peak around a week later represented an increase of almost five times compared to normal discharge at this point in the river.

Satellite images from 2010 help to understand the developing flood that followed the heavy rains upstream and resulted in large-scale inundation of the regions around the lower reaches of the Indus River (Figure 6.16). It is noteworthy that some of the worst flooding was related to the presence of the dams along the river. The Sukkur Barrage acted as a choke point on the river resulting in widespread flooding north of that city. River levels rose to such an extent that by August 8, 2010, the Indus burst its banks north of Sukkur, flooding substantial areas to the west of the main channel. A significant flow of water then moved toward the west before turning south as shown in Figures 6.16C and 6.16D. The large lake that formed as a result west of the Indus was present for many months after the flooding. Economic losses caused by the flooding that year were significant, with 570,000 ha of cropland being destroyed and more than 2,000 people killed (Singapore Red Cross, 2010). Around 20 million people are believed to have been displaced from their homes as a result of the 2010 floods. In the following year, further flooding resulted in the deaths of a further 434 people and the inundation of 690,000 ha of

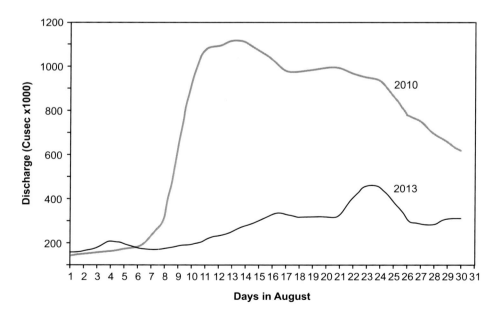

Figure 6.15 Diagram showing the contrasting discharge at Sukkur in Sindh province, southern Pakistan between a normal year, 2013 and the floods of 2010. (www.lbcawb.org/). Data from Left Bank Canals Area Water Board

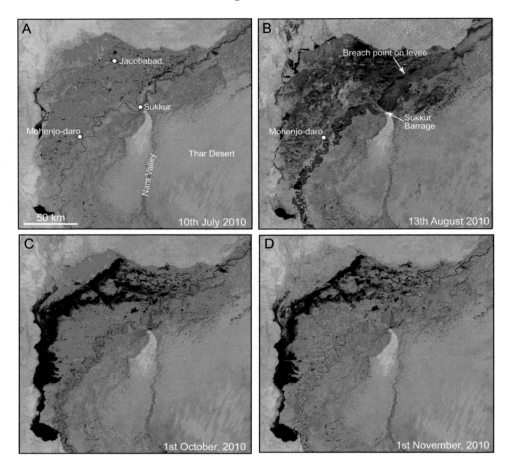

Figure 6.16 A series of satellite images taken by NASA's Moderate Resolution Imaging Spectroradiometer (MODIS) on their Aqua satellite. Images use a combination of infrared and visible light to increase the contrast between water and land. Water appears in varying shades of blue. Vegetation is green, and bare ground is pinkish brown. Clouds range in color from pale blue-green to bright turquoise. The images show the disastrous flooding around the main Indus, the effect of the damming in focusing where flooding occurred, as well as the effect of breaching the levees north of Sukkur resulting in flooding of areas west of the river.

arable land (Tariq and van de Giesen, 2012). Many of these losses related to breaching of the river levees as the result of damming impeding the ready flow of water through the system. Damage to structures, including road and railways as well as public buildings, was estimated to exceed US$4 billion (Hicks and Burton, 2010), and wheat crop damages alone were estimated to be over US$500 million. Total economic impact may have been as much as US$43 billion, which resulted in a significant decrease in gross domestic product (GDP) from the forecast growth of 4 percent to a net fall of around 3 percent of GDP (Looney, 2012). These losses continued for many months and disproportionately affected rural areas (Kirsch et al., 2012).

It is debatable whether any long-term lessons can be learned from the floods of 2010. Pakistan is essentially a rather dry country and one in which irrigation systems are fundamental to allowing agricultural production, because in a - normal year rainfall is insufficient to supply all the agriculture required to sustain the high population density. We reviewed the different types of irrigation systems that have been used in this region throughout history. Indeed, flooding has been seen in Europe as well in the past decade when rainfall overwhelms normal flood defenses, for example, on the Elbe and Danube in 2013 (Alfieri et al., 2016). Nonetheless, year 2010 showed that the same structures that are needed under normal circumstances can be very damaging when the monsoon becomes intense or storm-like and delivers very large quantities of water to the rivers over short time periods. Those measures that had been put in place did not significantly reduce the degree of damage in the wake of the 2010 flood (Tariq and van de Giesen, 2012). Whether other countries could have better coped with the conditions found in 2010 is questionable, although it is also clear that insufficient maintenance of the levee systems, coupled with some poor decisions about when and where to allow the levees to break did result in greater loss of life and property them might otherwise it been expected (Syvitski and Brakenridge, 2013). The Federal Flood Commission (FFC), which had been founded in 1977 following earlier flooding, had been charged with a number of projects that would have alleviated the severity of flood related damage, but poor leadership and corruption had contributed to making this body relatively ineffective in planning for the type of events seen in 2010 (Mustafa and Wrathall, 2011).

For future purposes, a pressing question is how often such events occur. Historical records of discharge do not extend far enough in the past to answer this and do not account for changes in frequency linked to climate change. However, data from the offshore Indus delta provides a sedimentary record of flooding activity in the river before around 1950 when the damming of the river prevented further major floods reaching the ocean (Clift et al., 2014). Sand layers at the head of the Indus Canyon have been interpreted as annual monsoon flood discharge proxies, many of which are thin over the past 150 years or so. However, seven to eight sand layers are somewhat thicker and two are much bigger than the rest, suggestive of deposition by a significant discharge event. The frequency of major flooding events thus appears to be on the order of one every

50–100 years. Whether that same rate of large-scale flooding can be extrapolated into the future is questionable given changes in global temperature that increase total rainfall and the storminess of the summer monsoon rains (Murakami et al., 2017).

6.5.4 The Indus Basin Project

Much of the large-scale dam construction in Pakistan has in recent years been linked to a large-scale initiative known as the "Indus Basin Project" (Wescoat et al., 2000). The project is encapsulated by a number of large dams, particularly the Mangla Dam on the Jhelum River and the Tarbela Dam located on the Indus River close to where the Indus reaches the floodplains. This project represents the Pakistani response to the "Indus Waters Treaty" that was signed between India and Pakistan in 1960. The treaty became important in formalizing the distribution of water resources in western India and Pakistan because of the Pakistani dependence on rivers that originate in India, but provide much of the water resources to this country (Alam, 2002). In its simplest form, the treaty allows India to use the eastern tributaries, Ravi, Beas, and Sutlej rivers before their entrance into Pakistan, whereas the western tributaries, Indus, Jhelum, and Chenab rivers are reserved for use by Pakistan (Sahni, 2006). Nonetheless, the very existence of an international border makes a fully coordinated response to changes in course and climate hard to achieve.

Tarbela Dam, in particular, is noteworthy in being the largest earth-filled dam in the world and as measured by its potential volume is the second-largest dam structure worldwide (Ahmad et al., 1993). This structure was completed in 1974 and is an important source of hydroelectricity, as well as water resources in northern Pakistan. One potential complexity of placing a dam across the mainstream of the Indus at the point where it leaves the mountain front is that the river is heavily laden with sediment, reflecting the erosive nature of the mountains north in the gorge, as well as the high stream power operating in the steep slopes upstream of the dam (Ahmed et al., 2018). Surveys of the reservoir show that the northern end of the lake formed behind the dam is marked by a major delta (Tate and Farquharson, 2000). Because of the sediment load of the Indus the reservoir has been filling with sediment since the start of operations (Figure 6.17)(Palmieri et al., 2001). What is evident is that the reservoir filled rapidly after the initial blocking of the river and constructed a significant delta within the reservoir before 1979. The cross section shows the operational limits of the reservoir and illustrates that by 1999 much of the usable space behind the dam had been filled by sediment. This meant that at minimum levels there was very little open water left behind the barrage. Initial specifications for the dam suggested that the reservoir would be full of sediment by around 2030, which would render it unusable, but the actual infilling was initially much faster than expected. However, since 1999 the rate of additional sediment storage within the reservoir has fallen substantially so that only small amounts of additional material have been added to the delta since that time. As a result, predictions now suggest that Tarbela Dam will be serviceable as an economic and water control structure at least until 2060 at present rates (White, 2005), although exactly why this decrease has happened is debated (Khan and Tingsanchali, 2009). Despite the potential complexities it is abundantly clear that damming will continue to play a central role in water management for arid countries such as Pakistan regardless of future variations in monsoon precipitation.

Whether competition for water rights remains peaceful remains to be seen, as disputes certainly exist. For example, the Wular Barrage project involves the construction by India of a barrier on the Jhelum River downstream of Wular Lake in Kashmir (Figure 6.18). The purpose of this barrier, apart from power generation and provision of water supplies, is to make the river navigable during the dry period between late October and mid-February. The barrier has geostrategic importance because of its location close to the Line of Control (Condon et al., 2009). Moreover, the Jhelum River rises in the Kashmir Valley and is supplied by Wular Lake, which is located about 25 km north of Srinagar at about 1581 m above sea level. For Pakistan the strategic importance is that its control provides India with the means to exert pressure on Pakistan because a dam on that site has the

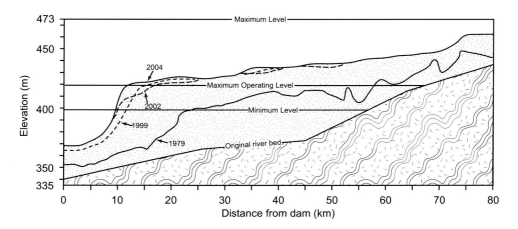

Figure 6.17 Diagram showing the progressive infilling of the Tarbela Reservoir following its completion in 1974. Note that sedimentation rates have slowed markedly in the recent past. Redrawn from the original by Khawaja and Sanchez (2009).

potential to disrupt the entire system of the "Triple Canals Project," a long-lived system of waterways developed under colonial rule that transfers water from the Jhelum to the Ravi in order to allow sufficient water to flow and maintain the navigational access to the Ravi and associated canals (Shakir et al., 2010; Zawahri, 2009). Because the Mangla Dam in Pakistan is also fed by the Jhelum, its long-term viability would also potentially be at stake. The Pakistani government fears that the control of the water flow by India to the construction of the Wular Barrage could deal a crippling blow to the agricultural economy of Pakistan.

6.6 Damming on the Yangtze

East Asia, too, has experienced major problems over the past century related to monsoon variability. Major floods affected the Yangtze River valley in 1954 and 1956, as well as in 1959 (Yu et al., 2009b), which was partly used as a natural explanation for the devastation caused by the Great Leap Forward that started in that year. The floods in 1954 particularly struck Hubei Province and killed about 33,000 people (Yin and Li, 2001a). However, the area of China that suffered from flooding during the Great Leap Forward was considerably smaller than those in

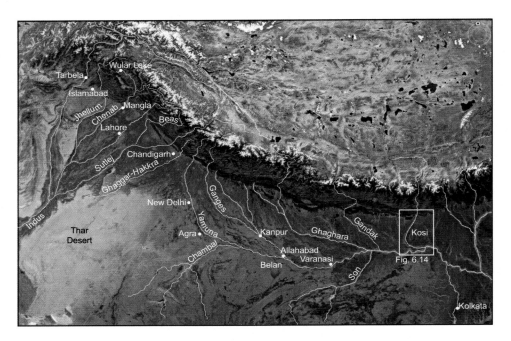

Figure 6.18 Satellite image of the Himalayan foreland basin showing the major tributaries of the Ganges and Indus River systems as well as geographic features mentioned in the text. Image from Google Earth.

decades to come (Zhang et al., 2016). Monsoon-related flooding reached its worst level in 40 years in the summer of 1998 (Zong and Chen, 2000). At that time, the Yangtze (Changjiang) experienced massive flooding of parts of its middle reaches, resulting in 3,704 dead, 15 million homeless and $26 billion in economic loss. Other sources report a total loss of 4150 people, and 180 million people having been affected (Spignesi, 2004). An area of 25 million acres (100,000 km^2) had to be evacuated, with 13.3 million houses damaged and/or destroyed. Although the rainfall was high, the degree of flooding was disproportionate in comparison. Human activities made the damage much worse than previously experienced because of large-scale reclamation of lakes and fluvial islands in the middle basin of the Yangtze (Zong and Chen, 2000). This process substantially reduced the floodwater storage and drainage capacity resulting in more ready overspilling from the main channel. Furthermore, deforestation in the upper catchment has accelerated soil erosion (Chen et al., 2001; Yin and Li, 2001b), resulting in a large amount of sediment being deposited in reservoirs whose storage capacity was thus reduced. Between 1970 and 2001 areas experiencing very rapid erosion have doubled in size (Yin and Li, 2001b), reducing capacity further. This deforestation is likely due to efforts aimed at increasing cultivated land during the Great Leap Forward and the Cultural Revolution, as well as the large amounts of forest and tree cutting that took place during Mao's efforts to construct backyard furnaces (Shapiro, 2001). Moreover, strengthening flood defenses to protect people living on the floodplain, largely in the form of levees, has raised the water level and encouraged silting up of the main channel during floods so that when large rainfall events occur there is breaching and overspilling to a much greater degree than had historically been the case. All these processes happened in the context of higher summer rainfall in the middle Yangtze basin after 1960, largely in the form of storms that rapidly cause river levels to rise, and likely driven by higher temperatures (Jiang et al., 2007). Gauging data from the lower reaches shows that the frequency of flooding has increased since the 1950s and that peak discharge has also increased since the 1860s (Yu et al., 2009a).

Not surprisingly, the Chinese government has taken action to prevent further damaging losses, principally in the form of damming to generate storage capacity, which has the additional benefit of providing power generation capacity. In 1919, the first president of the Republic of China, Sun Yat Sen, declared the intention to build a large dam across the Yangtze, yet little of substance happened on the project until the 1980s. At that time planning began for a large dam in the Three Gorges, a proposal that was controversial because of the famous natural beauty of the region, high biodiversity, and significant number of people that would have to be relocated to allow for the reservoir, totaling 632 km^2 when the reservoir is full

(Jackson and Sleigh, 2000). Nonetheless, construction finally began in 1994 with the main wall of the dam finished in 2009 and the last of generator turbines coming online in 2012 (Huang and Yan, 2009). The Three Gorges Dam is one of the largest on Earth, with a length of 2,335 m. The top of the dam is 185 m above sea level and 181 m above the rock foundation. The reservoir itself now extends 660 km upstream of the dam wall. Total costs exceeded $24 billion, which is expected to take 10 years to recoup from power generation.

Although the dam does bring many benefits, it has been noted that there are significant risks and costs associated with this as well. There are some increased hazards related to potential landsliding from the steep gorge sides into the reservoir (Cojean and Caï, 2011) and some suggestion that filling the reservoir has increased the seismic activity of faults in this region some of which pose a direct threat to the dam itself (Stone, 2008; Zhang et al., 2017). Because of the construction of the reservoir much of the sediment coming from the upper reaches is now trapped behind the dam and eventually this is beginning to fill the reservoir, much as seen in the Tarbela, the Pakistani example discussed earlier. After completion around 60 percent of the sediment reaching the reservoir from the upper reaches has been trapped (Xu and Milliman, 2009). Moreover, the river downstream of the Three Gorges, as well as associated lakes, are now starved of sediment (Zhou et al., 2016), and this also carries some risks in terms of increased flooding risk should the dam be unable to control the water flux. The flood plains of the lower reaches of the river were previously built up and strengthened by the continual arrival of silty material, whose flux is now strongly reduced. For coastal cities such as Shanghai this is potentially a serious problem because the land beneath them continues to subside, while at the same time the region is experiencing modest rates of sea level increase (Liu et al., 2007a; Zhongyuan and Stanley, 1998). This is causing large-scale loss of wetlands (Yang et al., 2006a), which are both economic resources and as we reviewed earlier, also traps of carbon, while helping enhance coastal resilience to changing climate. Furthermore, the low sediment load of the Yangtze downstream of the dam has resulted in the river incising and eroding its own channel, causing large-scale export of sediment from the river (Dai and Liu, 2013).

6.7 Sediment in the Yellow River

The delta coast of the Yellow River is prone to rapid progradation that is then modulated by the activity of tides and waves at the coast. Figure 6.19 shows how the water and sediment load in the modern Yellow River varies along its entire course. The amount of water entering the river increases sharply in its uppermost regions as a result of precipitation against the northern edge of the Tibetan Plateau, where orographic rainfall is focused. However, at the same time sediment loads in the river remain relatively modest as a result of the flow passing over strong bedrock units in the absence of very high stream power. Downstream of Lanzhou the river enters a relatively arid region and the amount of water reduces as the flow is used for agricultural purposes, while very little is added from local runoff. However, as the river moves around its northernmost bend within the region of the Chinese Loess Plateau, the amount of sediment rises sharply as silt from the Loess Plateau is reworked into the river (Nie et al., 2015). This process has been exacerbated by farming practices breaking up the soil and liberating more sediment, which is responsible for giving the river its name. Recent evidence suggests that humans had already begun this process in the deep past (Kidder et al., 2012; Zhuang and Kidder, 2014). At the same time there is a moderate increase in the amount of water discharge, which then remains relatively constant in the lower reaches. Sediment flux also falls within the lower reaches as a result of sedimentation especially upstream of Huayuankou (Ren and Shi, 1986). This reflects the fact that the river is close to its maximum carrying capacity and that as the regional

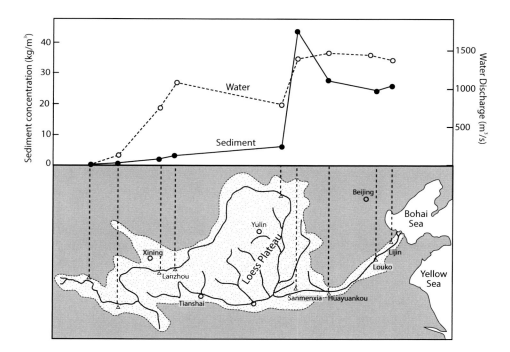

Figure 6.19 The Yellow River basin showing a 30-year average of water discharge and sediment concentration data for nine stations along the river (1950–1979). Note the abrupt increase of sediment concentration after the river flows through the Loess Plateau. Redrawn from Ren and Shi (1986).

slope shallows entering the lower reaches the river is able to deposit sediment within the floodplains as it loses stream power.

Although the sediment load of the Yellow River is famously high, it is noteworthy that it has dropped in recent years. Wang et al. (2007b) showed that in the short period between 2000 and 2005 CE the load fell to only 14 percent of the previous 1.08 Gt/year. This change was largely a function of human activities, especially the installation of dams and reservoirs in the upper reaches, as well as improvements in soil conservation. Furthermore, the carrying capacity has decreased 30 percent since the 1950s as result of weakened precipitation (Xu, 2018) and the removal of water from the river in the lower reaches for agricultural uses. While the older sediments of the Loess Plateau continue to be the source of the vast majority of the sediment to the delta, as demonstrated by isotopic fingerprinting of ancient particulate organic carbon (Tao et al., 2015), erosion from this source fell from the 1970s to 1990s because of better terracing, as well the construction of dams and reservoirs. Further reductions were achieved by large-scale vegetation restoration projects that were initiated in the 1990s (Wang et al., 2015). Reducing soil erosion and loss in the Loess Plateau is critical to maintaining its use as a productive agricultural area. Furthermore, silting of the lower reaches results in build-up of the levees along the river. Unless the river course is allowed to avulse this can result in significant elevation of the river above the flood plains and cause major disruption if these were to be broken during a flooding event, as shown by the example from Pakistan in 2010.

6.8 Impacts on Farming Systems in South Asia

Below, we use the example of the South Asian data to examine how changes in the monsoon and policy have intersected to produce changes in agricultural economies in South Asia. Agriculture provides work and livelihood for around 70 percent of the population of South Asia, as well as accounting for around 24 percent of the GDP and 16 percent of export earnings (Gadgil and Gadgil, 2006). Understanding how changes in the mean state of climate may have impacted the agricultural sector is thus essential. The influence of climate, has, however, to be understood in context of a number of transformations in agricultural practice that took place at the same time, which we reviewed earlier. The "Green Revolution" (late 1960s to 1970s): a series of agricultural innovations that focused on introducing dwarf crops of wheat and maize, expanding the use of chemical fertilizers, and introducing new methods of irrigation, substantially changed the dynamics of farming in India starting in the mid-late 1960s.

While the "Green Revolution" has been widely touted in the popular media as savings millions from starvation, due to its supposed increases in yield a recent analysis of production data suggests that these measures actually did little to increase overall yield. For instance, Stone (2019) re-analyzed production data from India and found that "Total food-grain production, which had been growing at 2.8% annually, slowed to a 1.9% growth rate during the Green Revolution years, a 32% drop." Rather he argues that there has been a steady linear rise in grain production with no signs of a revolution in food production (Stone, 2019). Rather, as we discuss subsequently, these innovations in technology had devastating effects on farmer alienation and displacement as well as on erosion.

First, we look at some of the impacts that changing precipitation may have had on farming. Farming is heavily dependent on the delivery of moisture at the right intensity and time in order to allow crops to thrive. In particular, soil moisture is a key variable that is affected not only by monsoon intensity but also by the timing of moisture delivery, air and ground temperatures, as well as the nature of the groundcover that controls evaporation.

6.8.1 Modelling the Indian Monsoon

The trends in monsoon precipitation described earlier, coupled with a general warming trend through the twentieth century, have affected growing conditions. Drought indices (e.g., Palmer Drought Severity Index (PDSI))(Palmer, 1965) can be strongly influenced by how climate models treat evapo-transpiration methods and this can result in overestimation of the extent and intensity of droughts if not properly accounted for. The intensified hydrological cycle associated with a warmer planet causes more precipitation so that predicting moisture in soils can be complex and recent studies have argued that twenty-first-century climate may be more prone to drought than the past century has been (Dai, 2013). Soil moisture estimated from land surface modeling can used as an alternative to observed and remotely sensed soil moisture values, which have been widely applied in drought prediction (Mishra and Singh, 2010; Sheffield et al., 2004).

Mishra et al. (2014) performed a land surface modeling study of India for the period 1950–2007. The model was based on daily precipitation data with a 0.25° spatial resolution that were gridded to provide regional coverage. The land cover issue was dealt with using a Variable Infiltration Capacity (VIC) model that simulates energy and water fluxes within each grid cell (Liang et al., 1996). In this model, multiple vegetation types can be represented in the same grid cell using a mosaic scheme. Land cover classes are represented using an estimate of the root fraction, canopy resistance, and leaf area to determine their influence in governing evaporation. The study particularly examined conditions during the two major growing seasons, the *Kharif* (planted in the spring and harvested during and after the monsoon) and the *Rabi* (October to April). As described in Chapter 3, the *Kharif* season (rice, millets, maize, cotton, sunflower, and soybean) are more monsoon dependent and we focus on those results here.

Modeling of the period 1950–2008 showed that precipitation has significantly declined during the May–June (*Kharif* sowing) period in northeastern India, and in the southern peninsular India, as outlined earlier (Figure 6.20A). However, at the same time, temperatures decreased over the Indo-Gangetic Plain but rose in southern India and in the far northwest, again consistent with observations from the aerosols, as described earlier (Figure 6.20D and E). The net result of this was higher soil moisture in the northeast, western Ghats, and Kashmir, but lower soil moisture in western India since 1950 (Figure 6.20G and H). When the same approach was applied to the winter *Rabi* season significant declines in soil moisture were found in western India and the Indo-Gangetic Plains . In contrast, some regions in southern India showed significant increases between 1950 and 2008 (Mishra et al., 2014), underlining the divergence of twentieth-century climatic change across the subcontinent.

6.8.2 Impacts of Droughts

This environmental model was effective at identifying the timing of three major droughts since 1950, that is, in 1972, 1987, and 2002. The drought of 2009 postdated that study period and could not be considered, although in that year All-India rainfall average for the monsoon season was reduced 23 percent compared to the long-term average, one of the most serious in recent times (Neena et al., 2011). The 1987 drought affected both growing seasons, but especially the monsoonal *Kharif*, and was particularly damaging, affecting 285 million people (Weisman, 1987). In 2002 the average rainfall deficit reached 21 percent and is recorded as the third most severe since the start of the twentieth century, only being eclipsed by 1972 and 1917 (Bhat, 2006). Conditions were especially bad in western India. Food grain production dipped by 29 million tonnes to 183 million tonnes, compared to 212 million tonnes in 2001 (Gadgil and Gadgil, 2006). Production of rice fell drastically to 75.72 million tonnes (2002–2003), as against 93.08 million tonnes during the previous year (Government of India, Ministry of Agriculture). The impact of the drought of 2002–2003 on hydroelectric power generation led to a decline of 13.9 percent (Panwar et al., 2011; Sarkar, 2011), with even more modest drought years resulting in around 5 percent loss of output (van Vliet et al., 2016). All of these effects had a major impact on the Indian GDP.

Figure 6.21 shows that in the modern era, grain production fell by as much as 20 percent in dry years and rose by almost 10 percent in wet years (Gadgil and

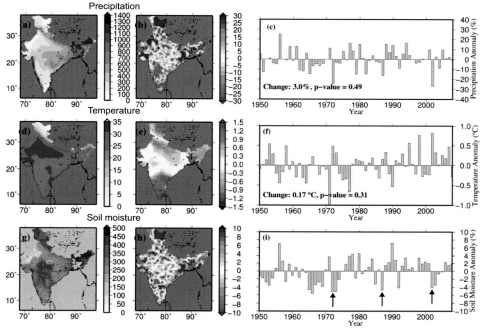

Changes in precipitation, temperature and soil moisture since 1950.

Figure 6.20 Changes (trend slope multiplied by total number of years) in precipitation, temperature, and model simulated total column soil moisture during the
Summer Monsoon (Kharif sowing) season for the period of 1950–2007. (A, D, and G)
Mean seasonal precipitation, temperature, and model simulated total column soil
moisture for the retrospective base period (1961–1990), (B, E, and H). Change in
precipitation (percent), air temperature (°C), and total column soil moisture (percent)
between 1950 and 2008, (C, F, and I) all India averaged seasonal anomalies of
precipitation (percent), air temperature (°C), and total column soil moisture (percent)
during the period of 1950–2007. Stippling indicates statistically significant changes at
5 percent level. In (C, F, and I), changes were estimated for all India averaged
anomalies of precipitation, air temperature, and simulated total column soil moisture.
Anomalies for precipitation, temperature, and soil moisture were estimated with
respect to mean of the retrospective base period (1961–1990). Black arrows indicate
years of extreme drought. Figure modified from Mishra et al. (2014).

Gadgil, 2006). The link between grain production and monsoon strength is not
entirely linear and shows some scatter, but there is a clear tendency to improving
yield with heavier rains. After the onset of the green revolution, improvements in
yield dropped sharply in years of higher than average precipitation, so than
a 15 percent increase in rainfall generated <1 percent improvement in yield in
1983, while in 1975 a slightly smaller additional rainfall caused >7 percent better

yield. However, in years of weaker than normal monsoon these new farming methods had much less effect, with the driest years still resulting in yields almost 10 percent below normal (Gadgil and Gadgil, 2006). The technology of the Green Revolution has thus not proved more resilient to changes in rainfall than the traditional systems of farming that were used prior to its introduction.

6.8.3 Rainfall Control on Agricultural Production

Attempts to estimate the impact of a weakening in the monsoon suggests that a 10 percent decrease in summer rainfall it would be worth more than 2 percent reduction in GDP and that would rise to more than 5 percent if the monsoon was 20 percent weaker than usual (Gadgil and Gadgil, 2006; Gadgil and Rupa Kumar, 2006; Krishna Kumar et al., 2004). Unfortunately, production does not rise to the same degree in the event that monsoon rains are heavier than normal, although generally speaking more monsoon rain tends to increase production, with 10 percent extra rainfall resulting in slightly more than 1 percent increase in Indian GDP (Gadgil and Gadgil, 2006). This is not surprising because during a drought water is a limiting resource, but this is not the case in normal or surplus rainfall years. Variations in monsoon precipitation not surprisingly impact the summer *Kharif*

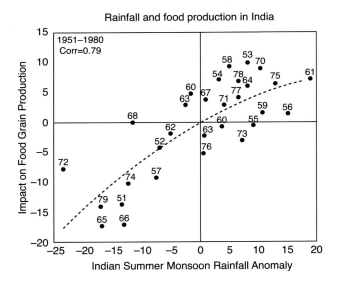

Figure 6.21 Plot showing the relationship between monsoonal rainfall and food production in India. IFGP is negative for all droughts (with values up to −20 percent) and positive for ISMR anomaly larger than 10 percent (with values up to +10 percent). Numbers represent years in the twentieth century. Reproduced with permission from Gadgil and Gadgil (2006).

crop much more than they do the winter *Rabi* (Prasanna, 2014). Since monsoon rainfall has generally become less frequent but more intense during the twentieth century this has had a negative impact on agricultural output. Auffhammer et al. (2012) estimated that yield fell by 1.7 percent as a result and a further 4 percent were lost due to warmer nights and weaker rainfall at the end of the *Kharif* growing season. This study concluded that these losses reduced by one fifth the increases gained from improved farming technology.

There is no simple linear relationship between monsoonal rainfall and agricultural productivity because not all plants are equally susceptible to damage by dry conditions, and growing conditions are also very important. Agricultural practices are also influential, as shown when we compare regular farms and intensively cultivated research stations where pesticides and fertilizers are used to enhance production. What is clear is that during years of heavy rainfall crops of all varieties cultivated by research stations tend to generate much higher yields, whereas in normal fields there is little additional benefit when rainfall is greater than normal (Figure 6.22) (Sivakumar et al., 1983). This is because of the consistent use of fertilizers in research stations compared to farms, where fertilizers are often not used when the monsoon is strong and despite the fact that the soil has become increasing depleted as a result of intensive cultivation over long periods of time. Regardless of growing conditions yields for all crops are at a minimum during years of weakest monsoon rains.

Of course, increases in agricultural yield of key commodity crops like wheat, rice and maize have been due not to changes in the monsoon alone but to focus on exclusively cultivating these crops to the detriment of some of the other crops, such as millets and smaller grains (Pingali, 2012)(Figure 6.23A). In order to better quantify the impact that climate change has had on yield it is better to try to subtract the influence of changes in yield due to the introduction of dwarf varieties and chemical fertilizers from the raw data in order to isolate the effects of weak and strong monsoon years (Lobell and Field, 2007). Clearly there is significant variation in crop yields over the past 60 years and many of these changes are related to abnormal monsoon years. Mishra et al. (2014) employed this approach to estimate the degree of correlation between crop yields and precipitation. They also considered the influence of soil moisture anomalies and the areal extent of drought-stricken regions during the summer monsoon planting and harvesting season since 1950. Precipitation and soil moisture anomalies showed positive correlations with crop yields (Figure 6.23C and D), while drought intensity, as defined from the degree of soil moisture, is negatively correlated, as might be expected (Figure 6.23E).

Care needs to be exercised when trying to understand how changes in precipitation affect crop yields compared to rising temperatures. Uncertainties linked to temperature are higher largely because there continues to be greater uncertainties

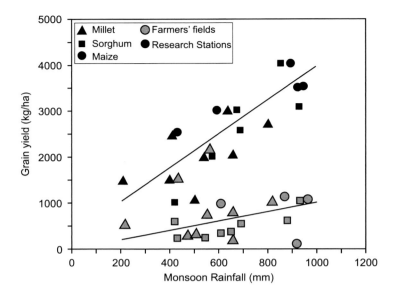

Figure 6.22 Relationship between rainfall during the monsoon and the yield of maize, sorghum and millet in 15 dryland locations in India (Gadgil and Gadgil, 2006). Data show that yields are higher in well managed agricultural stations in wet years because of the application of fertilizer and pesticides compared to farmer fields.

related to future temperature prediction, compared to precipitation. This is a reflection of the fact that precipitation has shown greater variability in the documented past and therefore the impacts are better known (Stocker et al., 2013). Lobell and Burke (2008) analyzed data concerning potential climate change from 94 combinations of crops and different regions, especially in areas where the population density is high, including South and East Asia. They concluded that uncertainties linked to temperature variability were a greater source of uncertainty when assessing the impact of climate change than those linked to changes in precipitation in most areas. That is not to say that they considered precipitation to be unimportant. In particular, in South Asia they noted that millets had experienced decreases in yield on the order of 10 percent as a result of changes in precipitation that were only one standard deviation in magnitude compared to the long-term average. In this respect, millet, a traditionally rain-fed crop, is recognized to be particularly sensitive to precipitation (Eyshi Rezaei et al., 2014). Not surprisingly, rice in South Asia and wheat in West Asia, as well as to a lesser extent soybean in China (Chen et al., 2016b), also showed sensitivity to changes in precipitation, as well as high degrees of uncertainty concerning their future response to changes in monsoon intensity.

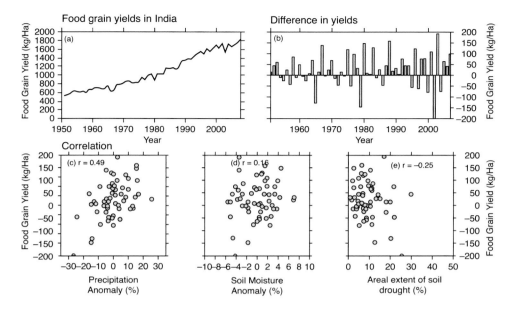

Figure 6.23 (A) Total food grain yields in India between 1950 and 2008, (B) first difference of total food grain yields, (C–E) correlation between total food grain yields and precipitation anomaly (percent), temperature anomaly (°C), and areal extent of soil moisture drought during Summer Monsoon season (KHARIF_SOW). Figure reproduced from Mishra et al.(2014).

Reduced rainfall during weak monsoon years has been associated with times of famine in India. Modeling the changing soil moisture linked to lower rainfall has resulted in the identification of seven major drought periods (1876–1882, 1895–1900, 1908–1924, 1937–1945, 1982–1990, 1997–2004, and 2011–2015) (Mishra et al., 2019). The five largest famines in India prior to 1900 (1873–1874, 1876, 1877, 1896–1897, 1899) correlate with times of high soil aridity, with the most deadly droughts being associated with times of strong El Niño. After 1900, and especially since Indian independence, soil drought conditions did not generally result in widespread famine as a result of better government policy in handling food distribution. The famine of 1943 stands out as being caused primarily by British colonial policy during World War II rather than because of monsoon weakness.

6.8.4 Impacts of the Green Revolution

Despite being heralded as saving the world from overpopulation, it is increasingly realized that the technology and policies of the Green Revolution itself has had impacts on smallholder farmers that may be more far reaching than changes in the monsoon (Shiva, 2016; Siegel, 2018; Stone, 2019; Subramanian, 2015). For instance, in the Thar Desert, Gagné (2013) describes how deep plowing and deforestation associated with the Green Revolution have created an erosional crisis. Green Revolution consultants encouraged farmers in the area to cut down trees that had grown in the area for hundreds of years in order to widen fields to accommodate the tractors designed to mechanize and improve yield; however, this turned many once fertile fields to sand. Unfortunately, the architects of the Green Revolution did not realize that the roots of these trees played an important role in maintaining soil cohesion and preventing erosion and desertification, but also maintained soil moisture (Gagné 2013). An ecological and erosional crisis has since characterized the area.

The traditional forms of farming we reviewed in Chapter 3 used complex methods of crop rotation, involving planting nitrogen-fixing legumes, such as alfalfa, lentils, chickpeas, and a variety of locally domesticated beans including mung beans. In addition to promoting erosion, the Green Revolution's focus on monoculture farming meant that soil fertility was severely damaged throughout India as farmers were told to no longer plant essential nitrogen-fixing crops in order to focus on increasing the yield of a very narrow array of crops (Shiva, 2016). For instance, Shiva (2016) notes that the area in which wheat has grown doubled during the years of the Green Revolution, whereas the area in which nitrogen-fixing legumes were cultivated was reduced by half (Shiva, 2016).

The varieties of crops promoted by the Green Revolution also require that farmers be able to afford and apply synthetic fertilizers in order to ensure their growth. The introduction of this program targeted the largest, wealthiest farms who could afford to invest in the chemical fertilizers that were necessary to ensure that the dwarf crops introduced could grow properly. This left many smaller farms behind and struggling to be competitive (Siegel, 2018; Subramanian, 2015).

One of the key conclusions we have learned is that farmers throughout history have dealt with uncertain climatic conditions through increasing the diversity of species and cultivars planted. The Green Revolution focus on crops like maize, cotton, and rice, which have value as international commodities, has eroded crop biodiversity in the region, thus weakening the response of South Asian farmers to variation in *Kharif* rainfall. Maize, cotton, and rice, particularly those types bred for the Green Revolution, have substantial water requirements, and assume that water can be regularly supplied, making the need for monsoon rain all the more critical. One the other hand, the millets we reviewed in Chapter 3, while not commonly traded on the commodity market, are far more tolerant of drought and low water conditions. As Stone (2002) notes: "The grain in the overflowing granaries is from Green Revolution plants highly dependent on irrigation; since 1950 the percentage of the wheat crop under irrigation has risen from 34% to 86% and that of the rice crop from 32% to 51%." This trend comes at the expense of most sustainable crops. Figure 6.24 compares long-term patterns in production of the more heavily irrigated crops (rice and wheat) with those in the production of the drought-tolerant crops of pearl millet and sorghum, which had for millennia played a major role in Indian subsistence.

6.9 Synthesis and Summary

Monsoon intensity has changed significantly over the past 100 years across monsoonal Asia and continued global warming is considered likely to accelerate this trend. Observational and modeling results now demonstrate that northern and western India and Pakistan have experienced drying over the last half of the twentieth century, as has northern China. Rainfall has fallen around 15 percent over the past 100 years across the North China and Indo-Gangetic plains. In contrast, southern India and southern China have experienced stronger summer monsoons, that increased by similar amounts. In South Asia at least some of this drying is the result of increased atmospheric pollution and the concentration of aerosols. This has resulted in significantly decreased runoff from rivers into the ocean, most notably in the already arid Indus basin and an earlier and weaker end to the monsoon rain season. In East Asia the westerly jet has migrated southward causing drying in northern China, but more risk of flooding to the south of the country. At the same time atmospheric reorganization has pushed the pathways of typhoons away from southern China and toward the northeast, increasing the impact along the eastern coast of China, Korea, and southern Japan. Rainfall in

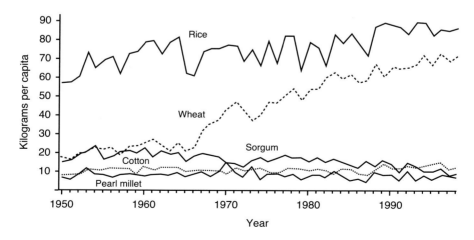

Figure 6.24 Per capita crop production in India, 1950–1999. Data from Indian Ministry of Agriculture (www.agricoop.nic.in). Redrawn from Stone (2002).

southern China has been maintained and even increased along the coast as a result of more typhoons being sourced within the South China Sea itself. In southernmost China, a lot of the increased rainfall is related to these locally sourced typhoons and is less useful for agriculture, coming in large volumes over short time periods.

Change of land use across India, especially deforestation, does not mean that changes in precipitation immediately result in reduced soil moisture and thus in an agricultural drought, although there is evidence that this has more strongly affected the northern Indian plains, western India and Pakistan than other parts of the subcontinent. There are positive links between monsoon rainfall and agricultural yield, but more recently changing practices mean that years of extra rain did not result in much additional yield, only that dry years can result in >15 percent reduction in food production. This is particularly the case for Green Revolution crops.

Although total production has increased with the population, the size of the budget now means that China and India can only partly make good any climatically induced deficits through importing food. While the Green Revolution in South Asia and collective agriculture in China could not have been two more different movements philosophically, these farming policies both lowered the resilience of farmers in Asia to changing climatic conditions by reducing the biodiversity available on farms and by promoting the planting of water intensive varieties of crops. These policies have also exacerbated human impacts on the landscape, such as tree cutting, in an attempt to mechanize farming mobilized top soil accelerating erosion and damaging soil fertility in already fragile ecosystems. Rather than improving the situation under a changing climate, these measures have failed to take into account the millennia of traditional knowledge on farming and landscape management that have made smallholder farmers resilient to changes in their rivers, rainfall, and temperature. Finally, the lessons of these two agricultural experiments have taught us that the greatest damage is done when farmers lose their autonomy and are unable to apply the expert knowledge they possess to manage their land. In the recent and well-documented history we review in this chapter, changes in the monsoon have had less of impact on human welfare than the decisions people (and particularly higher-level organizations such as governments) have made about farming. Changes in the monsoon can, however, add the final nail in the coffin to low resilience systems.

7

Future Monsoon Predictions

7.1 Monsoon and Global Climate Change

Today, monsoonal Asia is still largely agrarian: 58% of Asia's population lives in rural areas and 81% are dependent on farming for their livelihood (Porter et al., 2014). In Chapter 1, we outlined climatic and geomorphological factors that impact human agriculture and possible human adaptive responses to these processes. In particular we note the importance of changes in temperature (shifts in growing degree days, and extended length of the growing season), changes in precipitation, changes in river load and channel, and changes in sea level. We have reviewed how throughout history humans have used a variety of different strategies to adapt to these changes. It is clear that our global climate is experiencing changes not seen since the end of the Little Ice Age and at higher rates and magnitudes than seen at that time. If such changes continue, then these will have a decisive impact on agriculture worldwide, but especially in monsoonal Asia. In this chapter, we consider how changes in global climate are likely to affect the Asian monsoon over the next couple of centuries and what impact this will have on farming systems.

In considering these problems we rely mostly on predictions based on climate models (generally integrated General Circulation Models (GCMs)) that are ground-truthed by trying to reproduce modern or recent historical climatic conditions observed throughout the region. When that testing is successful then these models can be used to make future projections, based on predicted changes in the most important forcing factors. In particular, we make reference to recent attempts to synthesize the results of different competing climatic models and published by the Intergovernmental Panel on Climate Change (IPCC)(Stocker et al., 2013). Although these models do not all make the same predictions, there is a significant amount of overlap and it is in this area of agreement that we focus here. Most of the predictions with significant statistical confidence are based on competing GCMs that all make the same prediction. This means that it less likely that each prediction is just the result of a quirk or flaw in any given model.

It is apparent from our understanding of past changes in monsoon intensity and their timing in relation to changes in global climate that the intensity and range of summer rains and temperatures in Asia are intimately linked with other climatic phenomena, both globally and within the Asia-Pacific region. Processes of particular concern are ENSO, the Indian Ocean Dipole, the extent of solar forcing, and the degree of polar glaciation (Clemens et al., 2018; Kathayat et al., 2016). These links are not always very straightforward, because although over timescales of thousands of years we can say that summer monsoon rains are associated with warmer, interglacial periods and that weak rains are linked with colder, glacial times, it is not clear that it is the glaciation itself that is causing this variation (Tiwari et al., 2005). Indeed, Indian monsoon strength appears to be more closely related to the amount of solar energy that is being received in the mid-latitudes (Sagawa et al., 2014; Tiwari et al., 2005), which itself is linked, albeit indirectly, to the degree of Northern Hemispheric Glaciation. In contrast, records of discharge from the Yangtze River, which is heavily controlled by the intensity of the East Asian Summer Monsoon rainfall indicate that this is more sensitive to greenhouse gas concentrations and the extent of high-latitude ice than to direct insolation forcing (Clemens et al., 2018). Thus, although the extent of Northern Hemispheric Glaciation has remained approximately stable since around 10,000 years BP, at the start of the Holocene, the strength of the summer rains has decreased as solar energy delivery has reduced. Nonetheless, a common feature of many climate models is a prediction that monsoon rains will become stronger as the planet heats up, whether that be for natural or man-made reasons.

7.1.1 Evidence for Global Warming

There is very little doubt that global climate has been heating in the recent historical past and that much of this has been driven by the burning of fossil carbon fuels by human society. That is not to say that everywhere has been getting hotter, only that the global average has been increasing. Determining the impact that a continuation of that trend will have on monsoon strength is a key concern. Figure 7.1 shows a compilation of data that charts the change in surface temperature from 1901 to 2012. This particular compilation shows the result of three different data sets and while differences are present between them, it is also apparent that regardless of their origin long-term warming has been occurring over the past 100 years. This warming is by no means continuous, with warming being much stronger in the more recent past, averaging 0.72°C since 1951. The Hadley Centre/Climatic Research Unit gridded surface temperature data set 4 (HadCRUT4) does not have data for many parts the world, but where this is available warming has been recorded almost everywhere except to the south of Greenland (Figure 7.1). This result is moreover consistent with compilations from the Merged Land–Ocean Surface Temperature Analysis (MLOST) and Goddard Institute for Space Studies Surface Temperature Analysis (GISTEMP) data sets, giving confidence that this is a real development.

Some of the heating trend observed is related to urban heat island effects (i.e., the spread of urban centers that retain more heat than undeveloped land (Kumar et al., 2017; Li et al., 2017c). Nonetheless, this has been discounted as the primary driver of the global pattern, and a long-term heating during the twentieth century is now relatively undisputed as an observation. Indeed, the decadal averaged rate of warming is now the highest seen in the past 1,000 years (Smith et al., 2015). Although it is undeniable that there has been a slowing of this trend between ca. 2005 and 2015 (Lewandowsky et al., 2016), it is important to realize that other processes may affect global temperatures on this timescale, such as ENSO or the Northern Atlantic Oscillation. Some physical oceanographers have argued that some of the heat that appears to be "missing" has been stored at depth in the

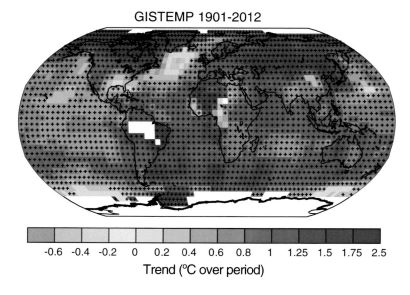

GISTEMP 1901-2012

Trend (°C over period)

-0.6 -0.4 -0.2 0 0.2 0.4 0.6 0.8 1 1.25 1.5 1.75 2.5

Figure 7.1 Change in surface temperature over the period 1901–2012 as determined by linear trend from the Goddard Institute for Space Studies Surface Temperature Analysis (GISTEMP) data set. White areas indicate incomplete or missing data. Trends have been calculated only for those grid boxes with greater than 70% complete records and more than 20% data availability in the first and last 10% of the time period. Black plus signs (+) indicate grid boxes where trends are significant (i.e., a trend of zero lies out- side the 90% confidence interval). Figure modified from Figure TS.2 of Stocker et al. (2013). Published by Cambridge University Press with permission of IPCC.

Pacific Ocean (Kosaka and Xie, 2013). However, more recent studies that incorporate both new and re-analyzed data now demonstrate that the Pacific Ocean is less important in this process than hitherto recognized. Instead, it is deeper levels in the Atlantic and Southern Oceans that have acted as the sink of significant amounts of heat. These basins act as heat sinks because of the salinity anomaly that allows dense water to sink in these areas (Chen and Tung, 2014).

Whether the mechanism of storing heat deep in the oceans will lead to long-term thermal stability is another matter because this same study predicted that similar cooling periods driven by this process have historically lasted between 20 and 35 years, after which renewed surface warming has occurred. A similar response might also be anticipated in the future. Furthermore, concerns that the present phase of heating is just part of a natural cycle seem less serious because although regional heating patterns have been noted in the past (e.g., the Roman and Medieval warm periods), these did not show the same extent or rapid rate of heating as seen in the twentieth century (Wilson et al., 2016), suggesting that something more significant is occurring. Premodern cycles of warming and cooling at centennial scales tend to be linked to variations in solar insolation and/or fluctuations in ocean currents, especially the North Atlantic circulation (Denton and Broecker, 2008), which is not a viable explanation for the recent trend.

Climate models have been used to predict future continental temperatures across Asia. In southwest Asia and the Persian Gulf region summer temperatures are expected to routinely exceed 45°C and even approach 60°C in extreme events by the end of the twenty-first century in the case of unrestricted global CO_2 emissions (Pal and Eltahir, 2015). Similar projections for central Asia indicate significant increases of +3° and +7°C compared to temperatures the late twentieth century (Ozturk et al., 2017), again depending on future emission levels. Higher emission estimates result in projected twenty-first century increases of +4°C in Sichuan, southwest China (Bannister et al., 2017), but higher estimated increases of +6°C have been made for Xinjiang, northwest China, with lower values of ~3.3°C in southeast China (Wang and Chen, 2014). Warming is expected to disproportionately affect the winters rather than the summer season in East Asia, a feature shared by climate predictions in South Asia (Revadekar et al., 2012). In northeast India and Bangladesh, mean annual temperatures are predicted to rise ~2.5°C by the mid-twenty-first century and ~4.5°C by 2100 (Dash et al., 2012). This same work suggested that warming in the Himalaya to the north of this region would be ~3.5°C and >6°C over the same intervals, consistent with regional predictions for the greatest warming to occur in and around the Tibetan Plateau (Hijioka et al., 2014). In northwest India (Rajasthan) the region of the Thar Desert is predicted to warm by 2.0°C–2.8°C from south to north by the mid-twenty-first century and by

2.7°C–4.3°C by 2100 in the event of continued CO_2 emissions (Sharma et al., 2018). This rise is accompanied by an increase in the number of extreme heat days. Further south in India, forward modelling of temperatures using the HadCM3 climate model, ground truthed with recent observational data indicate rises of ~3° C in central India and 2.7°C in the extreme south by the mid-twenty-first century and rises of 4.4°C and 3.8°C by the 2080s (Patwardhan et al., 2018). As far as Indochina is concerned modelling indicates modest warming over the first half of the twenty-first century regardless of whether emissions rise slowly or quickly but that in high emission models temperatures continue to rise after 2055 at a rate of 0.52°C per decade until the end of the century, and beyond (Zhu et al., 2019), with total temperature rises being similar to those seen in southern and central India (Hijioka et al., 2014).

7.1.2 Global Warming and Farming

Global warming will not impact farming systems across Asia in a uniform fashion, and while some countries or regions might stand to temporarily gain, others might lose substantial portions of their production based on the predicted changes in rainfall and temperature. While mid to high latitude regions might stand to initially gain, low latitude areas could be severely impacted. While in Chapter 1, we introduced the concept of growing degree days (GDDs) to refer to units of heat required for a crop to develop, killing degree days (or damaging heat units) are known to impact yield (Butler and Huybers, 2015). For instance, for maize, photosynthesis is inhibited at 38°C and at 45°C many plants experience protein denaturing (Taiz and Zeiger, 2002). For wheat, an important crop grown across the region we discussed, temperatures should range between 12°C and 22°C during anthesis and grain filling to prevent malformation and subsequent reductions in yield. Rice is also at risk for having its yield lowered by extreme temperatures (Mohammed and Tarpley, 2009; Tian et al., 2010). Phenological changes in the timing of flowering, but also increases in plant disease and reduction in yield have been associated with a 1°C increase in temperatures in Japan (Sugiura et al., 2012). The duration of extreme heat also has an impact on plants, with longer periods of heat exposure having a higher impact than shorter ones (Balla et al., 2019). However, temperatures such as these are regularly being reached across South and Southeast Asia under the context of global warming. For example, in June 2019 an intense heatwave affected northern India, with major potential damage to crops across the breadbasket of the subcontinent (Figure 7.2). Some regions experienced temperatures surpassing 45°C for almost three weeks. On June 10, Delhi reached its hottest day on record for the month, reaching 48°C. The map also shows major,

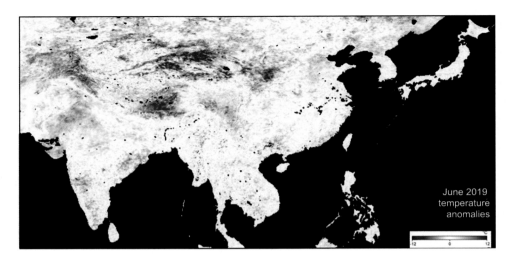

Figure 7.2 Map of temperature anomalies across Asia in June 2019. Temperature anomalies are compared to the average conditions during that period between 2001 and 2010. Places that are warmer than average are red, places that were near-normal are white, and places that are cooler than average are blue. Images by NASA Earth Observations (NEO) from MODIS data.

albeit lesser, month-long heat anomalies in the north China plain, parts of Indochina and southwest China. Extreme heating is noted in central Tibet.

Farmers could shift planting dates to take advantage of lengthened growing seasons in a warming world; however, this may not be helpful in regions such as South Asia where temperature was never a limiting factor in what could be grown. In other areas such as the Tibetan Plateau and high latitude Asia for instance, new crops could be grown and the area farmed might be expanded, especially given the predicted high degree of future temperature rise in that area (Hijioka et al., 2014). This, however, will benefit only some Asian countries while others may risk losing entire areas of cultivation for specific crops. In particular, some crops like apples, and some varieties of wheat and barley require specific cool periods to be success-fully grown. Today, South Asia produces 15% of global wheat, with 81% of this wheat being consumed locally in the region. Current studies suggest that warming temperatures will impact South Asia in a very geographically specific fashion. By 2050, Ortiz et al. (2008) predict that the Indo-Gangetic Plains will experience large reductions in wheat yield due to heat stress and drought, similar to that seen in 2019 (Figure 7.2). Other parts of South Asia may benefit from this. In high altitude parts of Pakistan, wheat yield is projected to increase by 40% under the B2 climate scenario; however, in low lying regions that are arid and semiarid yields will decrease (Hussain and Mudasser, 2007; Iqbal et al., 2010).

We have seen that throughout history either switching varieties of the same crop, switching to new types of species, or diversifying the types of crops used has been a key way of mitigating changes in temperature to either cooler or warmer conditions and of adapting to lower or higher rainfall. Throughout the last 2,000 years, humans have largely had to adapt to cooler and not to warmer conditions, making future adaptations to warmer conditions all the more challenging. Some of the landraces we reviewed are adapted to hotter and more arid conditions, such as millets, may prove useful in an increasingly warming future.

7.2 Hydrologic Response to Warming

A warming climate impacts the water cycle, and in turn the precipitation in different regions, including that of the Asian monsoon. As the planet heats up so rates of evaporation might be anticipated to increase and this in turn drives predictions of a global increase in precipitation, which most models are projected to increase gradually over the twenty-first century (Dai et al., 2017; Hijioka et al., 2014; Stocker et al., 2013). It is important to realize that a general increase in precipitation (and temperature) does not necessarily mean that any particular region is going to become much wetter (or hotter). In fact, models also predict some drying over limited regions, especially in the mid-latitudes and in subtropical arid and semiarid regions where evaporation outstrips precipitation (Ham et al., 2018). Figure 7.3 shows the predicted long-term changes anticipated for the period 2081–2100 based on the Representative Concentration Pathway (RCP8) climate model (Stocker et al., 2013). This model predicts an increase in precipitation across much of the monsoon region, but, in particular, in the area of modern-day Bangladesh and the northern Bay of Bengal. The increase is somewhat lower in southern China and Indochina. Uncertainties in the model are sufficient to make the amount of the rainfall increase unclear, at least at the time of writing.

Higher temperatures also lead to greater evaporation, so it is not a given that more precipitation would necessarily result in high humidity or soil moisture. Figure 7.3B shows the prediction of net change in precipitation after accounting for changes in evaporation. In this model the northern Bay of Bengal still shows a significant moistening trend, as does peninsular India, although to a lesser extent in this latter region. The model is less clear for much of East Asia, and especially China where predictions are within the error of the uncertainties. Indeed, when relative humidity (Figure 7.3C) and soil moisture (Figure 7.3D) are also simulated, a divergent trend between South and East Asia is predicted, with a rough increase in

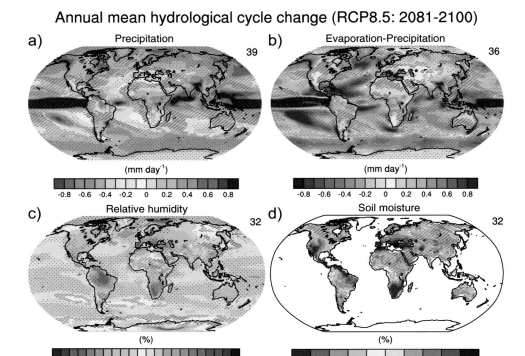

Annual mean hydrological cycle change (RCP8.5: 2081-2100)

a) Precipitation 39

b) Evaporation-Precipitation 36

(mm day^{-1})

-0.8 -0.6 -0.4 -0.2   0   0.2   0.4   0.6   0.8

(mm day^{-1})

-0.8 -0.6 -0.4 -0.2   0   0.2   0.4   0.6   0.8

c) Relative humidity 32

d) Soil moisture 32

(%)

-10   -8   -6   -4   -2   0   2   4   6   8   10

(%)

-10   -7.5   -5   -2.5   0   2.5   5   7.5   10

Figure 7.3 Annual mean changes in (A) precipitation (*P*), (B) evaporation-precipitation, (C) relative humidity and d) soil moisture for 2081–2100 relative to 1986–2005 under the Representative Concentration Pathway RCP8 model. The number of Coupled Model Intercomparison Project Phase 5 (CMIP5) models to calculate the multi-model mean is indicated in the upper right corner of each panel. Hatching indicates regions where the multi-model mean change is less than one standard deviation of internal variability. Stippling indicates regions where the multi-model mean change is greater than two standard deviations of internal variability and where 90% of models agree on the sign of change. Figure is modified from Figure 1 of Section TFE1 of Stocker et al. (2013) published by Cambridge University Press with permission of IPCC.

humidity and moisture in South Asia but increased average aridity in East Asia by 2081–2100. In central China, in the region of the modern Yangtze Valley, this predicted desiccation is strong enough to exceed two standard deviations in the model, making the prediction more robust than in other parts of the monsoon region. This prediction has been confirmed by more recent analysis of climate data collected since the 1980s (Park et al., 2017). This prediction is considered relatively robust and is of significant societal concern given the number and density of people now inhabiting that region.

7.2.1 Monsoons in Asia and beyond

GCMs allow us to predict and compare changes in the Asian monsoon with a number of different monsoon systems worldwide. Differences between the South and East Asian systems, as well as other continental precipitation cycles, are predicted from several sets of climate models (Figure 7.4). As noted earlier, climate models generally predict higher average precipitation for most projected future climate states, and this in part is reflected in an increase in summer precipitation during the twenty-first century (Stocker et al., 2013). This trend exists in spite of the fact that monsoon circulation is predicted to decrease over the same time (Turner and Annamalai, 2012). Nonetheless, most continental monsoon climate models tend to predict a clear increase in precipitation in both South and East Asia compared to other areas, so that this can be considered as a robust result. Figure 7.4 shows the range of predictions both for seasonal average precipitation (Pav), the standard deviation of precipitation (Psd), the amount of precipitation in the wettest five days of each season (R5d), and the duration of the summer rains. South and East Asian monsoons show moderately higher future precipitation, and in particular a longer monsoon in the near future. Whether such high precipitation will be delivered in an even fashion throughout the monsoon season in a way that would favor efficient agriculture is however an open issue. In particular, the storminess of the monsoon is open to significant uncertainties. Some climate models indicate a migration of the monsoon front in East Asia (Intertropical Convergence Zone, ITCZ), causing drying of northern China, but greater flooding and storminess in the Yangtze River valley and further south (Yu et al., 2004a).

7.2.2 Soil Moisture, Vegetation Changes, and Impacts of the Monsoon on Farming

In semiarid areas, such as large parts of South Asia or northern China, reductions in rainfall or changes in its timing will impact farming systems. In areas experiencing reductions in rainfall, expanding irrigation networks might be one solution, however, as we explained earlier, changes in river runoff driven by climate change may involve drying of the rivers that feed these networks. Too much rain can also have adverse impacts. Unusually heavy rainfall events, which are projected to increase with changes in the monsoon, can also affect farmland located in floodplains through increased flooding risk, but also in the uplands through increased landslides due to additional rainfall.

It is been known for some time that vegetation has feedbacks on environmental and climatic conditions in a given area as a result of the vegetation controlling the degree of evaporation (Dubbert and Werner, 2019). Although

Climate Models and Precipitation Predictions

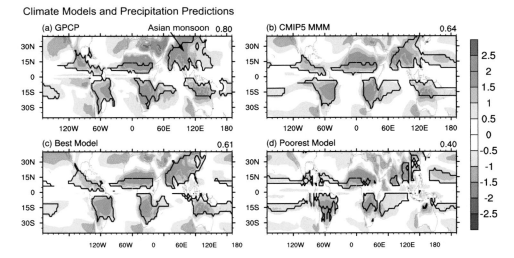

Figure 7.4 Monsoon precipitation intensity (shading, dimensionless) and monsoon precipitation domain (lines) are shown for (A) observation-based estimates from Global Precipitation Climatology Project (GPCP), (B) the CMIP5 multi-model mean, (C) the best model, and (D) the worst model in terms of the threat score for this diagnostic. These measures are based on the seasonal range of precipitation using hemispheric summer (May through September in the Northern Hemisphere (NH)) minus winter (November through March in the NH) values. The monsoon precipitation domain is defined where the annual range is >2.5 mm day^{-1}, and the monsoon precipitation intensity is the seasonal range divided by the annual mean. The threat scores (Wilks, 1995) indicate how well the models represent the monsoon precipitation domain compared to the GPCP data. The threat score in panel (A) is between GPCP and CMAP rainfall to indicate observational uncertainty, whereas in the other panel it is between the simulations and the GPCP observational data set. A threat score of 1.0 would indicate perfect agreement between the two data sets. See Wang and Ding (2008), Wang et al. (2011a), and Kim et al. (2011) for details of the calculations. Figure reproduced from Figure 9.32 of Flato et al. (2013) published by Cambridge University Press with permission of IPCC.

vegetation is itself controlled by the amount and timing of precipitation, the relationship is not entirely linear. The complexities of this linkage have been investigated by a number of studies to see how vegetation and soil moisture can vary as a result of long-term global warming. Mishra et al. (2014) used a GCM to see how observed past changes in soil moisture have related to changes in vegetation and then used these results to make predictions for the remaining part of the twenty-first century. As with the models discussed earlier, this study ground-truthed their model by reproducing past variations before trying to apply this method to predictions of future climate change. A particular emphasis was placed on the role that different plants play in controlling soil moisture. The conditions for drought in agricultural regions vary largely based on the precipitation and soil moisture, with the availability of moisture countering losses due to evaporation, which are linked to vegetative cover (Sheffield et al., 2004). Such predictions are complicated by changing plant physiology that further reduces evaporation via transpiration (Lemordant et al., 2018). Increasing atmospheric CO_2 at the leaf surface decreases the density of stomata on the leaf surface. Furthermore, individual openings are reduced and, therefore, less water is transpired per unit leaf area (de Boer et al., 2011).

A synthesis of available data indicates that there has been little long-term trend in the occurrence of drought since 1950 (Sheffield et al., 2004). However, a projection based on current Indian meteorological data input into the CMIP5 climate model predicts increasing precipitation between the present day and the end of the twenty-first century (Li et al., 2015). There is significant extra precipitation predicted in central and eastern India and a relative lack of change in the west (Figure 7.5). These predictions contrast with the standard GCM by also incorporating information about vegetation types. Models for the summer season, during the sowing of the rainy season (*Kharif*) crops, including millet, sorghum, maize, and sugarcane, are factored in to try and estimate their potential influence on soil moisture. Figures 7.5D–F show that accompanying the increasing precipitation over the next hundred years increasing temperature is also predicted, which in turn results in greater evaporation. This means that higher precipitation does not always result in greater soil moisture. Nonetheless, general long-term trend to greater soil moisture is predicted, with the important exception of north and northwestern India and Pakistan (Figure 7.5G–I). A significant increase in soil moisture is predicted for the southern part of peninsular India and especially the region of the Deccan Plateau by the end of the twenty-first century.

Predictions of drought at four stages over the next 100 years were made after accounting for the vegetation effect (Figure 7.6) (Mishra et al., 2014). Models are shown for the four different seasons. The model "*Kharif_*sow" represents the early summer months of the peak monsoon when these crops are planted. The early autumnal months are represented by the "*Kharif*" model. The *Rabi* crop sowing months (October–December) and the *Rabi* periods (January–April) when the wheat and the barley are harvested are represented by the other two models. Because of rising future temperatures, the model predicts an increase in the occurrence of drought, especially toward the middle part of the twenty-first century, with the worst occurrences of drought being present in western and northwestern India (Mishra et al., 2014), especially in the critical *Kharif* sow period. This will particularly be the case for water intensive *Kharif* crops like rice and maize.

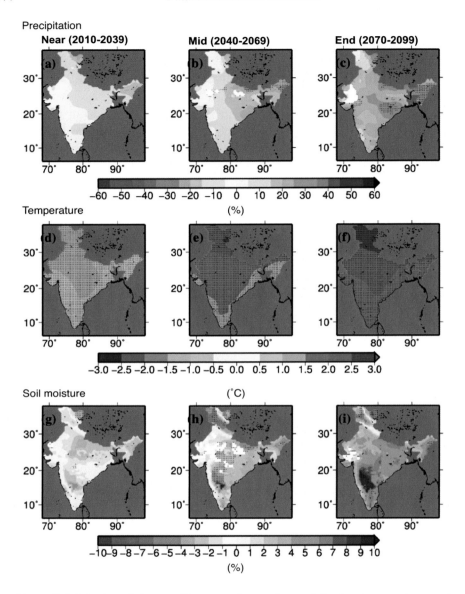

Figure 7.5 Multi-model ensemble mean changes in precipitation, temperature, and soil moisture in the Near (2010–2039), Mid (2040–2069), and End (2070–2099) twenty-first century climate during the Summer Monsoon (KHARIF_SOW) season: (A–C) precipitation (%), (D–F) temperature (°C), and (G–I) soil moisture (%). Stippling indicates that changes are robust. Changes were estimated for each individual GCM with respect to the historic base period (1961–1990) from the climate models and then mean was taken to estimate multi-model ensemble mean change. Figure reproduced from Mishra et al. (2014).

Cultivating crops with lower water requirements such as millets may help mitigate this situation. However, it should be understood that the model also predicts that toward the end of the twenty-first century the frequency of strong droughts will have abated substantially and should mostly affect only the north and northwest parts of India and Pakistan, especially during the early summer (Figure 7.6D). The main reason for the reversal of the trend to drying is a projected increase in the amount of monsoon precipitation because of the intensifying hydrological cycle. It is worth considering the fact that the modeling of climate that has extended into the twenty-second century (Schewe and Levermann, 2012) indicates that this trend to wetter conditions may be relatively short-lived and therefore that India could go back into a more drought-prone state after the year 2100.

South Asia faces an uncertain future of monsoon precipitation and decreasing soil moisture, especially in northern India and Pakistan. However, the situation in East Asia is potentially more serious. Predictions based on long-term global warming indicate that East Asia and especially central China should experience reduced humidity and lower soil moisture by the end of the twenty-first century (Figure 7.3D). Despite the relative weakness of monsoon precipitation, the population of northern China continues to expand and already water shortage issues are acute along the Yellow River Valley and in the regions of the North China Plain (Li et al., 2017b). In order to understand the drying of China further, Gao et al. (2003) simulated the effects of land use change over the whole country, based on predictions of land use in the future. This approach resulted in a prediction of decreased mean annual precipitation over northwest China (Figure 7.7), a region that is already known for its arid and semi-arid climate. Although the global coupled atmosphere–ocean model employed did predict an increase of mean annual surface air temperature over some areas, it also predicted a decrease in temperature and rainfall in coastal areas. Summer daily maximum temperatures increase in many locations, while winter mean daily minimum temperatures decrease in eastern China but increase in northwest China. The net result is that upper soil moisture would decrease significantly across most of China by the end of the twenty-first century, with potentially serious agricultural and societal consequences. More recent modeling studies argue for wetter conditions across most of western China but still predict drier conditions across most of eastern China in the mid-twenty-first century (Yin et al., 2015). Changes in the monsoon since 1960 have already been estimated to reduce rice yields by 1.7% (Auffhammer et al., 2012).

The traditional farming systems we reviewed over the course of this book have several mechanisms for dealing with lower precipitation conditions. One of the key mechanisms involves planting of arid adapted crops. Millets domesticated both in East Asia (broomcorn and foxtail millet), as well as the myriad of millets

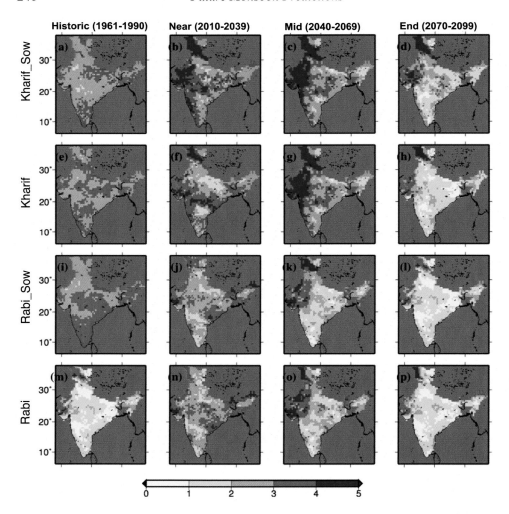

Figure 7.6 Frequency of severe, extreme, and exceptional soil moisture droughts during the historic base period (1961–1990) and in the Near (2010–2039), Mid (2040–2069), and End (2070–2099) term projected future climate. Drought frequency was estimated for each grid cell as the number of years when soil moisture percentiles fell below 10. Figure reproduced from Mishra et al. (2014).

domesticated in South Asia, together with those introduced from Africa all require substantially less water than the types of crops to whom these millets are losing ground under current scenarios of industrial farming, that is, wheat and maize. Unlike monocropped industrial farms that often only plant a single grain crop, traditional farming systems can allow for a wider range of bet-hedging strategies: plants to be intercropped (planted at the same time) or rotated throughout the year in order to increase the probability of harvest survival.

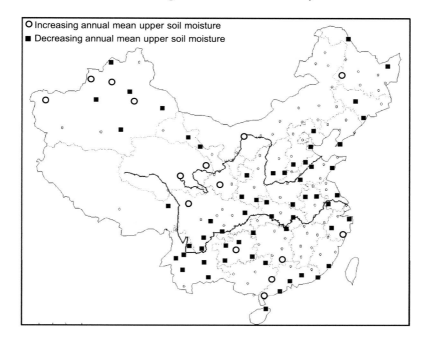

Figure 7.7 Predicted effects of land use change on annual mean upper soil moisture in China. Reproduced from Gao et al. (2003).

The traditional farming systems we reviewed also have more adapted mechanisms for preserving soil moisture than the types of industrial farming scenarios that are being promoted in monsoonal Asia. Traditional fields generally contain barriers of trees or bushes whose role is to provide shade, but often also nutrition in the form of fruits and nuts. These barrier plants around a field or settlements have the added benefit of retaining soil moisture. They thus will likely prove a critical strategy for reducing the losses in soil moisture that are likely to take place in the context of global warming.

7.3 Modeling Future Monsoon Intensity

Making robust predictions about future climate and the environment is often complex, and different climate models that account for different forcing parameters in different ways result in predictions that are not always in accord with one another. There is, however, general agreement about short-term variations in monsoon intensity being largely driven by connections with the Madden–Julian Oscillation (Madden and Julian, 1972; Madden and Julian, 1994) and with ENSO (Yun and Timmermann, 2018). These links are generally not simple or linear. While the South Asian monsoon has become less correlated with ENSO since the 1970s the same is not true in East Asia (Wu et al., 2012c). The recently developed concept of a "global monsoon" moreover emphasizes the links between monsoons in different continents and large-scale tropical circulation (Wang and Ding, 2008). The popular multi-model ensemble CMIP5 (Sperber et al., 2013) generally reproduces the observed spatial patterns of precipitation but somewhat underestimates the extent and intensity of rainfall, especially over Asia (Figure 7.8). This shortcoming is in spite of the fact that this model is generally quite good at estimating mean climate, seasonality, and intra-seasonal variations. Figure 7.8 compares observations with both the CMIP5 model and both the best and worst climate models within the considered set. The former, in particular, fails to capture the precipitation patterns across Asia, or their intensity, whereas Model CMIP5 does at least provide a reasonable simulation for current monsoon precipitation intensity. This lends confidence to predictions based on this suite of GCMs.

Difficulties in deriving realistic simulations of the South Asian Monsoon have been attributed to a number of issues, specifically a bias in many models toward colder sea surface temperatures over the Arabian Sea than actually measured (Levine and Turner, 2012), a weaker meridional temperature gradient (Joseph et al., 2012), or unrealistic development of the Indian Ocean Dipole (Achuthavarier et al., 2012; Boschat et al., 2012). These problems result in predictions that involve insufficient moisture being brought inland and therefore underestimate the degree of precipitation over peninsular India. Difficulties in predicting future monsoon in East Asia often involve issues with the complex topography and in determining the intensity of extreme weather events, largely in the form of greater flooding in central China (Wu et al., 2016b).

7.3.1 Impact of CO_2 Levels

Although the CMIP5 model is an improvement on previous predictions, it is still vulnerable to controls that are not yet well defined. As well as increasing temperature, one well-accepted feature of recent environmental change is the increase of carbon dioxide in the atmosphere, which on geological timescales is associated with warmer temperatures, reflecting the "greenhouse gas" properties of this compound (Crowley and Berner, 2001; Stuiver and Reimer, 1993). Simulations can be used to predict how precipitation in monsoonal Asia might be affected by a doubling of CO_2 in the atmosphere (Figure 7.9) (Turner and Annamalai, 2012). The results shown here are derived from the earlier CMIP3 model and represent the mean of the best four models. Although most of the predictions indicate an increase in future monsoon precipitation, it is noteworthy that there is a lot of variability both in strength and in

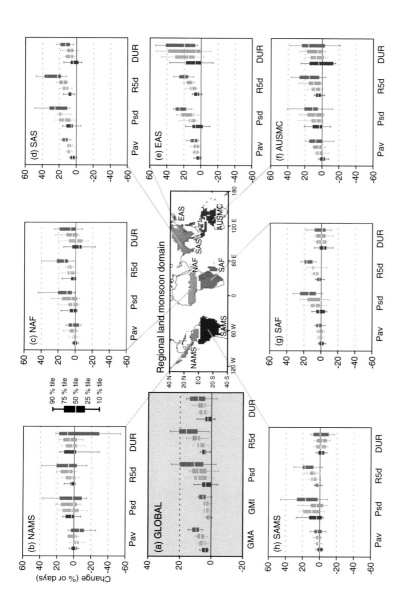

Figure 7.8 Future change in monsoon statistics between the present-day (1986–2005) and the future (2080–2099) based on CMIP5 ensemble from Model RCP2.6 (dark blue; 18 models), RCP4.5 (blue; 24), RCP6.0 (yellow; 14), and RCP8.5 (red; 26) simulations. (A) GLOBAL: Global monsoon area (GMA), global monsoon intensity (GMI), standard deviation of inter-annual variability in seasonal precipitation (Psd), seasonal maximum 5-day precipitation total (R5d) and monsoon season duration (DUR). Regional land monsoon domains determined by 24 multi-model mean precipitation in the present-day. (B)–(H) Future change in regional land monsoon statistics: seasonal average precipitation (Pav), Psd, R5d, and DUR in (B) North America (NAMS), (C) North Africa (NAF), (D) South Asia (SAS), (E) East Asia (EAS), (F) Australia-Maritime continent (AUSMC), (G) South Africa (SAF), and (H) South America (SAMS). Units are % except for DUR (days). Box-and-whisker plots show the 10th, 25th, 50th, 75th, and 90th percentiles. All the indices are calculated for the summer season (May to September for the Northern, and November to March for the Southern Hemisphere) over each model's monsoon domains. Figure reproduced Figure TS.24 from Stocker et al. (2013). Published by Cambridge University Press with permission of IPCC.

spatial range. The models that best simulated the monsoon cycle and variability, with particular reference to the connection with ENSO, confirm patterns of future stronger precipitation, especially in the Indian Ocean, and in the south of peninsular India, as well as the eastern Tibetan Plateau. Rather less increase is predicted over central and southern China, and no increase at all over most of northern China. Indeed, drying is expected in some areas of East Asia, in the region of the Loess Plateau. Almost none of the models predict drier conditions in most of South Asia, although the timing of monsoon onset does appear to be sensitive to CO_2 levels. One of the models that is characterized by a high land-sea thermal contrast predicts that the monsoon in South Asia would initiate earlier in the year under conditions of higher CO_2 compared to the present (Annamalai and Liu, 2005).

Increasing moisture supply in a warmer, more CO_2-rich world results in stronger extreme precipitation events, despite the fact that mean precipitation increases

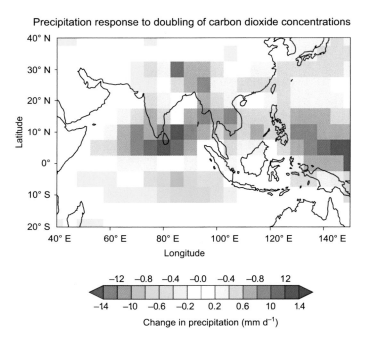

Figure 7.9 Precipitation response to doubling of carbon dioxide concentrations. Mean summer (June–September) precipitation projections in the 1% per year increasing carbon dioxide experiment (1pctto2x) of the CMIP3 multi-model database after doubling of carbon dioxide concentrations relative to control conditions, showing a subset of four models judged to reasonably simulate the monsoon seasonal cycle, interannual variability and the teleconnection between monsoon rainfall and ENSO (Annamalai et al., 2007). Figure is modified from Turner and Annamalai (2012).

more slowly because of energy constraints. It follows that the frequency of convection must decrease, or the strength of more moderate rainfall events must decline if both these predictions are true (Chou et al., 2009). Several models forecast that the number of wet days in the monsoon season of South Asia might decrease in the near future under the influence of higher CO_2 concentrations, even as the total amount of precipitation is projected to increase. Whether these predictions can be trusted is open to question because most of the extreme rainfall events that are modeled do not penetrate far inland, and thus do not really resemble monsoon depressions originating in the Bay of Bengal in the way that we understand this today.

7.3.2 Monsoon Breaks

The most unreliable part of future climate predictions relates to our ability to predict monsoon breaks, i.e., periods during the monsoon season when no rain is falling. A number of climate models predict that the duration and number of dry days during the monsoon might increase in a global warming situation (Joseph and Simon, 2005). This in turn is a product of predictions that show a decrease in the number of long duration rainfall events (Dash et al., 2009). That said, it is noteworthy that this conclusion is certainly not shared by all models (Rajeevan et al., 2006), and this appears to be because monsoon breaks are sensitive to a number of variables, which are not well reproduced in many modern models, making them hard to forecast. A relative lack of high-quality observations from the atmosphere and oceans over monsoon-influenced regions makes it hard to constrain the model physics (Annamalai and Sperber, 2016). Simulations based on the CMIP5 model, forced with higher CO_2 and global temperatures indicate that monsoon breaks should become more common and longer in the future (Sharmila et al., 2015). The consequences of more frequent monsoon breaks are potentially very serious because they are highly disruptive to agriculture and can result in large-scale crop failures. Improving predictions about how monsoon breaks will react to global climate change is a research and economic priority.

7.3.3 Pollution and the Monsoon

Adding to the potential complexity of predicting the future monsoon is the realization that clouds and aerosols may also play an important role in controlling rainfall intensity and may in fact overwhelm the influences of global warming and CO_2 increase. In particular, it has been noted that emissions of fossil fuel related pollutants have increased rather sharply in South Asia during the last century. Ramanathan et al. (2005) reported that between 1930 and 2005 emissions of black carbon and fossil fuel-derived sulfur dioxide increased six times in India and this contributed to large increases in the atmospheric concentrations of these and other aerosols. The presence of significant volumes of such materials in the atmosphere has the effect of reducing the amount of solar energy capable of reaching the ground surface and this in turn results in a feedback that reduces the influence of global warming (Li et al., 2016). Despite the fact that the last century has seen an increase in global surface temperatures, it is noteworthy that South Asia has been marked by a reduction in surface solar radiation, surface evaporation and even summer monsoon rainfall, which runs against the predictions of the models outlined earlier, but that is explicable in terms of increased aerosols.

In their study, Ramanathan et al. (2005) argued that the mismatch between observation and modeling is largely linked to the influence of atmospheric brown clouds caused by pollution. They do this based on a number of climate models that were designed to simulate the influence of the atmospheric brown clouds. Figure 7.10 shows the results of three such models and compares them with rainfall observations within peninsular India. The first of these models, entitled "GHGs+SO4_2050", makes a prediction in which output of greenhouse gases and sulfates is predicted to increase as a result of increasing industrialization and without any future reduction driven by mitigation or policy change (Dai et al., 2001). In that model only the direct effects of sulfates are considered. In addition, two other models were analyzed that account for brown clouds. Model "ABC_1998" builds on the simple model by adding in the effect of increasing brown clouds from 1930 until 1998. However, this model ignores further increases in pollution after that time. Model "ABC_2050" assumes that outputs in brown cloud pollutants will increase until 2050, following a "business-as-usual" scenario and extrapolates historical trends from the present day until 2050.

What is clear is that climate models that account for the influence of aerosols and pollutants differ significantly from those that ignore them. In particular, the influence of aerosols is seen to reduce the amount of precipitation and particularly so for those models that predict increasing aerosol production beyond 1998. These models are also in accord with the observed rainfall data (Bollasina et al., 2011), whereas those that do not account for aerosol production overpredict the amount of precipitation in recent historical times. It is the influence of aerosols in reducing solar heating of the surface that is the root cause of the overprediction in precipitation in models that do not account for pollution.

The strong reduction in precipitation predicted because of aerosols and brown clouds in the atmosphere increased particularly in the years up until 2020. Figure 7.11 explores this issue further by showing the number of years of drought per decade to see if this might become more common as a result of greater pollution

Observed and simulated summer rainfall for India

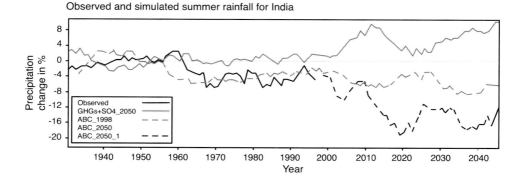

Figure 7.10 Time series of observed and simulated summer (June to September) rainfall for India from observations and PCM simulations. The results show the percent deviation of the rainfall from the 1930–1960 average. Observed rainfall data were obtained from Parthasarathy et al. (1995). The data are smoothed by an 11-year running mean averaging procedure. Modified from Ramanathan et al. (2005).

(Ramanathan et al., 2005). The model is compared with observations up until the end of the twentieth century. Predictions for the next 50 years are based on models that both do and do not account for brown cloud formation. For the purposes of this work a drought represented a time when rainfall decreases by at least 10% from the climatological average. We see that there is actually significant variation within the twentieth century, with the frequency of drought increasing until 1990. The modeling predicts that after accounting for increasing aerosols since 1998 that the number of droughts suffered by India would increase to more than four years per decade and after 2040 there could be as many as six droughts per decade. This influence is in stark contrast to the wettening trend predicted by CGMs driven by only temperature and CO_2 levels.

Such a steep deterioration in the climate of South Asia would have very serious implications for farming and society if this actually came to pass. Furthermore, the predictions appear to be statistically robust and have been reproduced by more recent modelling studies that have attempted to separate the influence of seawater warming and aerosol increases on the South Asian monsoon (Patil et al., 2019). The good agreement between the model output and historical observations suggests that the influence of brown cloud aerosols in controlling monsoon intensity has been underestimated and that such predictions need to be taken seriously. Greenhouse gases and atmospheric brown clouds seem to have competing effects on the monsoon. Without accounting for brown clouds or aerosols climate models often make predictions that are not only incorrect in their magnitude today but are wrong

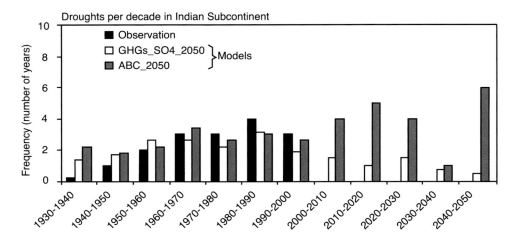

Figure 7.11 Simulated and observed frequency of droughts per decade in the Indian subcontinent. Drought is defined to occur when the summer rainfall decrease exceeds 10% of the climatological average, defined here as the average summer rainfall for 1930–1960. Reproduced from Ramanathan et al. (2005).

even in the direction of change. It is only when accounting for the influence of aerosols that we are able to get close to simulating observations and in making firm predictions for the next 100 years of future climate change.

Naturally, there are uncertainties in our ability to predict the influence of future South Asian aerosol influences because we are unable to know what future output is likely to be. This is dependent on both industrialization of the Indian subcontinent, controlled in turn by the global economy, and on the future climate change mitigation policies that may or may not be followed by the Indian government. It is also worth noting that the modeling undertaken in the Ramanathan et al. (2005) study did not involve any aerosol forcing in other parts of the tropics, which may also have a knock-on effect on the intensity of the South Asian monsoon. What is clear is that aerosols and pollutants may have an important negative feedback on the monsoon intensity compared to what might be expected when only making predictions based on increased global temperatures and higher CO_2 values.

East Asia too suffers from significant industrial pollution and in this region there are also predicted impacts on monsoon intensity. Measurements made in the hilly country down-wind of Xi'an in north central China show that particulate air pollution causes a decrease in precipitation. Rosenfeld et al. (2007) showed that when the weather was hazy due to emissions in Xi'an city rainfall in the hill country decreased by an astonishing 30%–50%. Observations across wider areas however show a more complex relationship between pollution and rain. Analysis of a storm system in 2008 in the Beijing region showed that industrial pollution had the effect of increasing total rainfall by 17% during that event. It was noteworthy than this increase was achieved because of heavier rainfall at the peak of the storm and despite reduced precipitation during the onset and dissipation of the storm (Guo et al., 2014). A similar result was derived from a study of a March 2009 storm in the region of Guangzhou in southern China in which precipitation was enhanced by 16% and lightning by 50% under the influence of heavy pollution (Wang et al., 2011d). Modelling suggests that the presence of pollution particles over southeast China changes the cloud concentration nuclei (CCN) content, which strongly affects the timing of storms that involve deep convective clouds, but not stratus clouds. Increasing CCN increases cloud droplet numbers and mass concentrations, and thus decreases raindrop number concentration, so delaying the start of precipitation (Fan et al., 2012a). Pollution not only affects regular monsoon precipitation but also the intensity of tropical cyclones, despite the observation that these reduced in frequency in China over the latter decades of the twentieth century. A synthesis of data from over mainland China spanning from 1980 to 2014 indicated that greater pollution was associated with more cyclone rain days but less light rain (Yang et al., 2018). Not only did the number of rain days increase, but also the intensity of the precipitation during the start and peak of many cyclonic storms.

It has further been noted that there is an important feedback in the aerosol-precipitation cycle in which dust caused by soil erosion may be advected into the atmosphere and in turn cause an increase in the heaviest precipitation, which then drives further soil erosion (Casazza et al., 2018). Future predictions for the monsoon in East Asia thus critically depend on the degree of industrial and agricultural pollution with potentially serious effects in terms of flash flooding and erosion.

7.3.4 Impact of the Walker Circulation

An alternative approach to making accurate future monsoon predictions was attempted focusing on the global mean temperature and the strength of the Pacific Walker Circulation during the spring season. This analysis attempted to understand the influence of the latter on precipitation in South Asia (Schewe and Levermann, 2012). The onset of monsoon precipitation in spring is initiated by the development of a tropospheric temperature contrast between land and ocean, and by the associated convergence of moist air over the continent (West et al., 2018). The Schewe and Levermann (2012) study used five millennial climate simulations, which reproduced precipitation well during a typical Indian summer monsoon season. Their model is based on the presumption that during the rainy season the heat budget is dominated by the moisture-advection feedback. Essentially, the release of latent heat by precipitation reinforces the land-ocean tropospheric temperature contrast, and in so doing stabilizing the circulation that brings in more moist air from the ocean. This in turn maintains precipitation. The Worcester Polytechnic Institute-Earth System Model (WPI-ESM) was able to accurately reproduce the precipitation on a day-to-day basis spanning the years 800 to 2005 CE (Figure 7.12). This ability to replicate observations gives confidence that the model is robust and may have predictive power. When the model was run based on estimates of rising global CO_2 concentrations and temperatures a rather different result was derived for the period 2151–2200 CE, somewhat further into the future than the other models discussed earlier. A prediction of monsoon change over the next 200 years was based on CO_2 emissions continuing to grow at present rates until 2100, followed by constant CO_2 concentrations until 2200 CE. In particular, the end of the twenty-first century is predicted to be the time when precipitation would move to the lower values (Figure 7.12B) (Schewe and Levermann, 2012). What is apparent is that the number of days with heavy rainfall is anticipated to reduce

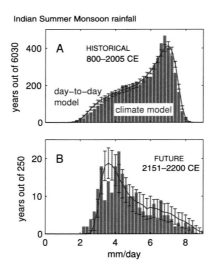

Figure 7.12 Indian Summer Monsoon (ISM) rainfall statistics. (A) Distribution of seasonal (JJA) mean ISM precipitation from the MPI-ESM climate model simulation ensemble over the period 800–2005 CE (gray bars; five simulations, 1206 years each) and from 6030 runs of the stochastic day-to-day model (blue line shows mean value, and error bars show ±1 standard deviation, from 100 realizations of the model). (B) Gray bars as in (A), but over the period 2151–2200 CE in a global warming scenario. Blue line and error bars as in (A), but from 250 runs and with a different set of model parameters. Modified from Schewe and Levermann (2012).

rather sharply, and although there are now a small number of extreme events with very high degrees of daily precipitation, the peak predicted precipitation is 4 mm/day and not the ca. 7 mm/day observed in the historical past.

This type of model implies significant drying in South Asia by the latter half of the twenty-second century, a result that is not at odds with predictions for higher precipitation driven by a warmer troposphere in the nearer future. This is because the WPI-ESM model specifically accounts for the Walker Circulation in the Pacific Ocean and loss in strength of the mean sea-level pressure linked to ongoing global warming. The rise of global temperatures into the twenty-first century was accompanied by a drop in the intensity of the Walker Circulation over the same time, slightly lagging the rising temperature (Figure 7.13). The result of this reduction is first to shift the mode of the Indian summer monsoon to lower values and then to reduce its mean intensity following the initial increase. Consequently, if this prediction has any validity an initial modest increase in South Asian monsoon precipitation is predicted for the early part of the twenty-first century, followed by a rather rapid decline until the end of that century. What is most concerning is the magnitude of this decrease, which if it were applied to the entire Indian subcontinent would see a substantial reduction from around 5.7 mm/day during the monsoon season to only around 4.5 mm/day, representing the establishment of what would be effectively permanent drought conditions.

7.4 Typhoons and Cyclones

High-energy storms affect coastal regions throughout East and South Asia and are linked to the temperature of the ocean, which provides the energy to drive these high magnitude climate events. It might therefore be expected that in the context of future global warming and heating of the oceans that there is potential for these storms to become more frequent and/or more intense compared to the present day (Kang and Elsner, 2016). Analysis of data from 1965 to 2015 shows that the frequency of typhoon genesis in the Philippine Sea is positively correlated with the sea surface temperature (SSTs) in that region but negatively related to SST offshore eastern Australia (Zhang et al., 2018a). Figure 6.7 shows the tracks of typhoons in the Western Pacific during 2011 as an example of a typical modern typhoon season. The map shows clearly that many of these storms originate within the Philippine Sea, in waters associated with the Western Pacific Warm Pool, where the greatest energy exists. From this origin many of the storms track toward the WNW, with many of these subsequently turning toward the north and even northeast, following the warm water of the Kuroshio Current as they evolve. The tracks are further driven by large-scale gyres in the atmosphere (Wu et al., 2012a).

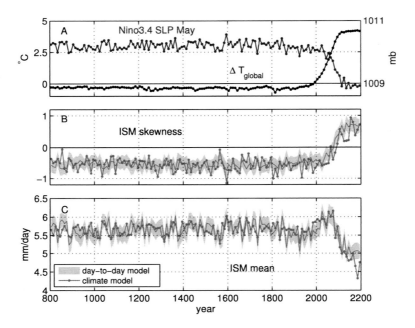

Figure 7.13 Application of the 'day-to-day model' to past and future Indian Summer Monsoon (ISM) variability. (A) Global mean surface temperature (black, in °C relative to the 1980–99 mean) and May mean sea-level pressure (MSLP) over the Nino3.4 region (blue, in mb) over the full period of the millennial MPI-ESM simulations and their continuations under a global warming scenario. Data have been averaged decadally (801–810, etc.) and over the five ensemble members, such that each point represents an average over 50 model years. (B) Skewness of the decadal frequency distribution of seasonal mean ISM rainfall in MPI-ESM (red). Each point is computed from a set of 50 model years (10 years, 5 ensemble members). The skewness computed from 50 runs each of the day-to-day model, with parameters set according to global mean temperature and May Nino3.4 MSLP as in (A), is shown in gray (thin line shows mean value, and shading shows ±1 standard deviation, from 100 realizations of the model). (C) As (B), but for the mean of the frequency distribution (in mm d−1). Modified from Schewe and Levermann (2012).

Consequently, storms entering the South China Sea may impact the south coast of China, but often head further west and strike Hainan island or northern Vietnam. Typhoons that change their track over the Philippines may also influence Taiwan, or head further toward the northeast and strike Japan, or even northern China and the Korean Peninsula. In contrast, the situation in the Indian Ocean shows substantially fewer such storm events and of those that do form these are restricted largely to the Bay of Bengal where they make landfall along the eastern coast of peninsular India and in particular in the delta region of the Ganges-Brahmaputra. Relatively few storms form within or enter the Arabian Sea and of those that do exist these largely parallel the west coast of India before making landfall in Gujarat (NW India) or close to the delta of the Indus River in Pakistan.

7.4.1 Cyclones in the Bay of Bengal

Climate models can be used to predict the frequency and track of storm events in this part of the Indian Ocean so that we can understand how they may respond to changing global conditions. Figure 7.14 shows simulations and observations for early and late monsoon storm tracks in the Bay of Bengal, following the work of Sarthi et al. (2014). That study employed a climate model known as PRECIS (Providing Regional Climates for Impacts Studies), a regional climate modeling system, which is a product of the UK Meteorological Office (www .metoffice.gov.uk/precis). The model was ground-truthed by comparing storm frequency and track data for the period between 1961 and 1990 and after achieving a reasonable simulation of the recent past this was used to predict future storm intensity. The purpose of that work was to understand future impacts on Indian Ocean typhoon intensity, assuming that global greenhouse gas emissions remain high and without any significant reduction efforts being implemented, i.e., effectively a worse-case scenario of global warming. The first step in this process was to use the emission prediction to calculate the sea surface temperatures of the Indian Ocean because these are the primary control on storm intensity (Emanuel et al., 2004). The model predicted changes in sea surface temperatures between the historical control period of May to September in 1961–1990 and the future climate state based on the high CO_2 prediction. During May the degree of future heating is greater than the present, and furthermore warm water is modeled to be concentrated along the eastern coast of India, whereas in September the amount of warming is lower in magnitude but is more widespread across the central and southern parts of the Bay of Bengal. This is important because most Bay of Bengal cyclones initiate in the east of the basin and particularly during the post-monsoon periods (Balaguru et al., 2014).

When comparing simulations and observations the PRECIS climate model generally overestimated the number of storms, except for those during May. Moreover, the simulated direction of storm tracks during May is not entirely in accord with observations, but predicts cyclones veering too far to the west and not enough northward. However, simulations for the summer months, including September, provide a much better match, although again the simulation tended to overestimate the number of storms in any particular year. The predictive future model shows, if anything, a decline in tropical cyclones during May in a future high emission world, but for other months a significant increase is predicted

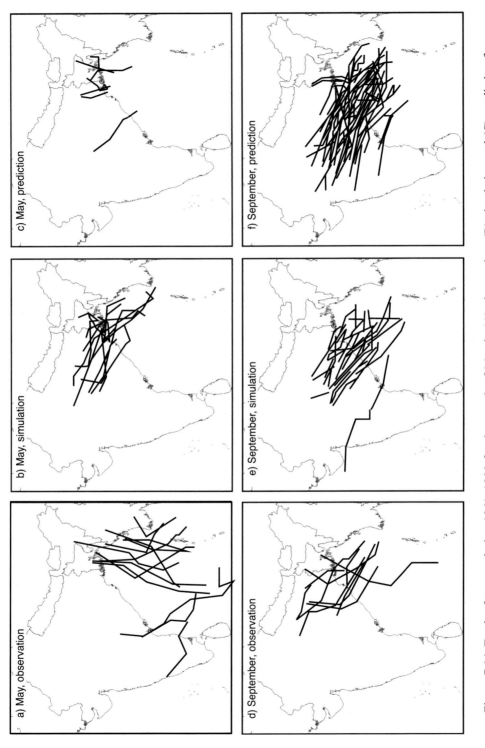

Figure 7.14 Track of storms during 1961–1990 for the month of May in (A) observation; (B) simulation and (C) prediction for 2071–2100 and (D) observation; (E) simulation and (F) prediction for September. Figure redrawn and modified from Sarthi et al. (2014).

a) May, observation

b) May, simulation

c) May, prediction

d) September, observation

e) September, simulation

f) September, prediction

(Figure 7.14F) with all storm tracks generally moving toward the northwest into peninsular India. One unexpected result of this model is that the simulation predicts fewer high intensity storms by 2071–2100 compared to the historical period 1961–1990. However, a drop in atmospheric pressure over the ocean in the future predictions also suggests that those storms that are generated over the Bay of Bengal would be stronger than those observed during the recent past. Increases in sea surface temperature and upper ocean heat content since 1981 have made the Bay of Bengal more conducive to cyclone intensification, while enhanced convective instability has also made the atmosphere more favorable to such phenomena (Balaguru et al., 2014).

7.4.2 Arabian Sea Cyclones

As far as the Arabian Sea is concerned it should be noted that this water mass is also subject to long-term heating and even in the present-day water temperatures are warm enough to allow the development of tropical cyclones. Their relative infrequency reflects the monsoon atmospheric circulation and associated vertical wind shear, which generally only allows such storms to form either before or after peak monsoon time (Evan and Camargo, 2011; Rao et al., 2008) (Figure 7.15). Recent modeling work now shows that the Arabian Sea monsoons may be affected by climate change and particularly by the increase of black carbon in the atmosphere linked to anthropogenic emissions, as described earlier. Evan et al. (2011) investigated this process and demonstrated that there has been an increase in the number of cyclones in the Arabian Sea since the 1970s, a phenomenon that they related to high emissions and increasing atmospheric carbon, which have weakened the monsoonal circulation through triggering a reduction of the southwesterly lower-level and easterly upper-level winds that define the monsoonal circulation over the Arabian Sea. The number of cyclones preceding the monsoon has reduced, but this mostly reflects the recent early onset of the southwest Monsoon, at least since 1998 (Wang et al., 2012a). If emissions were to continue to climb then this modeling predicts an intensification of storms in the region, as well as an increase in the numbers striking the northwest coast of India and the southern coast of Pakistan. This effect is independent of changes related to a warmer ocean. Further modelling now indicates that anthropogenically driven warming, rather than natural processes, may increase the frequency of cyclones in the Arabian Sea during the late stages of each monsoon season (Murakami et al., 2017).

Although the number of cyclones affecting the Arabian Sea in a warmer future climate state would still be relatively small, they contrast with those in the Western Pacific in almost always making landfall. This potentially makes changes in their frequency and intensity more dangerous than might be assumed.

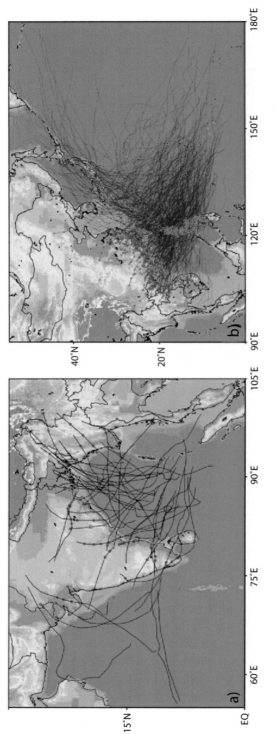

Figure 7.15 Cyclone and typhoon tracks and landfall location points for storms that make landfall at hurricane intensity (maximum 1-min sustained 64 kt) for the (A) Indian Ocean and (B) Western Pacific-East Asia. Each track line connects the six-hourly best-track positions, with red squares indicating a hurricane-force landfall location point and blue circles indicating overland observations of tropical storm strength (wind speed between 34 and 63 kt). Modified from Weinkle et al. (2012).

7.4.3 East Asian Typhoons

Climate modeling has also been employed to predict the impact of global warming on typhoon tracks in East Asia. Although the storms themselves originate over the warm temperatures of the Western Pacific Warm Pool, the direction that these storms take is largely controlled by regional scale wind circulation patterns (Qian and Huang, 2018). That this situation has been changing in the recent past is shown by data compilations for the period 1965–2003 that demonstrated an increase in the number of tropical cyclones. Over the same period typhoon activity in the South China Sea has decreased substantially, while typhoons appear to the making landfall in Taiwan on a more frequent basis (Wu et al., 2005). Between the 1960s and 1991–2011 typhoons have shown a greater tendency to curve north, reducing typhoon landfalls in southern China (Liu and Chan, 2019).

As in the Indian Ocean, predictive climate models have been developed and ground-truthed by trying to simulate known changes in sea surface temperatures since 1965 across eastern Asia (Wang et al., 2011c). The results of this study are in accord with the work of Sarthi et al. (2014) in South Asia in showing significant ocean heating in the recent past in the northern Indian Ocean, but further went on to show that strong wind flow toward the north in Eastern China might be responsible for the increasing diversion of typhoons away from the South China Sea and along the subtropical coast region over the past 40 years. A simulation was then attempted using the predicted climate change from the IPCC fourth assessment report (IPCC, 2007), and particularly focusing on those climate models that gave the best reconstructions of the recent climate changes in Eastern Asia.

The reduction in typhoon activity in that area does not mean that global warming is necessarily good for the regions around the South China Sea. In the past, oceanic and atmospheric conditions in that area have tended to dampen typhoon activity after they leave the influence of the Kuroshio Current and despite the warm water in South China Sea. This is because the surface layer of warm water is much thinner in the South China Sea than it is east of Luzon (Sun et al., 2017). Over interdecadal and longer timescales, however, sea-level rise and surface warming are reducing this buffering effect, driving stronger typhoons in this area that then do not weaken (Sun et al., 2017).

Figure 7.16 shows the predicted changes in large-scale steering wind flow for the period 2001 to 2040 based on the best climate predictions and in the context of warming sea surface temperatures. Figures 7.16A and 7.16B show changes in atmospheric circulation and in the strength of the steering flow between 1965 and 1998, respectively. Projections for the future (2001–2040) are guided by a sharp increase in sea surface temperatures in the equatorial Pacific during the first 40 years of the twenty-first century that is anticipated to continue into the future

(Figure 7.16C and D). The overall future wind pattern is relatively similar to the historic, but it is noteworthy that there is a cyclonic circulation, which is predicted to move from the southern coast of China to central and northern China during that time period. If this trend were actually to take place it would result in further decreases in the impact of tropical cyclones on the South China Sea, but more landfalls in eastern China, potentially affecting cities such as Shanghai and Qingdao, as well as parts of southwest Japan and the Korean Peninsula, rather than Vietnam, Guangzhou, and Hong Kong.

Overall the predictions for East Asia point to more destructive typhoons in the context of a warming climate. Knutson et al. (2010) noted that predictions are made harder because of the relative low frequency and great variability in intensity in the past record. They suggested that it remains uncertain whether past changes in tropical cyclone activity could be attributed to climate change. Future predictions based on analysis of historical tropical cyclone data (Emanuel, 2005) and theory (Emanuel, 1987) indicate an increasing intensity of hurricanes and typhoons with global warming, with a 2%–11% increase in intensity by 2100 projected.

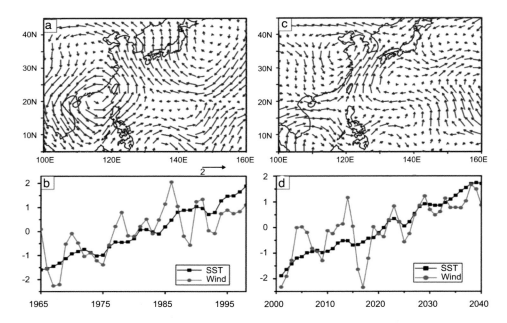

Figure 7.16 (A) Large-scale steering flows over the western North Pacific basin, and (B) the standardized SVD time series in the control runs during 1965–98. (C) and (D) As in (A) and (B), respectively, but for the five-model ensemble projection during the period 2001–2040 under the A1B scenario. Figure modified from Wang et al. (2011b).

A doubling of atmospheric CO_2 is estimated to result in a 40–50% increase in the destructive power of tropical cyclones (Emanuel, 1987). Modelling of the intensity of the strongest typhoons in the context of a 2°C warmer ocean suggested that the winds of the most intense future super-typhoon could attain wind speeds of 306–324 km/h (Tsuboki et al., 2015).

These high-velocity winds as well as the flooding associated with typhoons will have an obvious impact on farming strategies in future monsoon Asia. In particular, if typhoons arrive during the growing season (which in some regions is throughout the year), an entire region's crops may be destroyed.

7.5 Changes in River Runoff

The high mountainous regions of the Himalaya and Tibetan Plateau are often referred to as the "water towers of Asia" because of their central role in providing water through the river systems to many of the people living across this region (Bandyopadhyay, 2013). As noted earlier, water resources are increasingly valuable in this densely settled part of the world and are seen as fundamental to the continued vitality of regional economies. These regions are thus considered to be of high societal and strategic importance (Matthew, 2013). The water supplied from the plateau comes from a variety of sources, not only directly from the summer monsoon rains and the resultant runoff. Additional water is derived from the melting of glaciers, driven by long-term climate change, as well as the more seasonal melting of snows during the spring. In the western part of the Tibetan Plateau moisture is also brought from the westerly jet and precipitated as snow (Bao and You, 2019; Sato, 2009). In general, moisture declines toward the west in South Asia and toward the north in East Asia, with countries like Pakistan being more dependent on water supply from glacial melting than North China where no such supplies exist.

Glacial melt accounts for around 40% of the total runoff into the upper Indus basin, whereas it contributes only 11.5% to the upper Ganges (Lutz et al., 2014). Total glacial melt water supply from the glaciers of high Asia provides water for 136 million people, equivalent to the annual societal and industrial needs of Pakistan, Tajikistan, Turkmenistan, Uzbekistan, and Kyrgyzstan combined (Pritchard, 2017). Recently attempts have been made to model future water discharge in the major rivers draining the Tibetan Plateau in order to understand how fragile the supplies might be in the face of future climate change. Lutz et al. (2014) used an integrated cryospheric and hydrologic model to understand and predict future changes in the runoff from five large Himalayan rivers, the Indus, Ganges, Brahmaputra, Salween, and Mekong. In order to assess the impact of future climate change to the hydrology of the region the model was driven by the most recent collection of climate models (GCMs) vetted by the IPCC (Stocker et al., 2013). These were then ground-truthed against the reference period 1998–2007 before making predictions for 2050. In practice, this means an increase in temperature across the area of between 1.0 and 2.2°C, and changes in precipitation between −3.5 and +8.5%. Only the Indus basin shows a decrease in precipitation over this time period. The Indus is also exceptional in being the only part of the region in which glaciers are known to be growing, or are at least stable compared to retreating in other parts of the Himalaya (Gardelle et al., 2013; Hewitt, 2011). Nonetheless, all the climate models used in this study predicted that glaciers across the region would be in retreat by 2040, a result that is broadly supported by other modeling studies of large-scale glacial patterns for the next century (Radić et al., 2013).

Figure 7.17 shows the results of the modeling in terms of predicted runoff and the sources of this moisture for the period 2041–2050. What is particularly noteworthy is that the runoff volumes are approximately the same or slightly in excess of modern values. Only climate models that particularly emphasize dry and warm conditions predict a decrease. Even the Indus basin, which is the driest catchment considered, shows a maximum decrease that would be no more than 12% of the present value. Because the magnitude of future air temperature change is relatively well known, the uncertainty in predictions of melt-related discharge is similarly relatively low, resulting in relatively small uncertainties when predicting future discharge from the arid Indus basin. In contrast, the uncertainties related to the Mekong, Ganges, and Brahmaputra systems are much greater because of their dependency on precipitation predictions, which are less well constrained. However, it is worth noting that because high-volume discharge into the Indus is driven largely by ice melting this long-term trend is likely unsustainable beyond 2050, because the glaciers themselves will shrink and disappear. Until that time however the most important strategies aimed at alleviating climate change in this region would be best directed toward dealing with catastrophic weather events, such as flooding or severe droughts, rather than a total collapse in the water supply, which may be somewhat more distant.

Further inland the water supplies of central Asia are linked to snow and ice melt in the Tianshan. Warming of this region since 1960 has resulted in more precipitation falling as rain rather than snow, but there are significant differences in this balance along the length of the range. Catchments with a large proportion of glaciation show trends to increasing runoff, while river basins with less or no glacier development are affected by large variations in the observed runoff (Chen et al., 2016c). Because the extent of glaciation is reducing across the entire region the water storage capacity is reducing and this trend is predicted to continue through

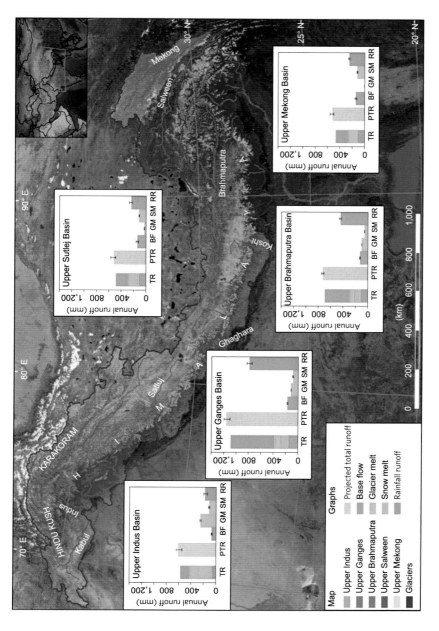

Figure 7.17 The upstream basins of Indus, Ganges, Brahmaputra, Salween, and Mekong. Bar plots show the average annual runoff generation (TR) for the reference period (1998–2007). The second column shows the mean projected annual total runoff (PTR) for the future (2041–2050 RCP4.5) when the model is forced with an ensemble of four GCMs. In the subsequent columns, PTR is split into four contributors (BF: baseflow, GM: glacier melt, SM: snow melt, RR: rainfall runoff). Error bars indicate the spread in model outputs for the model forced by the ensemble of four GCMs. Figure reproduced from Lutz et al. (2014).

the twenty-first century, with the greatest decreases after 2040. These trends particularly affect the central and eastern parts of the Tianshan range because of water supply from the westerly jet at the western end. Because of their importance to the arid regions of western China, Turkmenistan, Kyrgyzstan, and Kazakhstan melting and glacial loss in this region is potentially disastrous for the local communities.

7.6 Predictions of Future Sea-Level Rise

7.6.1 Controls on Sea-Level Rise

Along with changes in rainfall and temperature the present anthropogenically driven climate change has a profound impact on global, and therefore Asian, sea level. This rise in sea level is largely driven by melting of the polar ice caps, although the effects are complicated over different timescales. This is because of a variety of processes, most critically the gravitational attraction of the ice sheets themselves (Clark and Lingle, 1977; Mitrovica et al., 2001). This means that as the ice cap reduces the average level of global sea level increases, but the sea level around the ice cap also falls because of the removal of the ice that attracted water toward it due to its gravitational pull. As a result, sea level will rise less around the melting ice than in more distal regions. Removal of the ice also causes a flexural rebound that allows the crust under the ice to rise after melting (Ponte et al., 2018). Furthermore, because the continental lithosphere is elastic, uplift on the edge of the continent will result in deepening of the continental margin which provides space for water to be accommodated (Milne and Mitrovica, 2008). However, this flexural effect is relatively slow and takes several thousand years to have a major influence because of the high viscosity of the mantle, so that the immediate impact of polar melting is dominated by the amount of ice melted from the continent and added to the ocean and the location of that ice loss.

During the last interglacial period (115,000–130,000 years BP) sea level may have been 6–9 m higher than present and this difference is mainly attributable to a greater melting of the Antarctic polar cap. The dynamics of the ice sheet can be modeled under the influence of a warming climate. DeConto and Pollard (2016) examined the influence that shrinking Antarctic ice shelves have on melting. These buttress the continental ice, preventing its rapid loss. After their removal the potential for accelerating melting increases. As a result, DeConto and Pollard (2016) predicted around 1 m of average global sea-level rise driven by Antarctic melting by 2100, but as much as 15 m by 2500. This model was based on climate predictions with CO_2 emissions continuing to grow in an unrestricted mode, causing greater atmospheric warming. Moreover, they note that the ice loss is likely to continue for a long time, even if warming ceased because of the increasing role of ocean warming in driving melting and because the ocean has higher heat capacity than the atmosphere.

Uncertainties in sea-level predictions related to the stability of the major ice sheets are the single largest uncertainty in predicting future change. A recent reassessment of the effects of inter and intra-ice sheet processes predicted that a +2°C atmospheric temperature increase would cause a median 26 cm sea-level rise by 2100 (Bamber et al., 2019), but with a possible 95th percentile value of 81 cm. In the event that the average warming was +5°C, as estimated if emissions are not curtailed, then the predicted rise would be 51 cm and with a possible 95th percentile value of a 178 cm rise. If thermal expansion of the sea water is also factored in then the increase would be more than 2 m. The same study made a further projection to 2200 for the +5°C scenario and predicted a 7.5 m rise, largely caused by ice sheet collapse in both West and East Antarctica. (Bamber et al., 2019).

In the near future, the rate of sea-level rise is critical because it determines what sort of strategies can be employed to deal with this threat to coastal populations in Asia. In this respect modern satellite derived data can provide high-resolution global coverage that constrains how sea level has been rising and what is likely to happen in the near future. Chen et al. (2017) estimated a rate of sea-level rise from the sum of all observed melt contributions of 2.2 ± 0.3 mm/year in 1993, but found that this had risen to 3.3 ± 0.3 mm/year by 2014. These values appear robust because they are close to those derived from direct measurement of the rise, that is, 2.4 ± 0.2 mm/year (1993) to 2.9 ± 0.3 mm/year (2014), made by satellite observations. Dangendorf et al. (2017) compiled tide gauge data to estimate average rise rates of 1.1 ± 0.3 mm/year before 1990 but 3.1 ± 1.4 mm/year from 1993 to 2014. Although much of this effect is driven by Antarctic melting, Chen et al. (2017) note a statistically significant increase coming from the degradation of the Greenland ice sheet, which accounted for less than 5% of the global rise rate in 1993 but more than 25% by 2014. If that rate continues to increase in this way then future sea-level predictions would need to be adjusted appropriately.

7.6.2 Impact of Sea-Level Rise on Human Settlements and Farming

Sea-level rise is clearly a major hazard to those living on the coasts and impacts society in a number of ways, mostly negatively. The most obvious issue is land loss as a result of inundation of low-lying coastal areas. In particular, rising sea-level heightens the risks from inundation driven by cyclonic storm surges. The loss of coastal wetlands, for example, mangroves, which protect the coast from storms and nurture fisheries can have a negative economic effect, as does the progressive salinization of soil and water, rendering coastal flood plains infertile (Craft et al., 2009; Hossain, 2010). Because of their heavy concentrations of population and economic activity in low-lying coastal regions, South and East Asia are especially vulnerable to sea-level rise.

Attempts have been made to estimate regional sea-level rises, such as the study of Brown et al. (2016) which produced simulations with nine different projections of rising global temperatures until the 2090s (Figure 7.18). In the 2090s temperature rise with respect to 1961–1990 is estimated to fall in a range of 2.3°C–3.7 °C, with a mean rise of 2.9°C. This would result in sea-level rise ranges of 0.29–0.53 m, with a mean of 0.36 m. However, there were significant regional variations, with higher rises in the Mediterranean, but lower than average in East Asia. While rates in South and Southeast Asia are closer to the global average, impacts there are predicted to be some of the most severe because of the wide expanses of densely settled low elevation land.

Dasgupta et al. (2007) used satellite maps to identify coastal areas with low altitude above sea level and gentle gradients as being vulnerable to future sea-level rise, and assessed the probable consequences of continued rise for developing countries in Asia, including Brunei, Cambodia, China, Indonesia, D.P.R. Korea, Republic of Korea, Malaysia, Myanmar, Papua New Guinea, the Philippines, Thailand, and Vietnam. Dasgupta et al. (2007) estimated that a one-

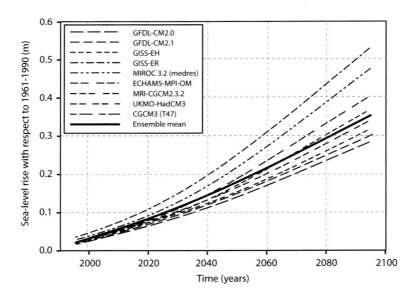

Figure 7.18 Global mean sea-level rise for the nine patterned CMIP3 models analyzed and the multi-model ensemble mean sea-level rise. Data is with respect to 1961–1990. Reproduced from Brown et al. (2016).

meter rise today would displace approximately 37 million people, while 60 million people would be impacted with a two-meter rise, and a three-meter rise would influence 90 million people, approximately the population of Vietnam (Figure 7.19). Future populations affected might be expected to be higher. Rising sea levels leave coasts more vulnerable to storm surges caused by cyclones that become stronger as ocean waters warm. Brecht et al. (2012) modeled the results of rising sea levels and stronger storms in order to assess the increasing vulnerability of major coastal cities to greater storm surges. They note that the geometry of the coast is also a key factor in determining how destructive a surge might be. Their analysis indicated that Asian cities were especially vulnerable to this process compared to many elsewhere. They highlighted Manila, Karachi, and Jakarta as being the urban centers with the worst increased risk during the twenty-first century, although others including Khulna and Chittagong in Bangladesh, Bangkok in Thailand, Kolkata in India, Ho Chi Minh City in Vietnam, and Yangon in Myanmar did not lie far behind.

Our ability to predict future storm flooding is critically dependent on an accurate topography of the region. Kulp and Strauss (2019) developed an improved digital elevation model that reduced the errors compared to older models and then made revised, improved predictions. This reassessment implies greater risk than previously recognized. While 110 million people were previously estimated to live below high tide ranges at the present day their improved topography implies that the true figure is closer to 190 million. Looking into the future and using a climate model in which emissions are not reduced they estimate that the number of people living in these potential flood zones rises to 340 million by 2050 and 630 million by 2100, many of them in monsoonal South and Southeast Asia.

The importance of waves and storm surges is often overlooked when considering the impact of sea-level rise. Vitousek et al. (2017) noted that the 10–20 cm global average rise commonly predicted by 2050 would have far reaching consequences, out of scale to this modest change. Global-scale estimates of increased coastal flooding due to sea-level rise do not consider elevated water levels driven by waves, and thus underestimate the potential impact. In their model, which factors in wave activity, they predict that storm surge flooding would more than double as a result of a 10–20 cm rise, but that these are primarily in low latitude regions, especially South and Southeast Asia.

A related focused study of the Red River delta region of Vietnam considered a range of rising sea-level rates and their impacts through 2050 (Neumann et al., 2015b). Higher sea levels are predicted to increase the frequency of the current 100-year storm surge (a storm surge of ~5 m), to once every 49 years. It is thus noteworthy that around 10% of the Hanoi region's GDP is vulnerable to

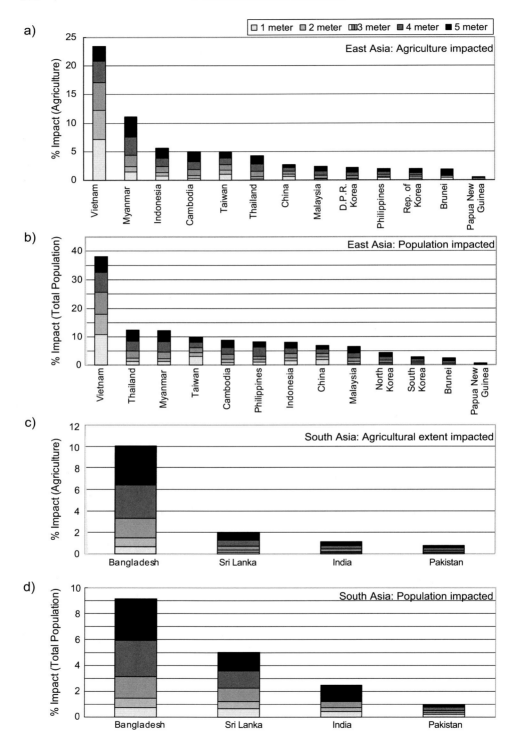

Figure 7.19 Estimates of proportion of agricultural land negatively impacted by sea-level rise of different magnitudes for different countries in (A) East Asia, (C) South Asia. Estimates of proportion of population impacted by sea-level rise for different countries in (B) East Asia (D) South Asia. Redrawn from Dasgupta et al. (2007).

permanent inundation due to sea-level rise as early as 2050. Furthermore, another 40% of this city would be vulnerable to damage by periodic storm surge of 5 m magnitude. The need for strengthened defenses is apparent and pressing in this densely settled (present population of 16 million) and economically important region.

The impacts on Ho Chi Minh City (current 13.5 million population) have also been assessed using hydrodynamic modeling based on IPCC scenarios of sea-level rise and socioeconomic change. Scussolini et al. (2017) predict that annual damages due to storm surges could grow by more than an order of magnitude by 2100, and potential casualties may grow 5- to 20-fold if no measures are taken to protect against rising waters. They further demonstrated that while elevating critical structures and areas of the city could be a more effective response than merely constructing a protective dike around the central zone that would have significant detrimental effects on the outlying parts of the city. Cost analysis suggests that because Ho Chi Minh City is the largest economic center in Vietnam that investment in flood defenses would have a very positive return.

A wider study combined modeling rising sea level with estimates of expanding populations in coastal area. Figure 7.20 shows the predicted increasing numbers of people that will be living within the 100-year storm surge zone across Asia and Africa. In this model by Neumann et al. (2015a), their "scenario C" was based on high-end estimates of increasing populations. Except for in western Asia the numbers of people at risk from the rising storm surges is seen to grow significantly from the present to 2030 and especially to 2060, far outstripping the numbers for Africa. We especially note that south central Asia shows the greatest percentage increase over this period with more than 350 million people in vulnerable situations by 2060, equivalent to the modern population of the United States.

7.6.3 Impact of Sea-Level Rise on Coastal Wetlands

Sea-level rise will impact farming and the humans who practice it by reducing the land available to farm as lands are submerged and taken out of production. Farmers and just generally people living in low-lying coastal parts of Asia, especially South and Southeast Asia, are probably at the most risk from climate change impacts in coastal Asia (Porter et al., 2014).

Sea-level rise is potentially destructive to wetland areas that protect coasts and play a crucial role in maintaining fisheries. Modelling of subsidence rates, tides, sediment supply rates and topography in the Yangtze delta region was used to assess the extended effects of sea-level rise using recent measured rates of sea-level rise

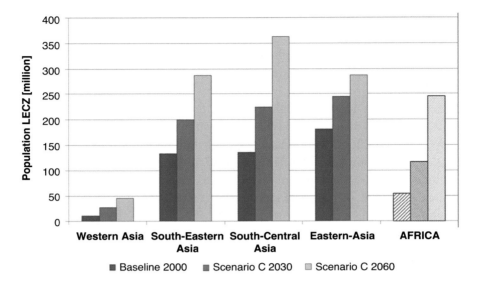

Figure 7.20 Low elevation coastal zone (LECZ) population in Asia in the year 2000 and projections for 2030/2060 per region. Included are totals of LECZ population in Africa for the baseline year 2000 and for 2030/2060. Reproduced from Neumann et al. (2015).

from this region (26 mm/year) and higher projected rates of 59 mm/year derived from the IPCC's A_1F_1 forward model (Cui et al., 2015). This model attempts to assess the state of the Yangtze wetlands in the 2030s, 2050s, and 2100s. By the 2100s, 3.0% of the coastal wetlands are predicted to be classified as being of low vulnerability based on current rise rates and 8.1% based on rates from the A_1F_1 forward model. 0.8% and 2.8% of wetlands will be rated as moderate vulnerability and 2.3% and 6.9% under high vulnerability based on the same rise rate models. Cui et al. (2015) noted the pressing need to improve coastal management and reduce subsidence caused by pumping water from aquifers if the wetlands were going to be preserved. The Yangtze River delta is already experiencing accelerated erosion driven by sea-level rise, which was estimated as 2.4 mm per year in 1981–2015 (Wang et al., 2018). However, this has been higher (3.1–5.0 mm/year) on the south side of the delta where the rate is also recorded to be accelerating. Given the location of Shanghai (population 34 million in metro area, $494 billion GDP in 2018) on this side of the delta the potential risks are enormous. Flooding and erosion have increased and represent a threat to the whole area in the near future. Already, the 100-year maximum tidal level has decreased to a 50-year return period and tidal bores are stronger and more frequent (Wang et al., 2018), requiring redesign of sea defenses, especially protective sea walls.

We have seen that migration has been a key strategy used throughout the prehistoric and historic period, for dealing with sea-level rise, and major shifts in river channel, temperature and precipitation. However, in our world today, characterized by nation states and national borders, large numbers of people needing to leave an area due to climatic change can become a refugee and humanitarian crisis (Williams, 2008). Current political rhetoric suggests that most countries are poorly equipped to show compassion and to open their borders to climate refugees (Kent and Behrman, 2018). Migration within nation state boundaries can also be destabilizing. Many of Asia's largest cities are located in coastal areas and many of these are at risk of experiencing the effects of rising sea level. It will be difficult for any one country to handle the inland migration of the many millions of people housed in any one of these.

We conclude that sea-level rise poses a particular risk to monsoonal Asia in the next century, given the high population numbers and densities in coastal regions. South and Southeast Asia are predicted to be the worst affected because of their low altitudes and because predicted sea-level rise and storms surges are especially destructive in this region, even under conditions of only +2°C average temperature rise. The need for improved coastal defenses in these regions is acute if the vitality of the major coastal cities and productivity of farmlands is to be preserved. The situation is especially serious for Vietnam, Bangladesh, Thailand, and Myanmar.

7.7 Synthesis and Summary

Asian monsoon precipitation is controlled by a number of climatic factors, many of which are predicted to undergo change in the relatively near future as a result of long-term increasing atmospheric CO_2 and warming of the planet. This trend is largely undisputed in its general pattern by a large majority of the scientific community, as summarized by the fifth report of the IPCC (Stocker et al., 2013). Higher average global temperatures, largely driven by anthropogenic emissions, are predicted to result in a generally stronger and longer summer monsoon rain season, especially in those regions adjacent to the Bay of Bengal and Bangladesh over the course of the twenty-first century. However, the increase in precipitation is not uniform and is predicted to reduce in Southwest Asia and northern China. At the same time extreme weather events will become more common. Because higher temperatures also result in greater evaporation, as well as heavier precipitation, increased moisture in the region is not guaranteed. Increasing evaporation until the year 2100 is predicted to result in long-term drying across North China, which is already threatened with significant water shortages. Going along with the higher temperatures, predictions of high CO_2 concentrations have their own effect on precipitation intensity. Like the temperature effect, combined climatic and land use models predict that high CO_2 will cause more precipitation across South Asia, but with no significant impact on precipitation patterns or intensity in China. Higher CO_2 levels are, however, associated with an earlier start to the monsoon rains and also an increase in the number of extreme precipitation events. In those areas where no significant increase in total precipitation is expected, the greater proportion of extreme precipitation events means that the rest of the monsoon season would be drier than the region currently experiences. This could be potentially very damaging for agriculture because heavy rains following drier conditions would result in stronger soil erosion and be generally less suitable for the cultivation of crops that require more continuous gentle watering.

Furthermore, additional problems for agriculture over the next 100 years are caused as a result of an increased number of monsoon breaks, whose duration is also predicted to increase. As well as dealing with global climate change the monsoonal climate of South and East Asia is also affected by pollution, which is already known to have resulted in a modest decrease in monsoon precipitation in the recent past. If some of the worst-case scenarios for increased industrialization and higher pollution in Asia come to pass then these would result in reduced precipitation across the area, which would cause a significant increase in the frequency of drought conditions by 2040 in South Asia. Pollution is predicted to increase storm intensity across East Asia. Although this process is partly offset by the heavier precipitation caused by higher temperatures, it remains to be seen which of these forcing factors might be more dominant over the long term. Further complexities in predicting the intensity of future monsoons involve the weakening of the Walker Circulation over the Pacific under the stress of a generally warmer planet. Again, the effect of the Walker Circulation appears to drive a trend to lower precipitation in the medium term, so that after an increase in summer rainfall during the early twenty-first century the prediction is for lower precipitation across India after 2150.

Changes in land use are also significant in influencing the net change in soil moisture, and thus environment. This is the critical variable when considering the impact of climate change on agriculture. As we reviewed in Chapter 6, traditional small-scale farming systems are often more effective at retaining soil moisture by allowing vegetation barriers to exist in between smaller fields and trees that are critical for retaining both moisture and soil cohesiveness to be part of the family farm.

When combining the influence of land use with the changing climate, models now predict that there would be more moisture across much of southern peninsular India through the twenty-first century, but with the important exception of northwest India and Pakistan, which would experience drying, even in the short term. Critically, predictions now suggest that there might be further drought even in South Asia, intensifying after 2100. The situation is even more serious for China based on similar predictions. Intense drought conditions are anticipated to affect much of the country, especially the north, in the context of a warming planet/high CO_2 emissions, together with continued land use patterns similar to the present day.

As well as the day-to-day rainfall of the summer monsoon season, changing global climate is anticipated to have an impact on major storm events, in particular the cyclones of the Indian Ocean and typhoons of the Pacific. Warming of the global ocean results in predictions that the Bay of Bengal will be influenced by more typhoons in the twenty-first century, but as this century draws to a close there would be a reduction in their frequency. Nonetheless, the peak intensity of those storms that do form is predicted to increase. The Arabian Sea is not heavily influenced by typhoons and those that are formed in this region appear to be especially sensitive to the impact of pollution. If pollution levels continue to rise then climate models predict more and stronger cyclones in this region, almost all of which make landfall in densely populated regions along the west coast of India and the deltaic regions of southern Pakistan. Recent increases in cyclone intensity in this basin suggest that these predictions may be reliable. Changes in large-scale atmospheric circulation across the Western Pacific influence how tropical typhoons will impact China in the near future. A general warming of the Western Pacific has resulted in changes in the large-scale gyres, which has already caused a decrease in cyclone activity in the South China Sea, with more and stronger storms diverging away from southern China and Vietnam, but more heavily influencing Taiwan, as well as coastal Eastern China, including the major industrial centers around Shanghai. This prediction suggests that typhoon intensity is likely to increase at least up until 2040 across eastern China.

The impact of increased typhoon activity is exacerbated by rising sea levels. Predictions of this are complicated depending on the extent of average temperature rise and thus to CO_2 emissions in the near future. Estimates commonly range around 0.29–0.53 m for 2100, with a mean of 0.36 m assuming a 2°C average atmospheric rise. Rises of ~25 cm are predicted to double the rate at which major tidal flooding of coastal lowlands are would occur. While 110 million people presently live below the high tide mark, this is estimated to increase to 340 million by 2050 and 630 million by 2100. These people disproportionately live in South and Southeast Asia, with Bangladesh and Vietnam, in particular, vulnerable to damaging effects because their major cities lie close to sea level, but clearly much productive farmland is also affected.

Finally, changes in the monsoon, as well as increasing rates of mountain glacial melting, have resulted in predictions of changing river runoff, which is an important supply of water to the increasing populations of the Asian monsoon region. Models predict that up until 2040 discharge in most of the major rivers that originate within the Tibetan Plateau might be relatively stable or possibly even increase. Of all fluvial basins the Indus is the most sensitive because of its high dependency on glacial and snow melt and this region is the one predicted to be most affected by reduced precipitation over the next century. Nonetheless, predictions now suggest that even in a worst-case scenario, runoff is unlikely to decrease more than 12% within the Indus. This would nevertheless impact many millions of people. It should also be noted that the loss of this apparent stability is driven by the melting of glaciers that compensate for the reduced monsoon precipitation supplied. Clearly the melting of the glaciers is not something that can permanently sustain water supplies across the area and so the long-term prognosis is not encouraging after the final reduction of these water sources within the context of a warming planet. What is clear is that there may be a number of significant changes in water supply, affecting agriculture both in South and especially East Asia that need to be addressed before the situation becomes critical or impossible to deal with in the shorter term. North China has problems with water supply even in the present day and predictions based on a number of variables suggest that the situation is likely to become more problematic in the future. Given the high population, as well as economic and strategic importance of the region this situation begs urgent action.

Bet-hedging, or not placing all of one's eggs in one basket, is perhaps the best strategy humans have developed over our history when faced with changing climate. As farming systems move more and more toward industrialized models focused on monocultures of single crops, we are reducing our resilience to what will be major changes in climate in our near future. What is clear from our review over the course of this book is that throughout history, Asian farmers within the monsoon region have developed some of the highest yield and most resilient forms of farming, that have not only supported high populations, but have adapted to changes in the mean state of climate. While future climate change may surpass anything experienced in the recent past, the biodiversity present in the crops and landraces grown by farmers and the wide range of ingenious techniques developed to preserve water and soil fertility are ones that may provide some of the most crucial lessons for us to draw on in dealing with the challenges our future holds. Ironically, most forward looking models for predicting how changing climate may impact crops have focused only on the cultivars employed by industrial agriculture (Kang et al., 2009; Lobell and Field, 2007) and few have examined the potential of these landrace crops. As recent history has shown, the loss of these place-based systems of knowledge through policies aimed at industrializing farming threatens humans' ability to adapt as much as changes in climate itself. As the UN's Food and Agriculture Organization has pointed out, it is crucial that the farming strategies we develop in the future serve not only the upper echelons of society, or to increase a country's GDP, but they must improve and serve rural livelihoods and promote equality, not only accumulation of wealth in the hands of a few (FAO et al., 2017). Given the state of future predictions, it is clear that humanity will need to develop as many adaptive strategies as it can and place as many eggs in as many baskets as possible. Many of these strategies have, however, already been developed. According importance and providing autonomy to those that practice these forms of farming may prove one of our greatest adaptive assets in an uncertain future.

References

Achuthavarier, D., Krishnamurthy, V., Kirtman, B. P., Huang, B. H., 2012. Role of the Indian Ocean in the ENSO-Indian summer monsoon teleconnection in the NCEP Climate Forecast System. *Journal of Climate*, 25: 2490–2508.

Achyuthan, H., Quade, J., Roe, L., Placzek, C., 2007. Stable isotopic composition of pedogenic carbonates from the eastern margin of the Thar Desert, Rajasthan, India. *Quaternary International*, 162–163: 50–60.

Acosta, R. P., Huber, M., 2017. The neglected Indo-Gangetic Plains low-level jet and its importance for moisture transport and precipitation during the peak summer monsoon. *Geophys. Res. Lett.*, 44: 8601–8610. DOI:10.1002/2017GL074440.

Adams, J., 1997. *Global Land Environments since the Last Interglacial*. Oak Ridge National Laboratory, TN, USA, p. 20.

Ahmad, N., Sufi, A., Hussain, T., 1993. *Water Resources of Pakistan*. Publisher Shahzad Nazir, Gulberg, Lahore, Pakistan.

Ahmed, M. F., Rogers, J. D., Ismail, E. H., 2018. Knickpoints along the upper Indus River, Pakistan: an exploratory survey of geomorphic processes. *Swiss Journal of Geosciences*, 111(1): 191–204. DOI:10.1007/s00015-017-0290-3.

Ahn, S.-M., 2010. The emergence of rice agriculture in Korea: archaeobotanical perspectives. *Archaeological and Anthropological Sciences*, 2(2): 89–98.

Alam, U. Z., 2002. Questioning the water wars rationale: a case study of the Indus Waters Treaty. *Geographical Journal*, 168(4): 341–353. DOI:10.1111/j.0016-7398.2002.00060.x.

Alfieri, L., Feyen, L., Salamon, P., Thielen, J., Bianchi, A., Dottori, F., Burek, P., 2016. Modelling the socio-economic impact of river floods in Europe. *Natural Hazards and Earth System Sciences*, 16(6): 1401–1411. DOI:10.5194/nhess-16-1401-2016.

Alizai, A., Carter, A., Clift, P. D., VanLaningham, S., Williams, J. C., Kumar, R., 2011a. Sediment provenance, reworking and transport processes in the Indus River by U-Pb dating of detrital zircon grains. *Global and Planetary Change*, 76: 33–55. DOI:10.1016/j.gloplacha.2010.11.008.

Alizai, A., Clift, P. D., Giosan, L., VanLaningham, S., Hinton, R., Tabrez, A. R., Danish, M., EIMF, 2011b. Pb isotopic variability in the modern and Holocene Indus River system measured by Ion Microprobe in detrital K-feldspar grains. *Geochimica et Cosmochimica Acta*, 75: 4771–4795. DOI:10.1016/j.gca.2011.05.039.

Alizai, A., Hillier, S., Clift, P. D., Giosan, L., 2012. Clay mineral variations in Holocene terrestrial sediments from the Indus Basin; a response to SW Asian Monsoon variability. *Quaternary Research*, 77(3): 368–381. DOI:10.1016/j.yqres.2012.01.008.

Allaby, M., Garratt, R., 2003. *Facts on File Dangerous Weather Series: Floods*. Facts on File, New York, p. 142.

Allan, R., Lindesay, J., Parker, D., 1996. *El Nino: Southern Oscillation and Climatic Variability*. CSIRO Publishing, Canberra, p. 416.

Alley, K. D., 2010. *The Goddess Ganga: Her Power, Mythos, and Worldly Challenges. Goddesses in World Culture*. Praeger, Santa Barbara, California, USA, pp. 33–48.

An, C.-B., Feng, Z., Tang, L., 2004. Environmental change and cultural response between 8000 and 4000 cal. yr BP in the western Loess Plateau, northwest China. *Journal of Quaternary Science*, 19: 529–535.

An, Z., 2000. The history and variability of the East Asian paleomonsoon climate. *Quaternary Science Reviews*, 19: 171–187.

An, Z. S., Wu, X. H., Wang, P. X., Wang, S. M., Dong, G. R., Sun, X. J., Zhang, D. E., Lu, Y. C., Zheng, S. H., Zhao, S. L., 1991. Changes in the monsoon and associated environmental changes in China since the last interglacial. In: Liu, T. S. (ed.), *Loess, Environment and Global Change*. Science Press, Beijing, pp. 1–29.

Anderson, E. N., 1988. *The Food of China*. Yale University Press, New Haven.

Annamalai, H., Hamilton, K., Sperber, K. R., 2007. The South Asian summer monsoon and its relationship with ENSO in the IPCC AR4 simulations. *Journal of Climate*, 20: 1071–1092.

Annamalai, H., Liu, P., 2005. Response of the Asian summer monsoon to changes in El Niño properties. *Quarterly Journal of the Royal Meteorological Society*, 131: 805–831.

Annamalai, H., Sperber, K. R., 2016. South Asian summer monsoon variability in a changing climate. In: de Carvalho, L. M. V., Jones, C. (eds.), *The Monsoons and Climate Change: Observations and Modeling*. Springer International Publishing, Cham, pp.25–46.

Anshari, G., Kershaw, A. P., van der Kaars, S., 2001. A late Pleistocene and Holocene pollen and charcoal record from peat swamp forest, Lake Sentarum Wildlife Reserve, West Kalimantan, Indonesia. *Palaeogeography, Palaeoclimatology, Palaeoecology*, 171 (3–4): 213–228.

Arimoto, R., 2001. Eolian dust and climate: relationships to sources, tropospheric chemistry, transport and deposition. *Earth-Science Reviews*, 54(1–3): 29–42.

Arnold, D., 1999. Hunger in the garden of plenty: the Bengal famine of 1770. In: Johns, A. (ed.), *Dreadful Visitations: Confronting Natural Catastrophe in the Age of Enlightenment*. Routledge, New York, pp. 81–111.

Ashraf, M., Majeed, A., 2006. *Water Requirements of Major Crops for Different Agro-Climatic Zones of Balochistan*. IUCN, Water programme, Balochistan Programme Office, Karachi.

Ashton, B., Hill, K., Piazza, A., Zeitz, R., 1992. Famine in China, 1958–61. In: Poston, D. L., Yaukey, D. (eds.), *The Population of Modern China*. Springer US, Boston, MA, pp.225–271.

Auffhammer, M., Ramanathan, V., Vincent, J. R., 2012. Climate change, the monsoon, and rice yield in India. *Climatic Change*, 111(2): 411–424. DOI:10.1007/s10584-011-0208-4.

Awan, T. H., Chauhan, B. S., Cruz, P. C. S., 2014. Growth plasticity of Junglerice (Echinochloa colona) for resource use when grown with different rice (Oryza sativa) planting densities and nitrogen rates in dry-seeded conditions. *Weed Science*, 62(4): 571–587.

Azad, S., Rajeevan, M., 2016. Possible shift in the ENSO-Indian monsoon rainfall relationship under future global warming. *Scientific Reports*, 6: 20145.

Bae, C. J., Kim, J. C., 2010. The late Paleolithic-Neolithic transition in Korea: current archaeological and radiocarbon perspectives. *Radiocarbon*, 52(2): 493–499. DOI:10.1017/S0033822200045525.

Bailey, I. W., Sinnott, E. W., 1916. The climatic distribution of certain types of angiosperm leaves. *American Journal of Botany*, 3(1): 24–39. DOI:10.2307/2435109.

Balaguru, K., Taraphdar, S., Leung, L. R., Foltz, G. R., 2014. Increase in the intensity of postmonsoon Bay of Bengal tropical cyclones. *Geophysical Research Letters*, 41(10): 3594–3601. DOI:10.1002/2014gl060197.

Balla, K., Karsai, I., Bónis, P., Kiss, T., Berki, Z., Horváth, Á., Mayer, M., Bencze, S., Veisz, O., 2019. Heat stress responses in a large set of winter wheat cultivars (Triticum aestivum L.) depend on the timing and duration of stress. *PLoS ONE*, 14(9): e0222639. DOI:10.1371/journal.pone.0222639.

Baltensperger, D. D., 2002. Progress with Proso, Pearl and other millets. In: Janick, J., Whipkey, A. (eds.), *Trends in New Crops and New Uses*. ASHS Press, Alexanderia, USA, pp. 100–103.

Bamber, J. L., Oppenheimer, M., Kopp, R. E., Aspinall, W. P., Cooke, R. M., 2019. Ice sheet contributions to future sea-level rise from structured expert judgment. *Proceedings of the National Academy of Sciences*, 116(23): 11195–11200. DOI:10.1073/pnas.1817205116.

Bamberg, A., Rosenthal, Y., Paul, A., Heslop, D., Mulitza, S., Rühlemann, C., Schulz, M., 2010. Reduced North Atlantic Central Water formation in response to early Holocene ice sheet melting. *Geophysical Research Letters*, 37(L17705). DOI:10.1029/2010GL043878.

Bandyopadhyay, J., 2013. Securing the Himalayas as the water tower of Asia: An environmental perspective. *Asia Policy*, 16: 45–50. DOI:10.1353/asp.2013.0042.

Bannister, D., Herzog, M., Graf, H.-F., Hosking, J.S., Short, C.A., 2017. An assessment of recent and future temperature change over the Sichuan Basin, China, Using CMIP5 Climate Models. *Journal of Climate*, 30(17): 6701–6722. DOI:10.1175/JCLI-D-16-0536.1.

Bao, Y., You, Q., 2019. How do westerly jet streams regulate the winter snow depth over the Tibetan Plateau? *Climate Dynamics*, 53: 353–370. DOI:10.1007/s00382-018-4589-1.

Bar-Yosef, O., 2011. Climatic fluctuations and early farming in West and East Asia. *Current Anthropology*, 52(S4): S175–S193.

Barber, D. C., Dyke, A., Hillaire, M. C., Jennings, A. E., Andrews, J. T., Kerwin, M. W., Bilodeau, G., McNeely, R., Southon, J., Morehead, M. D., Gagnon, J. M., 1999. Forcing of the cold event of 8,200 years ago by catastrophic drainage of Laurentide lakes. *Nature*, 400: 344–348.

Barker, R., 2011. The origin and spread of early-ripening champa rice: it's impact on Song Dynasty China. *Rice*, 4(3): 184.

Barnes, G., 2015. *Archaeology of East Asia*. Oxbow books, Oxford.

Barton, C. M., Ullah, I., Bergin, S. M., Mitasova, H., Sarjoughian, H., 2012. Looking for the future in the past: long-term change in socioecological systems. *Ecological Modelling*, 241: 42–53.

Barton, L., Brantingham, P. J., Ji, D. X., 2007. Late Pleistocene climate change and Paleolithic cultural evolution in northern China: implications from the Last Glacial Maximum. In: Madsen, D. B., Gao, X., Chen, F. H. (eds.), *Late Quaternary Climate Change and Human Adaptation in Arid China*. L. Elsevier, Amsterdam, pp. 105–128.

Barton, L., Newsome, S. D., Chen, F. -H., Wang, H., Guilderson, T. P., Bettinger, R. L., 2009. Agricultural origins and the isotopic identity of domestication in Northern China. *Proceedings of the National Academy of Sciences*, 106(14): 5523–5528. DOI:10.1073/pnas.0809960106.

Bayly, C. A., 1988. *Rulers, Townsmen and Bazaars: North Indian Society in the Age of British Expansion, 1770–1870*. CUP Archive, Cambridge.

Becker, J., 1998. *Hungry Ghosts: Mao's Secret Famine*. Holt Paperbacks, New York, p. 325.

Beinart, W., Coates, P., 1995. *Historical Connections: Environment and History.* Routledge, London.

Belcher, W. R., 2003. Fish exploitation of the Indus Valley tradition. In: Weber, S. A., Belcher, W. R. (eds.), *In Indus Ethnobiology: New Perspectives from the Field.* Lexington Books, Lanham, Maryland, pp. 95–174.

Belcher, W. R., Belcher, W. R., 2000. Geologic constraints on the Harappa archaeological site, Punjab Province, Pakistan. *Geoarchaeology*, 15(7): 679–713. DOI:10.1002/1520-6548(200010)15:7<679::aid-gea3>3.0.CO;2-9.

Bellwood, P., 2005. Asian farming diasporas? Agriculture, languages and genes in China and Southeast Asia. In: Stark, M. T. (ed.), *Archaeology of Asia.* Blackwell, Malden, MA, pp. 96–118.

Bestel, S., Crawford, G. W., Liu, L., Shi, J., Song, Y., Chen, X., 2014. The evolution of millet domestication, middle yellow river region, North China: evidence from charred seeds at the late Upper Paleolithic Shizitan Locality 9 site. *The Holocene*, 24(3): 261–265. DOI:10.1177/0959683613518595.

Bettinger, R. L., Barton, L., Morgan, C., 2010. The origins of food production in North China: a different kind of agricultural revolution. *Evolutionary Anthropology*, 19: 9–21.

Bettinger, R.L., Barton, L., Richerson, P.J., Boyd, R., Hui, W., Won, C., 2007. The transition to agriculture in Northwestern China. *Developments in Quaternary Sciences*, 9: 83–101.

Betzler, C., Eberli, G. P., Kroon, D., Wright, J. D., Swart, P. K., Nath, B. N., Alvarez-Zarikian, C. A., Alonso-García, M., Bialik, O. M., Blättler, C. L., Guo, J. A., Haffen, S., Horozai, S., Inoue, M., Jovane, L., Lanci, L., Laya, J. C., Mee, A. L. H., Lüdmann, T., Nakakuni, M., Niino, K., Petruny, L. M., Pratiwi, S. D., Reijmer, J. J. G., Reolid, J., Slagle, A. L., Sloss, C. R., Su, X., Yao, Z., Young, J. R., 2016. The abrupt onset of the modern South Asian Monsoon winds. *Scientific Reports*, 6: 29838. DOI:10.1038/srep29838.

Bhat, G. S., 2006. The Indian drought of 2002—a sub-seasonal phenomenon? *Quarterly Journal of the Royal Meteorological Society*, 132(621): 2583–2602. DOI:10.1256/qj.05.13.

Bi, X., Sheng, G., Liu, X., Li, C., Fu, J., 2005. Molecular and carbon and hydrogen isotopic composition of n-alkanes in plant leaf waxes. *Organic Geochemistry*, 36(10): 1405–1417. DOI:10.1016/j.orggeochem.2005.06.001.

Bigg, G. R., Clark, C. D., Hughes, A. L. C., 2008. A last glacial ice sheet on the Pacific Russian coast and catastrophic change arising from coupled ice–volcanic interaction. *Earth and Planetary Science Letters*, 265(3): 559–570. DOI:10.1016/j.epsl.2007.10.052.

Biome 6000 team, 2004. Biome 6000 version 4.2. BIOMES data. www.bridge.bris.ac.uk/resources/Databases/.

Blanford, H.F., 1884. On the connection of the Himalayan snowfall with dry winds and seasons of draughts in India. *Proceedings of the Royal Society, London*, 37: 3–22.

Boaretto, E., Wu, X., Yuan, J., Bar-Yosef, O., Chu, V., Pan, Y., Liu, K., Cohen, D., Jiao, T., Li, S., Gu, H., Goldberg, P., Weiner, S., 2009. Radiocarbon dating of charcoal and bone collagen associated with early pottery at Yuchanyan Cave, Hunan Province, China. *Proceedings of the National Academy of Sciences*, 106(24): 9595–9600. DOI:10.1073/pnas.0900539106.

Bollasina, M. A., Ming, Y., Ramaswamy, V., 2011. Anthropogenic aerosols and the weakening of the South Asian Summer Monsoon. *Science*, 334(6055): 502–505. DOI:10.1126/science.1204994.

Bookhagen, B., 2010. Appearance of extreme monsoonal rainfall events and their impact on erosion in the Himalaya. *Geomatics, Natural Hazards and Risk*, 1(1): 37–50. DOI:10.1080/19475701003625737.

Bookhagen, B., Burbank, D. W., 2006. Topography, relief, and TRMM-derived rainfall variations along the Himalaya. *Geophysical Research Letters*, 33:(L08405). DOI:10.1029/2006GL026037.

Boos, W. R., Kuang, Z., 2010. Dominant control of the South Asian monsoon by orographic insulation versus plateau heating. *Nature*, 463: 218–222. DOI:10.1038/nature08707.

Bordoni, S., Schneider, T., 2008. Monsoons as eddy-mediated regime transitions of the tropical overturning circulations. *Nature Geoscience*, 1: 515–519.

Bosboom, R. E., Dupont-Nivet, G., Houben, A. J. P., Brinkhuis, H., Villa, G., Mandic, O., Stoica, M., Zachariasse, W. -J., Guo, Z., Li, C., 2011. Late Eocene sea retreat from the Tarim Basin (west China) and concomitant Asian paleoenvironmental change. *Palaeogeography, Palaeoclimatology, Palaeoecology*, 299: 385–398.

Boschat, G., Terray, P., Masson, S., 2012. Robustness of SST teleconnections and precursory patterns associated with the Indian summer monsoon. *Climate Dynamics*, 38: 2143–2165.

Bouchery, P. 2010. Irrigation systems and religious interpretation of the local environment among the Hanis in Yunnan. In Lecomte-Tilouine, M. (ed.), *Nature, Culture and Religion at the Crossroads of Asia: 318–342*. Social Science Press, Delhi.

Bourassa, A. E., Robock, A., Randel, W. J., Deshler, T., Rieger, L. A., Lloyd, N. D., Llewellyn, E. J., Degenstein, D. A., 2012. Large volcanic aerosol load in the stratosphere linked to Asian monsoon transport. *Science*, 337(6090): 78–81. DOI:10.1126/science.1219371.

Bowman, D., 2019. *Principles of Alluvial Fan Morphology*. Springer, Dordrecht, p. 151.

Braconnot, P., Otto-Bliesner, B., Harrison, S., Joussaume, S., Peterchmitt, J. Y., Abe-Ouchi, A., Crucifix, M., Driesschaert, E., Fichefet, T., Hewitt, C. D., Kageyama, M., Kitoh, A., Laîné, A., Loutre, M. F., Marti, O., Merkel, U., Ramstein, G., Valdes, P., Weber, S. L., Yu, Y., Zhao, Y., 2007. Results of PMIP2 coupled simulations of the Mid-Holocene and Last Glacial Maximum – Part 1: experiments and large-scale features. *Clim. Past*, 3(2): 261–277. DOI:10.5194/cp-3-261-2007.

Brantingham, P. J., Kerry, K. W., Krivoshapkin, A. I., Kuzmin, Y. V., 2004. Time-space dynamics in the early upper Paleolithic of Northeast Asia. In: Madsen, D. B. (ed.), *Entering America: Northeast Asia and Beringia before the Last Glacial Maximum*. The University of Utah Press, Salt Lake City, pp. 255–284.

Brassell, S. C., Eglinton, G., Marlowe, I. T., Pflaumann, U., Sarnthein, M., 1986. Molecular stratigraphy: a new tool for climatic assessment. *Nature*, 320: 129–133. DOI:10.1038/320129a0.

Bray, F., 1986. *The Rice Economies: Technology and Development in Asian Societies*. Basil Blackwell, Oxford.

Bray, F., 1994. *The Rice Economies: Technology and Development in Asian Societies*. University of California Press, Berkeley, California, p. 254.

Bray, F., Needham, J., 1984. *Science and Civilisation in China: Volume 6, Biology and Biological Technology, Part 2, Agriculture*. Cambridge University Press, Cambridge.

Brecht, H., Dasgupta, S., Laplante, B., Murray, S., Wheeler, D., 2012. Sea-level rise and storm surges: high stakes for a small number of developing countries. *The Journal of Environment & Development*, 21(1): 120–138. DOI:10.1177/1070496511433601.

Breitenbach, S. F. M., Adkins, J. F., Meyer, H., Marwan, N., Kumar, K. K., Haug, G. H., 2010. Strong influence of water vapor source dynamics on stable isotopes in precipitation observed in Southern Meghalaya, NE India. *Earth and Planetary Science Letters*, 292(1): 212–220. DOI:10.1016/j.epsl.2010.01.038.

Briscoe, J., Qamar, U., Contijoch, M., Amir, P., Blackmore, D., 2006. *Pakistan's Water Economy: Running Dry*. Oxford University Press, Karachi.

Brown, S., Nicholls, R. J., Lowe, J. A., Hinkel, J., 2016. Spatial variations of sea-level rise and impacts: an application of DIVA. *Climatic Change*, 134(3): 403–416. DOI:10.1007/s10584-013-0925-y.

Brubaker, K. L., Entekhabi, D., Eagleson, P. S., 1993. Estimation of continental precipitation recycling. *Journal of Climate*, 6: 1077–1089.

Brumfiel, E., 1992. Distinguished lecture in archaeology: breaking and entering the ecosystem-gender, class, and faction steal the show. *American Anthropologist*, 94 (3): 551–568.

Burbank, D. W., Derry, L. A., France-Lanord, C., 1993. Reduced Himalayan sediment production 8 Myr ago despite an intensified monsoon. *Nature*, 364: 48–50.

Burns, S. J., Fleitmann, D., Matter, A., Neff, U., Mangini, A., 2001. Speleothem evidence from Oman for continental pluvial events during interglacial periods. *Geology (Boulder)*, 29(7): 623–626.

Bush, A. B., 2004. Modelling of late quaternary climate over Asia: a synthesis. *Boreas*, 33: 155–163.

Butler, E. E., Huybers, P., 2013. Adaptation of US maize to temperature variations. *Nature Climate Change*, 3(1): 68.

Butler, E. E., Huybers, P., 2015. Variations in the sensitivity of US maize yield to extreme temperatures by region and growth phase. *Environmental Research Letters*, 10(3): 034009. DOI:10.1088/1748-9326/10/3/034009.

Cai, Y. J., Tan, L., Cheng, H., An, Z., Edwards, R. L., Kelly, M. J., Kong, X., Wang, X., 2010. The variation of summer monsoon precipitation in central China since the last deglaciation. *Earth and Planetary Science Letters*, 291: 21–31. DOI:10.1016/j.epsl.2009.12.039.

Camoin, G. F., Montaggioni, L. F., Braithwaite, C. J. R., 2004. Late glacial to post glacial sea levels in the western Indian Ocean. *Marine Geology*, 206(1–4): 119–146. DOI:10.1016/j.margeo.2004.02.003.

Cao, S., Liu, X., Er, H., 2010a. Dujiangyan irrigation system – a world cultural heritage corresponding to concepts of modern hydraulic science. *Journal of Hydro-Environment Research*, 4(1): 3–13. DOI:10.1016/j.jher.2009.09.003.

Cao, X., Xu, Q., Jing, Z., Tang, J., Li, Y., Tian, F., 2010b. Holocene climate change and human impacts implied from the pollen records in Anyang, central China. *Quaternary International*, 227: 3–9.

Cao, Z. H., Ding, J. L., Hu, Z. Y., Knicker, H., Kagel-Knabner, I., Yang, L. Z., Yin, R., Lin, X. G., Dong, Y. H., 2006. Ancient paddy soils from the Neolithic age in China's Yangtze River Delta. *Naturwissenschaften*, 93(5): 232–236.

Cappucci, M., Freedman, A., 2019. *From Tropical Storm to Category 5 in 18 Hours: Super Typhoon Hagibis Intensifies at One of the Fastest Rates on Record*, Washington Post, Washington DC.

Cardenas, A., 1983. A pheno-climatological assessment of millets and other cereal grains in tropical cropping patterns, University of Nebraska, Master of Science. Department of Horticulture.

Casazza, M., Lega, M., Liu, G., Ulgiati, S., Endreny, T. A., 2018. Aerosol pollution, including eroded soils, intensifies cloud growth, precipitation, and soil erosion: A review. *Journal of Cleaner Production*, 189: 135–144. DOI:10.1016/j.jclepro.2018.04.004.

Castillo, C., 2011. Rice in Thailand: the archaeobotanical contribution. *Rice*, 4(3): 114–120. DOI:10.1007/s12284-011-9070-2.

Castillo, C., Fuller, D. Q., 2010. Still too fragmentary and dependent upon chance? Advances in the study of early Southeast Asian archaeobotany. In: Bellina, Bacus,

Pryce, Weissman Christie (eds.), *50 Years of Archaeology in Southeast Asia*. River Books, London, pp. 91–111.

Castillo, C. C., Bellina, B., Fuller, D. Q., 2016. Rice, beans and trade crops on the early maritime Silk Route in Southeast Asia. *Antiquity*, 90(353): 1255–1269. DOI:10.15184/aqy.2016.175.

Castillo, C. C., Fuller, D. Q., Piper, P. J., Bellwood, P., Oxenham, M., 2018. Hunter-gatherer specialization in the late Neolithic of southern Vietnam – The case of Rach Nui. *Quaternary International*, 489: 63–79. DOI:10.1016/j.quaint.2016.11.034.

Chakraborty, T., Kar, R., Ghosh, P., Basu, S., 2010. Kosi megafan: historical records, geomorphology and the recent avulsion of the Kosi River. *Quaternary International*, 227(2): 143–160.

Charles, C. D., Hunter, D. E., Fairbanks, R. G., 1997. Interaction between the ENSO and the Asian monsoon in a coral record of tropical climate. *Science*, 277: 925–928.

Chauhan, O. S., Vogelsang, E., Basavaiah, N., Kader, U. S. A., 2010. Reconstruction of the variability of SW monsoon during the past 3 Ka from the continental margin of the South Eastern Arabian Sea. *Journal of Quaternary Science*, 25(5): 798–807.

Chen, F., Xu, Q., Chen, J., Birks, H. J. B., Liu, J., Zhang, S., Jin, L., An, C., Telford, R. J., Cao, X., 2015a. East Asian summer monsoon precipitation variability since the last deglaciation. *Scientific Reports*, 5: 11186.

Chen, F. H., G. H. Dong, Zhang, D. J., Liu, X. Y., Jia, X., An, C. B., Ma, M. M., Xie, Y. W., Barton, L., Ren, X. Y., Zhao, Z. J., Wu, X. H., Jones, M. K., 2015b. Agriculture facilitated permanent human occupation of the Tibetan Plateau after 3600 B.P. *Science*, 349(6219): 248–250.

Chen, F. H., Shi, Q., Wang, J.M., 1999. Environmental changes documented by sedimentation of Lake Yiema in arid China since the Late Glaciation. *Journal of Paleolimnology*, 22: 159–169.

Chen, J., Rao, Z., Liu, J., Huang, W., Feng, S., Dong, G., Hu, Y., Xu, Q., Chen, F., 2016a. On the timing of the East Asian summer monsoon maximum during the Holocene—Does the speleothem oxygen isotope record reflect monsoon rainfall variability? *Science China Earth Sciences*, 59(12): 2328–2338. DOI:10.1007/s11430-015–5500-5.

Chen, S., Kung, J. K.-s., 2011. The Malthusian Quagmire: maize and population growth in China, 1500–1900. Hong Kong University of Science and Technology, working paper. https://pdfs.semanticscholar.org/46e6/c9e14a5182a8da14841c7f0f6cd02253b3b4.pdf.

Chen, S., Yu, P. -L., 2017. Intensified foraging and the roots of farming in China. *Journal of Anthropological Research*, 73(3): 381–412. DOI:10.1086/692660.

Chen, T., Xia, G., Liu, T., Chen, W., Chi, D., 2016b. Assessment of drought impact on main cereal crops using a standardized precipitation evapotranspiration index in Liaoning Province, China. *Sustainability*, 8(10): 1069. DOI:10.3390/su8101069.

Chen, X., Tung, K. -K., 2014. Varying planetary heat sink led to global-warming slowdown and acceleration. *Science*, 345(6199): 897–903 DOI:10.1126/science.1254937.

Chen, X., Zhang, X., Church, J. A., Watson, C. S., King, M. A., Monselesan, D., Legresy, B., Harig, C., 2017. The increasing rate of global mean sea-level rise during 1993–2014. *Nature Climate Change*, 7: 492. DOI:10.1038/nclimate3325.

Chen, Y., Li, W., Deng, H., Fang, G., Li, Z., 2016c. Changes in Central Asia's water tower: past, present and future. *Scientific Reports*, 6: 35458. DOI:10.1038/srep35458.

Chen, Y., Syvitski, J. P., Gao, S., Overeem, I., Kettner, A. J., 2012. Socio-economic impacts on flooding: a 4000-year history of the Yellow River, China. *Ambio*, 41(7): 682–698.

Chen, Z., Li, J., Shen, H., Zhanghua, W., 2001. Yangtze River of China: historical analysis of discharge variability and sediment flux. *Geomorphology*, 41(2): 77–91. DOI:10.1016/S0169-555X(01)00106-4.

Cheng, R., Dong, Z., 2010. Breeding and production of foxtail millet in China. In: He, Z., Bonjean, A.P.A. (eds.), *Cereals in China*. International Maize and Wheat Improvement Center, Mexico, pp. 87–97.

Chi, Z., 2002. The discovery of early pottery in China. *Documenta Praehistorica*, 29: 29–35.

Chou, C., Neelin, J. D., Chen, C. A., Tu, J. Y., 2009. Evaluating the "rich-get-richer" mechanism in tropical precipitation change under global warming. *Journal of Climate*, 99: 1982–2005.

Chun, C. K., 1961. Agrarian policy of the Chinese Communist Party. *Indian Journal of Agricultural Economics*, 16(902–2016-66960): 79–80.

Clark, J. A., Lingle, C. S., 1977. Future sea-level changes due to West Antarctic ice sheet fluctuations. *Nature*, 269(5625): 206.

Clark, M. K., House, M. A., Royden, L. H., Whipple, K. X., Burchfiel, B. C., Zhang, X., Tang, W., 2005. Late Cenozoic uplift of southeastern Tibet. *Geology*, 33(6): 525–528. DOI:10.1130/G21265.1.

Clarke, G., Leverington, D., Teller, J., Dyke, A., 2004. Paleohydraulics of the last outburst flood from glacial Lake Agassiz and the 8200 BP cold event. *Quaternary Science Reviews*, 23: 389–407.

Clemens, S. C., Holbourn, A., Kubota, Y., Lee, K. E., Liu, Z., Chen, G., Nelson, A., Fox-Kemper, B., 2018. Precession-band variance missing from East Asian monsoon runoff. *Nature Communications*, 9(1): 3364. DOI:10.1038/s41467-018-05814-0.

Clemens, S. C., Murray, D. W., Prell, W. L., 1996. Nonstationary phase of the Plio-Pleistocene Asian monsoon. *Science*, 274(5289): 943–948.

Clemens, S. C., Prell, W. L., 2003. A 350,000 year summer-monsoon multi-proxy stack from the Owen Ridge, Northern Arabian Sea. *Marine Geology*, 201: 35–51.

Clemens, S. C., Prell, W. L., Sun, Y., 2010. Orbital-scale timing and mechanisms driving Late Pleistocene Indo-Asian summer monsoons: Reinterpreting cave speleothem $\partial^{18}O$. *Paleoceanography*, 25(PA4207). DOI:10.1029/2010PA001926.

Clift, P. D., Webb, A. G., 2018. A history of the Asian monsoon and its interactions with solid Earth tectonics in Cenozoic South Asia. In: Searle, M.P., Treloar, P.J. (eds.), *Himalayan Tectonics: A Modern Synthesis*. Special Publications, Geological Society, London, 631–652.

Clift, P. D., Carter, A., Giosan, L., Durcan, J., Tabrez, A. R., Alizai, A., Van Laningham, S., Duller, G. A. T., Macklin, M. G., Fuller, D. Q., Danish, M., 2012. U-Pb zircon dating evidence for a Pleistocene Sarasvati River and Capture of the Yamuna River. *Geology*, 40(3): 212–215. DOI:10.1130/G32840.1.

Clift, P. D., Giosan, L., Henstock, T., Tabrez, A. R., 2014. Sediment storage and reworking on the shelf and in the Canyon of the Indus River-Fan System since the last glacial maximum. *Basin Research*, 26: 183–202. DOI:10.1111/bre.12041.

Clift, P. D., Hodges, K., Heslop, D., Hannigan, R., Hoang, L. V., Calves, G., 2008. Greater Himalayan exhumation triggered by Early Miocene monsoon intensification. *Nature Geoscience*, 1: 875–880. DOI:10.1038/ngeo351.

Cobo Castillo, C., 2018. The archaeobotany of Khao Sek. *Archaeological Research in Asia*, 13: 74–77. DOI:10.1016/j.ara.2017.05.002.

Coggan, M., 2008. Death toll rises from Indian floods. Australian Broadcasting Corporation, www.abc.net.au/news/2008-08-29/death-toll-rises-from-indian-floods/493582.

Cojean, R., Caï, Y. J., 2011. Analysis and modeling of slope stability in the Three-Gorges Dam reservoir (China) – The case of Huangtupo landslide. *Journal of Mountain Science*, 8(2): 166. DOI:10.1007/s11629-011-2100-0.

Colin, C., Siani, G., Sicre, M. -A., Liu, Z., 2010. Impact of the East Asian monsoon rainfall changes on the erosion of the Mekong River basin over the past 25,000 yr. *Marine Geology*, 271(1–2): 84–92. DOI:10.1016/j.margeo.2010.01.013.

Colinet, G., Koulos, K., Bozhi, W., Yongmei, L., Lacroix, D., Youbo, S., Chapelle, J., Fullen, M. A., Hocking, T., Bock, L., 2011. Agro-pedological assessment of the traditional Yuanyang rice terraces of Yunnan Province, China. *Journal of Resources and Ecology*, 2(4): 353–362

Condon, E., Hillmann, P., King, J., Lang, K., Patz, A., 2009. Resource disputes in South Asia: Water scarcity and the potential for interstate conflict, Workshop in International Public Affairs, p. 1. http://citeseerx.ist.psu.edu/viewdoc/download?doi=10.1.1.404.2127&rep=rep1&type=pdf.

Coningham, R., Young, R., 2015. *The Archaeology of South Asia: From the Indus to Asoka, c.6500 BCE–200 CE. Cambridge World Archaeology.* Cambridge University Press, Cambridge.

Conklin, H., 2008. Ethnoecological approach to shifting agriculture, ethnoecology and the defense of swidden agriculture. *Transactions of the New York Academy of Sciences*, 17: 133–142. DOI:10.1111/j.2164-0947.1954.tb00402.x

Conklin, H.C., 1969. An ethnoecological approach to shifting agriculture. In: Vayda, A.P. (ed.), *Environment and Cultural Behavior.* The Natural History Press, New York, pp. 221–233.

Constantini, L., 1984. The beginning of agriculture in the Kachi Plain: the evidence of Mehrgarh. In: Allchin, B. (ed.), *South Asian Archaeology.* Cambridge University Press, Cambridge, pp. 29–33.

Contreras, D., 2015. Correlation is not enough. Building better arguments in the archaeology of human environment interactions. In: Contreras, D. (ed.), *The Archaeology of Human Environment Interactions.* The Natural History Press, New York, pp. 3–22.

Cook, E. R., Anchukaitis, K. J., Buckley, B. M., D'Arrigo, R. D., Jacoby, G. C., Wright, W. E., 2010. Asian monsoon failure and megadrought during the last millennium. *Science*, 328(5977): 486–489. DOI:10.1126/science.1185188.

Cosmo, N. D., Hessl, A., Leland, C., Byambasuren, O., 2018. Environmental stress and steppe nomads: rethinking the history of the Uyghur Empire (744–840) with paleo-climate data. *Journal of Interdisciplinary History*, 48(4): 439–463. DOI:10.1162/JINH_a_01194.

Courty, M. A., 1995. Late Quaternary environmental change and natural constraints to ancient landuse (Northwest India). In: Johnson, E. (ed.), *Ancient Peoples and Landscapes.* Museum of Texas Tech University, Lubbok TX, pp. 106–126.

Craft, C., Clough, J., Ehman, J., Joye, S., Park, R., Pennings, S., Guo, H., Machmuller, M., 2009. Forecasting the effects of accelerated sea-level rise on tidal marsh ecosystem services. *Frontiers in Ecology and the Environment*, 7(2): 73–78.

Crawford, G., 2012. Early rice exploitation in the lower Yangzi valley: What are we missing? *The Holocene*, 22(6): 613–621. DOI:10.1177/0959683611424177.

Crawford, G., Chen, X., Wang, J., 2006. 山东济南长清地月庄遗址发现的后李文化时期的碳化稻 Shandong Jinan Changqing Diqu Yuezhuang Yizhi Faxian de Houli Wenhua Shiqi de Tanhuadao (Discovery of fossilized crops from the Houli site of Yuezhuang in the Jinan Region of Shandong). *Dongfang Kaogu*, 3: 247–251.

Crawford, G., Lee, G. A., 2003. Agricultural origins in the Korean Peninsula. *Antiquity*, 77(295): 87–95.

Crawford, G., Underhill, A. P., Zhao, Z., Lee, G. A., Feinman, G., Nicholas, L., Luan, F., Yu, H., Fang, H., Cai, F., 2005. Late Neolithic plant remains from Northern China: Preliminary results from Liangchengzhen, Shandong. *Current Anthropology*, 46(2): 309–317.

Crawford, G. W., 1992. The transitions to agriculture in Japan. In Gebauer, Anne E., Price T Douglas (eds.), *Transitions to Agriculture in Prehistory*. Prehistory Press, Madison, WI, pp. 117–132.

Crawford, G. W., 2011. Advances in understanding early agriculture in Japan. *Current Anthropology*, 52(S4): S331–S345. DOI:10.1086/658369.

Crosby, A. W. J., 2003. *The Columbian Exchange: Biological and Cultural Consequences of 1492*. Praeger, Westport, CT.

Crowley, T. J., Berner, R. A., 2001. CO2 and climate change. *Science*, 292: 870–872. DOI:10.1126/science.1061664.

Crutzen, P. J., 2002. Geology of mankind. *Nature*, 415(6867): 23. DOI:10.1038/415023a.

Crutzen, P. J., Stoermer, E. F., 2000. The 'Anthropocene'. *Global Change Newsletter*, 41: 17–18.

Cui, L., Ge, Z., Yuan, L., Zhang, L., 2015. Vulnerability assessment of the coastal wetlands in the Yangtze Estuary, China to sea-level rise. *Estuarine, Coastal and Shelf Science*, 156: 42–51. DOI:10.1016/j.ecss.2014.06.015.

Cui, X., Graf, H. -F., Langmann, B., Chen, W., Huang, R., 2007. Hydrological impacts of deforestation on the Southeast Tibetan Plateau. *Earth Interactions*, 11(15): 1–18. DOI:10.1175/ei223.1.

Currie, B. S., Rowley, D. B., Tabor, N. J., 2005. Middle Miocene paleoaltimetry of southern Tibet: implications for the role of mantle thickening and delamination in the Himalayan orogen. *Geology (Boulder)*, 33(3): 181–184.

Currie, R. I., Fisher, A. E., Hargreaves, P. M., 1973. Arabian Sea upwelling. In: Zeitschel, B., Golach, S. A. (eds.), *The Biology of the Indian Ocean*. Springer, New York, pp. 37–52.

Curry, W. B., Ostermann, D. R., Guptha, M. V. S., Itekkot, V., 1992. Foraminiferal production and monsoonal upwelling in the Arabian Sea; evidence from sediment traps. In: Summerhayes, C. P., Prell, W. L., Emeis, K. C. (eds.), *Upwelling Systems: Evolution since the Early Miocene*. Special Publication. Geological Society, London, pp. 93–106.

d'Alpoim Guedes, J., 2011. Millets, rice, social complexity, and the spread of agriculture to the Chengdu Plain and Southwest China. *Rice*, 4(3–4): 104–113. 10.1007/s12284-011-9071-1.

d'Alpoim Guedes, J., 2015. Rethinking the spread of agriculture to the Tibetan Plateau. *The Holocene*, 25(9): 1498–1510.

d'Alpoim Guedes, J., 2016. Model building, model testing, and the spread of agriculture to the Tibetan Plateau. *Archaeological Research in Asia*, 5(2016): 16–23.

d'Alpoim Guedes J, Jin G, Bocinsky RK 2015. The impact of climate on the spread of rice to North-Eastern China: A new look at the data from Shandong Province. *PLOS ONE*, 10(6): e0130430. https://doi.org/10.1371/journal.pone.0130430.

d'Alpoim Guedes, J., Lu, H., Hein, A., Schmidt, A. H., 2015. Early evidence for the use of wheat and barley as staple crops on the margins of the Tibetan Plateau. *Proceedings of the National Academy of Sciences*, 112(18): 5625–5630.

d'Alpoim Guedes, J., Ming, J., He, K., Xiaohong, W.,Jiang, Z., 2013. Site of Baodun yields the earliest evidence for the spread of rice and foxtail millet agriculture to Southwest China. *Antiquity*, 87: 758–771.

d'Alpoim Guedes, J., Austermann, J., Mitrovica, J. X., 2016a. Changing sea level during Meltwater Pulse 1A and lost foraging opportunities for East Asian Hunter-Gatherers. *Geoarchaeology*, 31(4): 255–266. DOI 10.1002/gea.21542.

d'Alpoim Guedes, J., Bocinsky, R. K., 2018. Climate change stimulated agricultural innovation and exchange across Asia. *Science Advances*, 4(10): eaar4491. DOI:10.1126/sciadv.aar4491.

d'Alpoim Guedes, J., Butler, E., 2014. Modeling constraints on the spread of agriculture to Southwest China with thermal niche models. *Quaternary International*, 349(2014): 29–41.

d'Alpoim Guedes, J., Jin, G., Bocinsky, R. K., 2015. The impact of climate on the spread of rice agriculture to North-Eastern China: An example from Shandong. *PLOS-One*, 10 (6): e0130430.

d'Alpoim Guedes, J., Lu, H., Li, Y., Spengler, R. N., Wu, X., Aldenderfer, M. S., 2014. Moving agriculture onto the Tibetan plateau: The archaeobotanical evidence. *Archaeological and Anthropological Sciences*, 6(3): 255–269.

d'Alpoim Guedes, J., Manning, S. W., Bocinsky, R. K., 2016b. A 5,500-year model of changing crop niches on the Tibetan Plateau. *Current Anthropology*, 57(4): 517–522. 10.1086/687255.

d'Alpoim Guedes, J. A., Crabtree, S. A., Bocinsky, R. K., Kohler, T. A., 2016c. Twenty-first century approaches to ancient problems: Climate and society. *Proceedings of the National Academy of Sciences*, 113(51): 14483–14491. 10.1073/pnas.1616188113.

Dai, A., 2013. Increasing drought under global warming in observations and models. *Nature Climate Change*, 3: 52–58. DOI:10.1038/nclimate1633.

Dai, A., Rasmussen, R. M., Liu, C., Ikeda, K., Prein, A. F., 2017. A new mechanism for warm-season precipitation response to global warming based on convection-permitting simulations. *Climate Dynamics*, 55: 343–368.

Dai, A., Wigley, T. M. L., Boville, B. A., Kiehl, J. T., Buja, L. E., 2001. Climates of the twentieth and twenty-first centuries simulated by the NCAR Climate System Model. *Journal of Climate*, 14: 485–519.

Dai, Z., Liu, J. T., 2013. Impacts of large dams on downstream fluvial sedimentation: an example of the Three Gorges Dam (TGD) on the Changjiang (Yangtze River). *Journal of Hydrology*, 480: 10–18.

Dal Martello, R., Min, R., Stevens, C., Higham, C., Higham, T., Qin, L., Fuller, D. Q., 2018. Early agriculture at the crossroads of China and Southeast Asia: Archaeobotanical evidence and radiocarbon dates from Baiyangcun, Yunnan. *Journal of Archaeological Science: Reports*, 20: 711–721.

Dales, G. F., 1964. The mythical massacre at Mohenjo-daro. *Expedition*, 6: 36–43.

Dangendorf, S., Marcos, M., Wöppelmann, G., Conrad, C. P., Frederikse, T., Riva, R., 2017. Reassessment of 20th century global mean sea level rise. *Proceedings of the National Academy of Sciences*, 114(23): 5946–5951. DOI:10.1073/pnas.1616007114.

Dasgupta, S., Laplante, B., Meisner, C., Wheeler, D., Yan, J., 2007. *The Impact of Sea Level Rise on Developing Countries: A Comparative Analysis*, The World Bank, New York.

Dash, S. K., Kulkarni, M. A., Mohanty, U. C., Prasad, K., 2009. Changes in the characteristics of rain events in India. *Journal of Geophysical Research, Atmospheres*, 114 (D10109), DOI:10.1029/2008JD010572.

Dash, S. K., Sharma, N., Pattnayak, K. C., Gao, X. J., Shi, Y., 2012. Temperature and precipitation changes in the north-east India and their future projections. *Global and Planetary Change*, 98–99: 31–44. DOI:10.1016/j.gloplacha.2012.07.006.

Davies, H. R., 1909. *Yun-nan, the Link between India and the Yangtze*. Cambridge University Press, London, UK.

Davis, M., 2001. *Late Victorian Holocausts: El Niño Famines and the Making of the Third World*. Verso, London.

Day, J. W., Ramachandran, R., Giosan, L., Syvitski, J., Kemp, G. P., 2019. *Delta Winners and Losers in the Anthropocene, Coasts and Estuaries*. Elsevier, Dordrecht, pp. 149–165.

de Boer, H. J., Lammertsma, E. I., Wagner-Cremer, F., Dilcher, D. L., Wassen, M. J., Dekker, S. C., 2011. Climate forcing due to optimization of maximal leaf conductance in subtropical vegetation under rising CO_2. *Proceedings of the National Academy of Sciences*, 108(10): 4041–4046. DOI:10.1073/pnas.1100555108.

de Vries, B., 2006. In search of sustainability: What can we learn from the past? In: Hornborg, A., Crumley, C.L. (eds.), *The World System and the Earth System: Global Socioenvironmental Challenge and Sustainability Since the Neolithic*. Left Coast Press, Walnut Creek, CA, pp. 243–257.

Debroy, B., Debroy, D., 2011. *The Holy Vedas: Rigveda, Yajurveda, Samaveda, Atharvaveda*. BR Publishing Corporation, Delhi.

DeConto, R. M., Pollard, D., 2016. Contribution of Antarctica to past and future sea-level rise. *Nature*, 531: 591. DOI:10.1038/nature17145.

Demske, D., Tarasov, P. E., Wünnemann, B., Riedel, F., 2009. Late glacial and Holocene vegetation, Indian monsoon and westerly circulation in the Trans-Himalaya recorded in the lacustrine pollen sequence from Tso Kar, Ladakh, NW India. *Palaeogeography, Palaeoclimatology, Palaeoecology*, 279(3–4): 172–185.

Deng, Z., Qin, L., Gao, Y., Weisskopf, A. R., Zhang, C., Fuller, D. Q., 2015. From early domesticated rice of the Middle Yangtze Basin to millet, rice and wheat agriculture: Archaeobotanical Macro-Remains from Baligang, Nanyang Basin, Central China (6700–500 BC). *PLOS ONE*, 10(10): e0139885. DOI:10.1371/journal.pone.0139885.

Denton, G. H., Broecker, W. S., 2008. Wobbly ocean conveyor circulation during the Holocene? *Quaternary Science Reviews*, 27(21): 1939–1950. DOI:10.1016/j.quascirev.2008.08.008.

Dercourt, J., Ricou, L. E., Vrielynck, B., 1993. *Atlas Tethys, Paleoenvironmental Maps*. Gauthier-Villars, Paris, 307 pp.

Derry, L. A., France-Lanord, C., 1996. Neogene Himalayan weathering history and river $^{87}Sr/^{86}Sr$; impact on the marine Sr record. *Earth and Planetary Science Letters*, 142: 59–74.

Diamond, J., 2005. *Collapse: How Societies Choose to Fail or Succeed*. Penguin, New York, p. 592.

Diao, X., 2017. Production and genetic improvement of minor cereals in China. *The Crop Journal*, 5(2): 103–114. DOI:10.1016/j.cj.2016.06.004.

Dillon, S. L., Shapter, F. M., Henry, R. J., Cordeiro, G., Izquierdo, L., Lee, L. S., 2007. Domestication to crop improvement: Genetic resources for sorghum and saccharum (Andropogoneae). *Annals of Botany*, 100(5): 975–989. DOI:10.1093/aob/mcm192.

Dixit, Y., Hodell, D. A., Giesche, A., Tandon, S. K., Gázquez, F., Saini, H. S., Skinner, L. C., Mujtaba, S. A. I., Pawar, V., Singh, R. N., Petrie, C. A., 2018. Intensified summer monsoon and the urbanization of Indus Civilization in northwest India. *Scientific Reports*, 8(1): 4225. DOI:10.1038/s41598-018-22504-5.

Dixit, Y., Hodell, D. A., Petrie, C. A., 2014. Abrupt weakening of the summer monsoon in northwest India ~4100 yr ago. *Geology*, 42: 339-342. DOI:10.1130/G35236.1.

Dong, G., Yang, Y., Zhao, Y., Zhou, A., Zhang, X., Li, X., Chen, F., 2012. Human settlement and human–environment interactions during the historical period in Zhuanglang County, western Loess Plateau, China. *Quaternary International*, 281: 78–83. DOI:10.1016/j.quaint.2012.05.006.

Donner, L., Schubert, W., Somerville, R. (eds.), 2011. *The Development of Atmospheric General Circulation Models: Complexity, Synthesis and Computation*. Cambridge University Press, Cambridge.

Doose-Rolinski, H., Rogalla, U., Scheeder, G., Lückge, A., von Rad, U., 2001. High-resolution temperature and evaporation changes during the late Holocene in the northeastern Arabian Sea. *Paleoceanography*, 16(4): 358–367.

Dore, J., 1959. Response of rice to small differences in length of day. *Nature*, 183: 413–414.

Dreybrodt, W., Scholz, D., 2011. Climatic dependence of stable carbon and oxygen isotope signals recorded in speleothems: From soil water to speleothem calcite. *Geochimica et Cosmochimica Acta*, 75(3): 734–752. DOI:10.1016/j.gca.2010.11.002.

Drumond, A., Nieto, R., Gimeno, L., 2011. Sources of moisture for China and their variations during drier and wetter conditions in 2000–2004: a Lagrangian approach. *Climate Research*, 50(2–3): 215–225.

Duan, K., Yao, T., Thompson, L.G., 2004. Low-frequency of southern Asian monsoon variability using a 295-year record from the Dasuopu ice core in the central Himalayas. *Geophysical Research Letters*, 31(16): n/a-n/a. 10.1029/2004GL020015.

Dubbert, M., Werner, C., 2019. Water fluxes mediated by vegetation: emerging isotopic insights at the soil and atmosphere interfaces. *New Phytologist*, 221(4): 1754–1763. DOI:10.1111/nph.15547.

Duke, J., 1983. *Handbook of Energy Crops*. https://hort.purdue.edu/newcrop/duke_energy/dukeindex.html.

Durcan, J. A., Thomas, D. S. G., Gupta, S., Pawar, V., Singh, R. N., Petrie, C. A., 2019. Holocene landscape dynamics in the Ghaggar-Hakra palaeochannel region at the northern edge of the Thar Desert, northwest India. *Quaternary International*, 501: 317–327. DOI:10.1016/j.quaint.2017.10.012.

Dykoski, C. A., Edwards, R. L., Cheng, H., Yuan, D., Cai, Y., Zhang, M., Lin, Y., Qing, J., An, Z., Revenaugh, J., 2005. A high-resolution, absolute-dated holocene and deglacial asian monsoon record from Dongge Cave, China. *Earth and Planetary Science Letters*, 233(1–2): 71–86. DOI:10.1016/j.epsl.2005.01.036.

Eagle, R. A., Risi, C., Mitchell, J. L., Eiler, J. M., Seibt, U., Neelin, J. D., Li, G., Tripati, A. K., 2013. High regional climate sensitivity over continental China constrained by glacial-recent changes in temperature and the hydrological cycle. *Proceedings of the National Academy of Sciences*, 110(22): 8813–8818. DOI:10.1073/pnas.1213366110.

East, A. E., Clift, P. D., Carter, A., Alizai, A., VanLaningham, S., 2015. Fluvial–Eolian interactions in sediment routing and sedimentary signal buffering: An example from the Indus Basin and Thar Desert. *Journal of Sedimentary Research*, 85: 715–728. DOI:10.2110/jsr.2015.42.

Edmond, J. M., Huh, Y., 1997. Chemical weathering yields from basement and orogenic terrains in hot and cold climates. In: Ruddiman, W.F. (ed.), *Tectonic Climate and Climate Change*. Plenum Press, New York, pp.330–353.

Eglinton, G., Hamilton, R. J., 1967. Leaf epicuticular waxes. *Science*, 156: 1322–1334. DOI:10.1126/science.156.3780.1322.

Eglinton, T. I., Eglinton, G., 2008. Molecular proxies for paleoclimatology. *Earth Planetary Science Letters*, 275(1–16): 1–16. DOI:10.1016/j.epsl.2008.07.012.

Ehhalt, D. H., 1974. The atmospheric cycle of methane. *Tellus*, 26: 58–70.

Ehleringer, J. R., Rundel., P. W., 1989. Stable isotopes: History, units, and instrumentation. In: Rundel, P. W., Ehleringer, J. R., Nagy, K. A. (eds.), *Stable Isotopes in Ecological Research*. Springer Verlag, New York.

Ehrlich, P. R., 1968. *The Population Bomb*. Ballantine Books, New York, 201 pp.

Eiler, J. M., 2007. "Clumped-isotope" geochemistry—the study of naturally-occurring, multiply-substituted isotopologues. *Earth and Planetary Science Letters*, 262(3): 309–327. DOI:10.1016/j.epsl.2007.08.020.

Ellison, C. R. W., Chapman, M. R., Hall, I. R., 2006. Surface and deep ocean interactions during the cold climate event 8200 years ago. *Science*, 312(5782): 1929–1932. DOI:10.1126/science.1127213.

Elston, R. G., Guanghui, D., Dongju, Z., 2011. Late Pleistocene intensification technologies in Northern China. *Quaternary International*, 242(2): 401–415. DOI:10.1016/j. quaint.2011.02.045.

Emanuel, K., 2005. Increasing destructiveness of tropical cyclones over the past 30 years. *Nature*, 436(7051): 686–688. DOI:10.1038/nature03906.

Emanuel, K., DesAutels, C., Holloway, C., Korty, R., 2004. Environmental control of tropical cyclone intensity. *Journal of the Atmospheric Sciences*, 61(7): 843–858.

Emanuel, K. A., 1987. The dependence of hurricane intensity on climate. *Nature*, 326 (6112): 483–485. DOI:10.1038/326483a0.

Enderlin, E. M., Howat, I. M., Jeong, S., Noh, M. -J., van Angelen, J. H., van den Broeke, M. R., 2014. An improved mass budget for the Greenland ice sheet. *Geophysical Research Letters*, 41(3): 866–872. 10.1002/2013GL059010.

Enzel, Y., Ely, L. L., Mishra, S., Ramesh, R., Amit, R., Lazar, B., Rajaguru, S. N., Baker, V. R., Sandle, A., 1999. High-resolution Holocene environmental changes in the Thar Desert, northwestern India. *Science*, 284: 125–128.

Erickson, C. L., 1999. Neo-environmental determinism and agrarian 'collapse'in Andean prehistory. *Antiquity*, 73(281): 634–642.

Evan, A. T., Camargo, S. J., 2011. A climatology of Arabian Sea cyclonic storms. *Journal of Climate*, 24: 140–158.

Evan, A. T., Kossin, J. P., Chung, C. E., Ramanathan, V., 2011. Arabian Sea tropical cyclones intensified by emissions of black carbon and other aerosols. *Nature*, 479: 94–98. DOI:10.1038/nature10552.

Eyshi Rezaei, E., Gaiser, T., Siebert, S., Sultan, B., Ewert, F., 2014. Combined impacts of climate and nutrient fertilization on yields of pearl millet in Niger. *European Journal of Agronomy*, 55: 77–88. DOI:10.1016/j.eja.2014.02.001.

Fagan, B., 2004. *The Long Summer: How Climate Changed Civilization*. Basic Books, New York.

Fairbanks, R.G., 1989. A 17,000-year glacio-eustatic sea level record: influence of glacial melting rates on Younger Dryas event and deep-ocean circulation. *Nature*, 342: 637–642.

Fan, J., Leung, L. R., Li, Z., Morrison, H., Chen, H., Zhou, Y., Qian, Y., Wang, Y., 2012a. Aerosol impacts on clouds and precipitation in eastern China: Results from bin and bulk microphysics. *Journal of Geophysical Research: Atmospheres*, 117(D16). DOI:10.1029/2011JD016537.

Fan, L., Lu, C., Yang, B., Chen, Z., 2012b. Long-term trends of precipitation in the North China Plain. *Journal of Geographical Sciences*, 22(6): 989–1001. 10.1007/s11442-012–0978-2.

FAO, IFAD, UNICEF, WFP, WHO, 2017. *The State of Food Security and Nutrition in the World 2017*. Building resilience for peace and food security, Rome.

Farquhar, G. D., Ehleringer, J. R., Hubick, K. T., 1989. Carbon isotope discrimination and photosynthesis. *Annual Reviews of Plant Physiology and Molecular Biology*, 40: 503–537. DOI:10.1146/annurev.pp.40.060189.002443.

Feakins, S. J., deMenocal, P. B., Eglinton, T. I., 2005. Biomarker records of late Neogene changes in northeast African vegetation. *Geology*, 33(12): 977–980. DOI:10.1130/G21814.1.

Feakins, S. J., Sessions, A. L., 2010. Controls on the D/H ratios of plant leaf waxes in an arid ecosystem. *Geochimica et Cosmochimica Acta*, 74(7): 2128–2141. DOI:10.1016/j.gca.2010.01.016.

Feng, Z. -D., An, C. B., Wang, H. B., 2006. Holocene climatic and environmental changes in the arid and semi-arid areas of China: a review. *The Holocene*, 16: 119–130. DOI:10.1191/0959683606hl912xx.

Fick, S. E., Hijmans, R. J., 2017 Fick et al. WorldClim 2: new 1-km spatial resolution climate surfaces for global land areas. *International Journal of Climatology*, 37(12): 4302–4315. DOI:10.1002/joc.5086.

Fisher, M. H., 2016. *Mughal Empire, The Ashgate Research Companion to Modern Imperial Histories*. Routledge, Abingdon, UK, pp.181–206.

Flad, R., Chen, P., 2006. The archaeology of the Sichuan Basin and surrounding areas during the Neolithic period. In: Li, S., von Falkenhausen, L. (eds.), *Salt Archaeology in China: Ancient Salt Production and Landscape Archaeology in the Upper Yangtse Basin: Preliminary Studies*. Kexue Chubanshe, Beijing, pp.183–259.

Flad, R., Chen, P., 2013. *Ancient Central China: An Archaeological Study of Centers and Peripheries along the Yangzi River*. Cambridge University Press, Cambridge, 397 pp.

Flad, R. K., Yuan, J.袁., Li, S.李., 2007. Zooarcheological evidence for animal domestication in northwest China. In: Madsen, D.B., Chen, F.H., Gao, X. (eds.), *Late Quaternary Climate Change and Human Adaptation in Arid Chin*. Elsevier, Amsterdam, pp. 167–203.

Flam, L., 1981. *The Paleogeography and Prehistoric Settlement Patterns in Sind, Pakistan (ca. 4000–2000 BC). PhD Thesis*, University of Michigan, Ann Arbor, Michigan.

Flato, G., Marotzke, J., Abiodun, B., Braconnot, P., Chou, S.C., Collins, W., Cox, P., Driouech, F., Emori, S., Eyring, V., Forest, C., Gleckler, P., Guilyardi, E., Jakob, C., Kattsov, V., Reason, C., Rummukainen, M., 2013. Evaluation of Climate Models. In: Stocker, T.F. et al. (eds.), *Climate Change 2013: The Physical Science Basis. Contribution of Working Group I to the Fifth Assessment Report of the Intergovernmental Panel on Climate Change*. Cambridge University Press, Cambridge, United Kingdom and New York, NY, pp. 741–882.

Fleitmann, D., Burns, S. J., Mudelsee, M., Neff, U., Kramers, J., Mangini, A., Matter, A., 2003. Holocene forcing of the Indian monsoon recorded in a stalagmite from southern Oman. *Science*, 300(5626): 1737–1739.

Fuller, D., 2002. Fifty years of archaeobotanical studies in india: laying a solid foundation. In: Settar, S., Korisettar, R. (eds.), *Indian Archaeology in Retrospect. Archaeology and Interactive Disciplines*, Manohar, Delhi, pp.247–363.

Fuller, D., 2007. Contrasting patterns in crop domestication and domestication rates: recent archaeobotanical insights from the old world.*Annals of Botany*, 100(5): 903–924. 10.1093/aob/mcm048.

Fuller, D., 2012. Pathways to Asian civilizations: tracing the origins and spread of rice and rice cultures. *Rice*, 4(3): 78–92. 10.1007/s12284-011–9078-7.

Fuller, D., Castillo, C., Kingwell-Banham, E., Qin, L., Weisskopf, A., 2018. Charred pomelo peel, historical linguistics and other tree crops: approaches to framing the historical context of early Citrus cultivation in East, South and Southeast Asia. In: Zech-Matterne, V., Fiorentino, G. (eds.), *AGRUMED: Archaeology and History of Citrus Fruit in the Mediterranean Acclimatization, Diversifications, Uses*. Centre Jean Bérard, Naples, pp.31–50.

Fuller, D., Ling, Q., Harvey, E., 2008. Evidence for a late onset of agriculture in the Lower Yangtze Region and challenges for an archaeobotany of rice. In: Sanchez-Mazas, A., Blench, R., Ross, M., Lin, M., Pejros, I. (eds.), *Past Human Migrations in East Asia:*

Matching Genetic, Linguistic and Archaeological Evidence. Taylor and Francis, London, pp.40–83.

Fuller, D., Qin, L., 2009. Water management and labor in the origins and dispersal of Asian rice. *World Archaeology*, 41(1): 88–111.

Fuller, D., Qin, L., 2010. Declining oaks, increasing artistry, and cultivating rice: The environmental and social context of the emergence of farming in the lower Yangtze Region. *Environmental Archaeology*, 15(2): 139–159.

Fuller, D., Qin, L., Zheng, Y., Zhao, Z., Chen, X., Hosoya, L., Sun, G., 2009. The Domestication process and domestication rate in rice: spikelet bases from the Lower Yangtze. *Science*, 323(5921): 1607–1610.

Fuller, D. Q., Castillo, C., Weisskopf, A., 2016. Pathways of rice diversification across Asia. *Archaeology International*, 2016(19): 84–96.

Fuller, D. Q., Harvey, E., Qin, L., 2007. Presumed domestication? Evidence for wild rice cultivation and domestication in the fifth millennium BC of the Lower Yangtze region. *Antiquity*, 81(312): 316–331. DOI:10.1017/S0003598X0009520X.

Fuller, D. Q., Madella, M., 2001. Issues in Harappan archaeobotany: retrospect and prospect. In: Settar, S., Korisettar, R. (eds.), *Indian Archaeology in Retrospect. Protohistory*, vol. II. Manohar Publishers, New Delhi, pp.317–390.

Fuller, D. Q., Van Etten, J., Manning, K., Castillo, C., Kingwell-Banham, E., Weisskopf, A., Qin, L., Sato, Y. -I., Hijmans, R. J., 2011. The contribution of rice agriculture and livestock pastoralism to prehistoric methane levels: An archaeological assessment. *The Holocene*, 21(5): 743–759.

Gadgil, S., 2006. The Indian monsoon:1. *Variations in space and time. Resonance*, 11 (8): 8–21.

Gadgil, S., Gadgil, S., 2006. The Indian monsoon, GDP and agriculture. *Economic and Political Weekly*, 41(47): 4889–4895.

Gadgil, S., Rupa Kumar, K., 2006. *The Asian Monsoon – Agriculture and Economy, The Asian Monsoon*. Springer Berlin Heidelberg, Berlin, Heidelberg, pp. 651–683.

Gagné, K., 2013. Gone with the trees: Deciphering the Thar desert's recurring droughts. *Current Anthropology*, 54(4): 497–509. 10.1086/671074.

Gamble, S. D., 1933. *How Chinese Families Live in Peiping*. Funk and Wagnalls Company, New York and London.

Gansu Sheng Wenwu Kaogu Yanjiusuo, 2006. 秦安大地湾：新石器时代遗址发掘报告 *Qinan Dadiwan: Xinshiqi Shidai Yizhi Fajue Baogao (Qin'an Dadiwan: Excavation Report on the Neolithic Site)*. Wenwu Chubanshe, Beijing.

Gao, C., Robock, A., Ammann, C., 2008. Volcanic forcing of climate over the past 1500 years: An improved ice core-based index for climate models. *Journal of Geophysical Research: Atmospheres*, 113(D23111). DOI:10.1029/2008JD010239.

Gao, L.-z., 2003. The conservation of Chinese rice biodiversity: genetic erosion, ethnobotany and prospects. *Genetic Resources and Crop Evolution*, 50(1): 17–32. DOI:10.1023/A:1022933230689.

Gao, X., Luo, Y., Lin, W., Zhao, Z., Giorgi, F., 2003. Simulation of effects of land use change on climate in China by a regional climate model. *Advances in Atmospheric Sciences*, 20(4): 583–592.

Gardelle, J., Berthier, E., Arnaud, Y., Kääb, A., 2013. Region-wide glacier mass balances over the Pamir-Karakoram-Himalaya during 1999–2011. *Cryosphere*, 7: 1263–1286.

Garris, A. J., Tai, T. H., Coburn, J., Kresovich, S., McCouch, S. R., 2005. Genetic structure and diversity in Oryza sativa L. *Genetics*, 169: 1631–1638.

Gasse, F., 2000. Hydrological changes in the African tropics since the Last Glacial Maximum. *Quaternary Science Reviews*, 19: 189–211.

Gautam, R., Hsu, N. C., Lau, K. M., Tsay, S. C., Kafatos, M., 2009. Enhanced pre-monsoon warming over the Himalayan-Gangetic region from 1979 to 2007. *Geophysical Research Letters*, 36(7): n/a-n/a. 10.1029/2009GL037641.

Ge, B., Huang, C., Zhou, Y., 2010. OSL dating of the Jinghe River palaeoflood events in the late period of the Longshan culture. *Quaternary Sciences*, 2010(30): 422–429.

Gernet, J., 1996. *A History of Chinese Civilization*. Cambridge University Press, Cambridge, England.

Ghose, A. K., 1982. Food supply and starvation: A study of famines with reference to the Indian sub-continent. *Oxford Economic Papers*, 34(2): 368–389.

Ghosh, P., Adkins, J., Affek, H., Balta, B., Guo, W., Schauble, E. A., Schrag, D., Eiler, J. M., 2006. 13C–18O bonds in carbonate minerals: A new kind of paleothermometer. *Geochimica et Cosmochimica Acta*, 70(6): 1439–1456. DOI:10.1016/j.gca.2005.11.014.

Gilmartin, D., 2015. *Blood and Water: The Indus River Basin in Modern History*. University of California Press, Oakland, p. 341.

Giosan, L., Clift, P. D., Macklin, M. G., Fuller, D. Q., Constantinescu, S., Durcan, J. A., Stevens, T., Duller, G. A. T., Tabrez, A., Adhikari, R., Gangal, K., Alizai, A., Filip, F., VanLaningham, S., Syvitski, J. P. M., 2012. Fluvial landscapes of the Harappan civilization. *Proceedings of the National Academy of Sciences*, 109(26): 1688–1694. DOI:10.1073/pnas.1112743109.

Giosan, L., Orsi, W. D., Coolen, M., Wuchter, C., Dunlea, A. G., Thirumalai, K., Munoz, S. E., Clift, P. D., Donnelly, J. P., Galy, V., Fuller, D. Q., 2018. Neoglacial climate anomalies and the Harappan metamorphosis. *Climate of the Past*, 14(11): 1669–1686. DOI:10.5194/cp-14-1669-2018.

Giosan, L., Ponton, C., Usman, M., Blusztajn, J., Fuller, D.Q., Galy, V., Haghipour, N., Johnson, J. E., McIntyre, C., Wacker, L., Eglinton, T. I., 2017. Massive erosion in monsoonal central India linked to late Holocene land cover degradation. *Earth Surface Dynamics*, 5: 781–789. DOI:10.5194/esurf-5-781-2017.

Glennie, K. W., Singhvi, A., 2002. Event stratigraphy, paleoenvironment and chronology of SE Arabian deserts. *Quaternary Science Reviews*, 21: 853–869.

Goebel, T., 1999. Pleistocene human colonization of Siberia and the peopling of the Americas: An ecological approach. *Evolutionary Anthropology*, 8: 208–227.

Goldstein, M. C., Beall, C. M., 1991. Change and continuity in nomadic pastoralism on the western Tibetan Plateau. *Nomadic Peoples*, 28: 105–122.

Gong, D. Y., Ho, C. H., 2002. The Siberian High and climate change over middle to high latitude Asia. *Theoretical and Applied Climatology*, 72: 1–9.

Griffiths, M. L., Drysdale, R. N., Gagan, M. K., Zhao, J. -x., Ayliffe, L. K., Hellstrom, J. C., Hantoro, W. S., Frisia, S., Feng, Y. -x., Cartwright, I., Pierre, E. S., Fischer, M. J., Suwargad, B. W., 2009. Increasing Australian–Indonesian monsoon rainfall linked to early Holocene sea-level rise. *Nature Geoscience*, 2: 636–639. DOI:10.1038/ngeo605.

Grimes, J., 2004. *The Vivekacūḍāmaṇi of Śaṅkarācārya Bhagavatpāda: An Introduction and Translation*. Motilal Banarsidass Publisher, Delhi, p. 284.

Grossman, M. J., Zaiki, M., Nagata, R., 2015. Interannual and interdecadal variations in typhoon tracks around Japan. *International Journal of Climatology*, 35(9): 2514–2527.

Grossman, M. J., Zaiki, M., Oettle, S., 2016. An analysis of typhoon tracks around Japan using ArcGIS. *Papers in Applied Geography*, 2(3): 352–363.

Grove, J. M., 2012. *The Little Ice Age*. Routledge, Abingdon, UK.

Grove, R.H., 2007. The great El Niño of 1789–93 and its global consequences: Reconstructing an extreme climate event in world environmental history. *Medieval History*, 10: 75–98.

Gunnell, Y., Anupama, K., Sultan, B., 2007. Response of the south Indian runoff-harvesting civilization to northeast monsoon rainfall variability during the last 2000 years: instrumental records and indirect evidence. *Holocene*, 17(2): 207–215. DOI:10.1177/0959683607075835.

Guo, D. S., 1995. Hongshan and related cultures. In: Nelson, S. M. (ed.), *The Archaeology of Northeast China: Beyond the Great Wall*. Routledge, New York, pp. 21–64.

Guo, Q. Y., 1996. Climatic change and East Asian Monsoon. In: Shi, Y. F. (ed.), *Historical Climatic Changes of China (1): Climatic and Sea Level Change and Their Trend and Impact*. Shandong Science and Technology Press, Jinan, pp. 468–483 (in Chinese).

Guo, X., Fu, D., Guo, X., Zhang, C., 2014. A case study of aerosol impacts on summer convective clouds and precipitation over northern China. *Atmospheric Research*, 142: 142–157. DOI:10.1016/j.atmosres.2013.10.006.

Gupta, A. K., Anderson, D. M., Overpeck, J. T., 2003. Abrupt changes in the Asian southwest monsoon during the Holocene and their links to the North Atlantic Ocean. *Nature*, 421: 354–356.

Gupta, A. K., Das, M., Anderson, D. M., 2005. Solar influence on the Indian summer monsoon during the Holocene. *Geophysical Research Letters*, 32(L17703). DOI:10.1029/2005GL022685.

Gupta, A. K., Yuvaraja, A., Prakasam, M., Clemens, S. C., Velu, A., 2015. Evolution of the South Asian monsoon wind system since the late middle miocene. *Palaeogeography, Palaeoclimatology, Palaeoecology*, 438: 160–167. DOI:10.1016/j.palaeo.2015.08.006.

Gupta, H., Kao, S. -J., Dai, M., 2012. The role of mega dams in reducing sediment fluxes: A case study of large Asian rivers. *Journal of Hydrology*, 464: 447–458.

Gutaker, R. M., Groen, S. C., Bellis, E. S., Choi, J. Y., Pires, I. S., Bocinsky, R. K., Slayton, E. R., Wilkins, O., Castillo, C. C., Negrão, S., Oliveira, M. M., Fuller, D. Q., d'Alpoim Guedes, J. A., Lasky, J. R., Purugganan, M. D., 2019. Genomic history and ecology of the geographic spread of rice. *BioRxiv*: 748178. DOI:10.1101/748178.

Habib, I., 1999. *The Agrarian System of Mughal India: 1556–1707*. Oxford University Press, New Delhi.

Habu, J., 2008. Growth and decline in complex hunter-gatherer societies: A case study from the Jomon period Sannai Maruyama site, Japan. *Antiquity*, 82(317): 571–584.

Hairong, D., Yong, L., Chongjian, S., Svirchev, L., Qiang, X., Zhaokun, Y., Liang, Y., Shijun, N., Zeming, S., 2017. Mechanism of post-seismic floods after the Wenchuan earthquake in the upper Minjiang River, China. *Journal of Earth System Science*, 126 (7): 96. DOI:10.1007/s12040-017-0871-6.

Ham, Y. -G., Kug, J. -S., Choi, J. -Y., Jin, F. -F., Watanabe, M., 2018. Inverse relationship between present-day tropical precipitation and its sensitivity to greenhouse warming. *Nature Climate Change*, 8(1): 64–69. DOI:10.1038/s41558-017-0033-5.

Han, Z., Wu, L., Ran, Y., Ye, Y., 2003. The concealed active tectonics and their character-istics as revealed by drainage density in the North China plain (NCP). *Journal of Asian Earth Sciences*, 21(9): 989–998. DOI:0.1016/S1367-9120(02)00175-X.

Hanebuth, T., Stattegger, K., Grootes, P. M., 2000. Rapid flooding of the Sunda Shelf: a late-glacial sea-level record. *Science*, 288(5468): 1033–1035.

Hao, S., Xue, J., Cui, J., 2008a. 董虎林四好人墓葬中的果壳 Donghulin Sihaoren Muzangzhong de Guoke (Seeds Uncovered from Burial no. 4 at Donghulin). *Renleixue Xuebao*, 27: 249–255.

Hao, Z., Zheng, J., Ge, Q., 2008b. Precipitation cycles in the middle and lower reaches of the Yellow River (1736–2000). *Journal of Geographical Sciences*, 18(1): 17–25. 10.1007/s11442-008–0017-5.

Harinarayana, G., 1987. *Pearl Millet in Indian Agriculture, Proceedings of the International Pearl Millet Workshop, 7–11 April 1986, ICRISAT Center, Patancheru, India*. Patancheru. Research Institute for the Semi-Arid Tropics., Andhra Pradesh, India, pp. 5–17.

Harris, N. B. W., 2006. The elevation of the Tibetan Plateau and its impact on the monsoon. *Palaeogeography Palaeoclimatology Palaeoecology*, 241: 4–15.

Hayden, B., 2011. *Rice: The First Luxury Asian Food?, Why Cultivate? Anthropological and Archaeological Approaches to Foraging/Farming Transitions in Southeast Asia*. McDonald Institute Monographs, Cambridge, pp.75–93.

Held, I. M., Soden, B .J., 2006. Robust responses of the hydrological cycle to global warming. *Journal of Climate*, 19: 5686–5699.

Henan Sheng Wenwu Kaogu Yanjiusuo, 1999. 舞阳贾湖 *Wuyang Jiahu (Report on the site of Jiahu at Wuyang)*. Kexue Chubanshe, Beijing.

Henan Working Team no. 1 of IA CASS, 1983. 河南新郑沙窝里新石器时代遗址 Henan Xinzheng Shawoli Shidai Yizhi (Excavation of the Neolithic Site at Shawoli). *Kaogu*, 195: 1057–1065.

Henan Working Team No. 1 of IA CASS, 1984. Peiligang Yizhi Fajue Baogao (Excavation of the Neolithic Site at Peiligang). *Kaogu Xuebao*, 72: 23–52.

Herzschuh, U., 2006. Palaeo-moisture evolution in monsoonal Central Asia during the last 50,000 years. *Quaternary Science Reviews*, 25(1–2): 163–178.

Herzschuh, U., Tarasov, P., Wqnnemann, B., Hartmann, K., 2004. Holocene vegetation and climate of the Alashan Plateau, NW China, reconstructed from pollen data. *Palaeogeography Palaeoclimatology Palaeoecology*, 211: 1–17.

Herzschuh, U., Winter, K., Wünnemann, B., Li, S., 2006. A general cooling trend on the central Tibetan Plateau throughout the Holocene recorded by the Lake Zigetang pollen spectra. *Quaternary International*, 154–155: 113–121. DOI:10.1016/j. quaint.2006.02.005.

Hewitt, K., 2011. Glacier change, concentration, and elevation effects in the Karakoram Himalaya, upper Indus Basin. *Mt. Res. Dev. . Mountain Research and Development*, 31: 188–200.

Hicks, M. J., Burton, M. L., 2010. *Preliminary Damage Estimates for Pakistani Flood Events, 2010*, Ball State University, Muncie, IN.

Higham, C., 2002. *Early Cultures of Mainland Southeast Asia*. River Books, Bangkok, 375 pp.

Higham, C., 2005. East Asian agriculture and its impact. In: Scarre, C. (ed.), *The Human Past. World Prehistory and the Development of Human Societies*. Thames and Hudson, London, pp. 234–263.

Higham, C., 2013. Hunter-gatherers in southeast Asia: from prehistory to the present. *Human Biology*, 85(1): 21–44.

Higham, C. F. W., Xie, G., Lin, Q., 2011. The prehistory of a Friction Zone: first farmers and hunters-gatherers in Southeast Asia. *Antiquity*, 85: 529–543.

Hijioka, Y., Lin, E., Pereira, J. J., Corlett, R. T., Cui, X., Insarov, G. E., Lasco, R. D., Lindgren, E., Surjan, A., 2014. Asia. In: Barros, V. R. et al. (eds.), *Climate Change 2014: Impacts, Adaptation, and Vulnerability. Part B: Regional Aspects. Contribution of Working Group II to the Fifth Assessment Report of the Intergovernmental Panel on Climate Change*. Cambridge University Press, Cambridge UK, pp.1327–1370.

Hijma, M. P., Cohen, K. M., 2010. Timing and magnitude of the sea-level jump preluding the 8.2 kiloyear event. *Geology*, 38(3): 275–278. DOI:10.1130/G30439.1.

Hill, C. V., 2008. *South Asia: An Environmental History*. ABC-CLIO, Santa Barbara, California, p. 329.

Ho, C. -H., Baik, J. -J., Kim, J. -H., Gong, D. -Y., Sui, C. -H., 2004. Interdecadal Changes in Summertime Typhoon Tracks. *Journal of Climate*, 17(9): 1767–1776. 10.1175/1520–0442(2004)017<767:icistt>2.0.Co;2.

Ho, P.-t., 1959. Aspects of social mobility in China, 1368–1911. *Comparative Studies in Society and History*, 1(4): 330–359.

Hodges, K., 2003. Geochronology and thermochronology in orogenic systems. In: Rudnick, R. (ed.), *The Crust*. Elsevier-Science, Amsterdam, pp.263–292.

Hoffmann, T., Penny, D., Stinchcomb, G., Vanacker, V., Lu, X., 2015. Global soil and sediment transfers in the anthropocene. *Pages Magazine*, 23(1): 37.

Hosner, D., Wagner, M., Tarasov, P. E., Chen, X., Leipe, C., 2016. Spatiotemporal distribution patterns of archaeological sites in China during the Neolithic and Bronze Age: An overview. *The Holocene*, 26(10): 1576–1593.

Hossain, M. A., 2010. *Global Warming Induced Sea Level Rise on Soil, Land and Crop Production Loss, 19th World Congress of Soil Science, Soil Solutions for a Changing World*, Brisbane, Australia.

Hsu, R. C., 1982. *Food for One Billion: China's Agriculture Since 1949*. Routledge, New York, p. 156.

Hu, C., Henderson, G. M., Huang, J., Xie, S., Sun, Y., Johnson, K. R., 2008. Quantification of Holocene Asian monsoon rainfall from spatially separated cave records. *Earth and Planetary Science Letters*, 266: 221–232. DOI:10.1016/j.epsl.2007.10.015.

Hu, D., Böning, P., Köhler, C. M., Hillier, S., Pressling, N., Wan, S., Brumsack, H. -J., Clift, P. D., 2012. Deep sea records of the continental weathering and erosion response to East Asian monsoon intensification since 14ka in the South China Sea. *Chemical Geology*, 326–327: 1–18. DOI:10.1016/j.chemgeo.2012.07.024.

Hu, D., Clift, P. D., Böning, P., Hannigan, R., Hillier, S., Blusztajn, J., Wang, S., Fuller, D. Q., 2013a. Holocene evolution in weathering and erosion patterns in the Pearl River delta. *Geochemistry Geophysics Geosystems*, 14: 2349–2368. DOI:10.1002/ggge.20166.

Hu, L., Chao, Z., Gu, M., Li, F., Chen, L., Liu, B., Li, X., Huang, Z., Li, Y., Xing, B., Dai, J., 2013b. Evidence for a Neolithic Age fire-irrigation paddy cultivation system in the lower Yangtze River Delta, China. *Journal of Archaeological Science*, 40(1): 72–78. DOI:10.1016/j.jas.2012.04.021.

Huang, C., Pang, J., Zha, X., 2011. Extraordinary floods related to the climatic event at 4200 a BP on the Qishuihe River, middle reaches of the Yellow River, China. *Quaternary Science Reviews*, 2011(30): 460–468.

Huang, C.C., Zhou, Y., Zhang, Y., Guo, Y., Pang, J., Zhou, Q., Liu, T., Zha, X., 2017. Comment on "Outburst flood at 1920 BCE supports historicity of China's Great Flood and the Xia dynasty". *Science*, 355(6332): 1382.

Huang, H., Yan, Z., 2009. Present situation and future prospect of hydropower in China. *Renewable and Sustainable Energy Reviews*, 13(6): 1652–1656. DOI:10.1016/j. rser.2008.08.013.

Huang, P., Schaal, B. A., 2012. Association between the geographic distribution during the last glacial maximum of Asian wild rice, Oryza rufipogon (Poaceae), and its current genetic variation. *American Journal of Botany*, 99(11): 1866–1874.

Huang, R., Fan, X., 2013. The landslide story. *Nature Geoscience*, 6: 325. DOI:10.1038/ ngeo1806.

Hubei Sheng Wenwu Guanlisuo, 1981. 武安磁山Wu'an Cishan (The site of Cishan in Wu'an). *Kaogu Xuebao*, 1981(3): 303–338.

Huber, M., Goldner, A., 2012. Eocene monsoons. *Journal of Asian Earth Sciences*, 44: 3–23. DOI:10.1016/j.jseaes.2011.09.014.

Hunan Sheng Wenwu Kaogu Yanjiusuo, 2006. 彭头山与巴士垱 *Pengtoushan yu Bashidang (Pengtoushan and Bashidang)*. Science Press, Beijing.

Hunt, K. M. R., Curio, J., Turner, A. G., Schiemann, R., 2018. Subtropical westerly jet influence on occurrence of western disturbances and Tibetan Plateau Vortices. *Geophysical Research Letters*, 45(16): 8629–8636. DOI:10.1029/2018GL077734.

Huntington, K., Wernicke, B., Eiler, J., 2010. Influence of climate change and uplift on Colorado Plateau paleotemperatures from carbonate clumped isotope thermometry. *Tectonics*, 29(3): TC3005. DOI:10.1029/2009TC002449

Hurrell, J. W., 1995. Decadal trends in the North Atlantic Oscillation: Regional temperatures and precipitation. *Science*, 269(5224): 676–679. DOI:10.1126/science.269.5224.676.

Hussain, S. M., Paperno, R., Khatoon, Z., 2010. Length–weight relationships of fishes collected from the Korangi-Phitti Creek area (Indus delta, northern Arabian Sea). *Journal of Applied Ichthyology*, 26(3): 477–480. DOI:10.1111/j.14390426.2009.01374.x.

Hussain, S. S., Mudasser, M., 2007. Prospects for wheat production under changing climate in mountain areas of Pakistan – An econometric analysis. *Agricultural Systems*, 94(2): 494–501. DOI:10.1016/j.agsy.2006.12.001.

Ikawa-Smith, F., 2004. Humans along the Pacific Margin of North East Asia before the last glacial maximum. In: Madsen, D. B. (ed.), *Entering America: Northeast Asia and Beringia before the Last Glacial Maximum*. University of Utah Press, Salt Lake City, Utah, pp. 287–309.

Immerzeel, W. W., van Beek, L. P., Bierkens, M. F., 2010. Climate change will affect the Asian water towers. *Science*, 328(5984): 1382–1385.

Inam, A., Clift, P. D., Giosan, L., Tabrez, A. R., Tahir, M., Rabbani, M. M., Danish, M., 2007. The geographic, geological and oceanographic setting of the Indus River. In: Gupta, A. (ed.), *Large Rivers: Geomorphology and Management*. John Wiley and Sons, Chichester, UK, pp. 333–345.

Ingraham, N. L., Taylor, B. E., 1991. Light stable isotope systematics of large-scale hydrologic regimes in California and Nevada. *Water Resources Research*, 27: 77–90.

IPCC, 2007. *Contribution of Working Group I to the Fourth Assessment Report of the Intergovernmental Panel on Climate Change*, Cambridge University Press, Cambridge.

Iqbal, M. M., Goheer, A. M., Khan, A. M., 2010. Climate-change aspersions on food security of Pakistan. *Science Vision*, 15: 15–23.

Ivanochko, T. S., Ganeshram, R. S., Brummer, G. -J. A., Ganssen, G., Jung, S. J. A., Moreton, S. G., Kroon, D., 2005. Variations in tropical convection as an amplifier of global climate change at the millennial scale. *Earth and Planetary Science Letters*, 235(1–2): 302–314.

Jackson, S., Sleigh, A., 2000. Resettlement for China's Three Gorges Dam: socio-economic impact and institutional tensions. *Communist and Post-Communist Studies*, 33(2): 223–241. DOI:10.1016/S0967-067X(00)00005-2.

Jacques, F. M. B., Su, T., Spicer, R. A., Xing, Y., Huang, Y., Wang, W., Zhou, Z., 2011. Leaf physiognomy and climate: Are monsoon systems different? *Global and Planetary Change*, 76(1): 56–62. DOI:10.1016/j.gloplacha.2010.11.009.

Jacquot, M., Courtois, B., 1987. *Upland Rice*. The Tropical Agriculturalist. CTA, Paris.

Jain, V., Kumar, R., Kaushal, R. K., Gautam, T., Singh, S., 2018. The dynamic Kosi River and its tributaries. In: Singh, D. S. (ed.), *The Indian Rivers*. Springer, pp. 221–237.

James, E. W., Banner, J. L., Hardt, B., 2015. A global model for cave ventilation and seasonal bias in speleothem paleoclimate records. *Geochemistry, Geophysics, Geosystems*, 16(4): 1044–1051. DOI:10.1002/2014GC005658.

Jarrige, C., Jarrige, J. -F., Meadow, R. H., Quivron, G., 1995. *Mehrgarh: Field Reports 1974–1985, from Neolithic Times to the Indus Civilization*. Department of Culture and Tourism of Sindh, Karachi.

Jenny, Jean-Philippe et al., 2019. Human and climate global-scale imprint on sediment transfer during the Holocene. *Proceedings of the National Academy of Sciences*, 116 (46): 22972.

Jian, Z. M., Li, B. H., Pflaumann, U., Wang, P. X., 1996. Late Holocene cooling event in the western Pacific. *Science in China (Series D)*, 39: 542–550.

Jiang, L., Li, L., 2006. New evidence for the origins of sedentism and rice domestication in the Lower Yangzi River, China. *Antiquity*, 80: 355–361.

Jiang, T., Su, B., Hartmann, H., 2007. Temporal and spatial trends of precipitation and river flow in the Yangtze River Basin, 1961–2000. *Geomorphology*, 85(3): 143–154. DOI:10.1016/j.geomorph.2006.03.015.

Joint Typhoon Warning Center, 2002. *The Joint Typhoon Warning Center Tropical Cyclone Best-Tracks, 1945–2000*. US Navy, www.usno.navy.mil/JTWC/.

Jones, H., Civáň, P., Cockram, J., Leigh, F. J., Smith, L. M. J., Jones, M. K., Charles, M. P., Molina-Cano, J. -L., Powell, W., Jones, G., Brown, T. A., 2011a. Evolutionary history of barley cultivation in Europe revealed by genetic analysis of extant landraces. *BMC Evolutionary Biology*, 11(1): 320. 10.1186/1471–2148-11–320.

Jones, M., Hunt, H., Lightfoot, E., Lister, D., Liu, X., Motuzaite-Matuzeviciute, G., 2011b. Food globalization in prehistory. *World Archaeology*, 43(4): 665–675. DOI:10.1080/00438243.2011.624764.

Jordan, P., Zvelebil, M., 2009. *Ceramics Before Farming: The Dispersal of Pottery Among Prehistoric Eurasian Hunter-Gatherers, 53*. Left Coast Press, Walnut Creek, CA.

Joseph, P. V., Simon, A., 2005. Weakening trend of the southwest monsoon current through peninsular India from 1950 to the present. *Current Science*, 89: 687–694.

Joseph, S., Sahai, A. K., Goswami, B.N., Terray, P., Masson, S., Luo, J. J., 2012. Possible role of warm SST bias in the simulation of boreal summer monsoon in SINTEX-F2 coupled model. *Climate Dynamics*, 38: 1561–1576.

Kaleemullah, S., 2001. Thermogravimetric analysis of paddy straw. *Madras Agricultural Journal*, 88(10/12): 582–584.

Kamkar, B., Koocheki, A., Nassiri Mahallati, M., Rezvani Moghaddam, P., 2006. Cardinal temperatures for germination in three millet species (Panicum miliaceum, Pennisetum glaucum and Setaria italica). *Asian Journal of Plant Sciences*, 5: 316–319.

Kang, N. -Y., Elsner, J. B., 2016. Climate mechanism for stronger typhoons in a warmer world. *Journal of Climate*, 29(3): 1051–1057. 10.1175/jcli-d-15–0585.1.

Kang, Y., Khan, S., Ma, X., 2009. Climate change impacts on crop yield, crop water productivity and food security – A review. *Progress in Natural Science*, 19(12): 1665–1674. DOI:10.1016/j.pnsc.2009.08.001.

Karim, A., Veizer, J., 2002. Water balance of the Indus river basin and moisture source in the Karakoram and western Himalayas: implications from hydrogen and oxygen isotopes river water. *Journal of Geophysical Research*, 107(D18): 4362. DOI:10.1029/2000JD000253.

Kaspari, S., Hooke, R. L., Mayewski, P. A., Kang, S. C., Hou, S. G., Qin, D. H., 2008. Snow accumulation rate on Qomolangma (Mount Everest), Himalaya: Synchroneity with sites across the Tibetan plateau on 50–100 year timescales. *Journal of Glaciology*, 54: 343–352. DOI:10.3189/002214308784886126.

Kathayat, G., Cheng, H., Sinha, A., Spötl, C., Edwards, R. L., Zhang, H., Li, X., Yi, L., Ning, Y., Cai, Y., Lui, W. L., Breitenbach, S. F. M., 2016. Indian monsoon variability on millennial-orbital timescales. *Scientific Reports*, 6(1): 24374. DOI:10.1038/srep24374.

Katz, A., 1973. The interaction of magnesium with calcite during crystal growth at 25°C–90°C and one atmosphere. *Geochimica et Cosmochimica Acta*, 37(6): 1563–1586. DOI:10.1016/0016-7037(73)90091-4.

Keally, C. T., Taniguchi, Y., Kuzmin, Y. V., Shewkomud, I. Y., 2004. Chronology of the beginning of pottery manufacture in East Asia. *Radiocarbon*, 46(1): 345–351.

Kendall, R. A., Mitrovica, J. X., Milne, G. A., Törnqvist, T. E., Li, Y., 2008. The sea-level fingerprint of the 8.2 ka climate event. *Geology*, 36(5): 423–426. DOI:10.1130/G24550A.1.

Kennett, J. P., Shackleton, N. J., 1976. Oxygen isotopic evidence for the development of the psychrosphere 38 Myr ago. *Nature*, 260: 513–515.

Kenoyer, J. M., 1997. Trade and technology of the Indus valley: New insights from Harappa, Pakistan. *World Archaeology*, 29(2): 262–280. 10.1080/00438243.1997.9980377.

Kenoyer, J. M., 1998. *Ancient Cities of the Indus Valley Civilization*. American Institute of Pakistan Studies, Madison, WI.

Kent, A., Behrman, S., 2018. *Facilitating the Resettlement and Rights of Climate Refugees: An Argument for Developing Existing Principles and Practices*. Routledge, London, p. 190.

Khalid, B., Cholaw, B., Alvim, D. S., Javeed, S., Khan, J. A., Javed, M. A., Khan, A. H., 2018. Riverine flood assessment in Jhang district in connection with ENSO and summer monsoon rainfall over Upper Indus Basin for 2010. *Natural Hazards*, 92(2): 971–993. DOI:10.1007/s11069-018-3234-y.

Khan, N. M., Tingsanchali, T., 2009. Optimization and simulation of reservoir operation with sediment evacuation: a case study of the Tarbela Dam, Pakistan. *Hydrological Processes*, 23(5): 730–747. 10.1002/hyp.7173.

Khawaja, B. A., Sanchez, M., 2009. *Tarbela Dam: A Numerical Model for Sediment Management in the Reservoir, Coastal and Maritime Mediterranean Conference*. Centre Français du Littoral, Tunisia.

Kidder, T., Liu, H., Xu, Q., Li, M., 2012. The alluvial geoarchaeology of the Sanyangzhuang site on the Yellow River floodplain, Henan Province, China. *Geoarchaeology*, 27(4): 324–343.

Kidder, T. R., Liu, H., 2014. Bridging theoretical gaps in geoarchaeology: archaeology, geoarchaeology, and history in the Yellow River valley, China. *Archaeological and Anthropological Sciences*, 9: 1585–1602. 10.1007/s12520-014-0184-5.

Kidwai, S., Ahmed, W., Tabrez, S. M., Zhang, J., Giosan, L., Clift, P., Inam, A., 2019. The Indus delta—catchment, river, coast, and people. In: Wolanski, E., Day, J. W., Elliiott, M., Ramachandran, R. (eds.), *Coasts and Estuaries: the Future*. Elsevier, Amsterdam, pp.213–232.

Kim, H.-J., Takata, K., Wang, B., Watanabe, M., Kimoto, M., Yokohata, T., Yasunari, T., 2011. Global monsoon, El Nino, and their interannual linkage simulated by MIROC5 and the CMIP3 CGCMs. *Journal of Climate*, 24: 5604–5618.

Kim, H. J., Wang, B., Ding, Q. H., 2008. The global monsoon variability simulated by CMIP3 coupled climate models. *Journal of Climate*, 21: 5271–5294.

Kim, J. H., Wu, C. C., Sui, C. H., Ho, C. H., 2012. Tropical cyclone contribution to interdecadal change in summer rainfall over South China in the early 1990s. *Terrestrial, Atmospheric and Oceanic Sciences*, 23: 49–58. DOI:10.3319/TAO.2011.08.26.01(A).

Kim, M. -K., Lau, W. K. M., Kim, K. -M., Sang, J., Kim, Y. -H., Lee, W. -S., 2016. Amplification of ENSO effects on Indian summer monsoon by absorbing aerosols. *Climate Dynamics*, 46(7): 2657–2671. DOI:10.1007/s00382-015-2722-y.

Kirsch, T. D., Wadhwani, C., Sauer, L., Doocy, S., Catlett, C., 2012. Impact of the 2010 Pakistan floods on rural and urban populations at six months. *PLoS Currents*, 4: e4fdfb212d2432. 10.1371/4fdfb212d2432.

Kitamoto, A., 2011. *Digital Typhoon*. National Institute of Informatics, www.digital-typhoon.org/.

Kleiven, H. F., Kissel, C., Laj, C., Ninnemann, U. S., Richter, T. O., Cortijo, E., 2008. Reduced North Atlantic Deep Water coeval with the glacial Lake Agassiz freshwater outburst. *Science*, 319: 60–64.

Knutson, T. R., McBride, J. L., Chan, J., Emanuel, K., Holland, G., Landsea, C., Held, I., Kossin, J. P., Srivastava, A. K., Sugi, M., 2010. Tropical cyclones and climate change. *Nature Geoscience*, 3: 157. DOI:10.1038/ngeo779.

Kodama, D., 2003. Komakino stone circle and its significance for the study of Jomon social structure In: Habu, J., Savelle, J.M., Koyama, S., Hongo, H. (eds.), *Hunter-gatherers of the North Pacific Rim (Senri Ethnological Studies 63)*. National Museum of Ethnological Studies, Osaka, pp.235–261.

Kohler, T. A., Johnson, C. D., Varien, M., Ortman, S., Reynolds, R., Kobti, Z., Cowan, J., Kolm, K., Smith, S., Yap, L., 2007. Settlement ecodynamics in the prehispanic central Mesa Verde region. In: Kohler, T.A., Leeuw, S.v.d. (eds.), *The Model-Based Archaeology of Socionatural Systems*. School for Advanced Research Press, Santa Fe, NM, pp.61–104.

Kondolf, G., Rubin, Z., Minear, J., 2014. Dams on the Mekong: Cumulative sediment starvation. *Water Resources Research*, 50(6): 5158–5169.

Kosaka, Y., Xie, S. -P., 2013. Recent global-warming hiatus tied to equatorial Pacific surface cooling. *Nature*, 501: 403–407. DOI:10.1038/nature12534.

Kovach, M. J., Sweeney, M. T., McCouch, S. R., 2007. New insights into the history of rice domestication. *TRENDS in Genetics*, 23(11): 578–587.

Krishna Kumar, K., Rupa Kumar, K., Ashrit, R. G., Deshpande, N. R., Hansen, J. W., 2004. Climate impacts on Indian agriculture. *International Journal of Climatology*, 24(11): 1375–1393. 10.1002/joc.1081.

Krishnamurthy, V., Goswami, B. N., 2000. Indian monsoon–ENSO relationship on inter-decadal timescale. *Journal of Climate*, 13(3): 579–595. 10.1175/1520–0442(2000) 013<0579:imeroi>2.0.Co;2.

Krishnan, R., Ramesh, K. V., Samala, B. K., Meyers, G., Slingo, J. M., Fennessy, M. J., 2006. Indian Ocean-monsoon coupled interactions and impending monsoon droughts. *Geophysical Research Letters*, 33(L08711). DOI:10.1029/2006GL025811.

Krishnaswamy, J., Vaidyanathan, S., Rajagopalan, B., Bonell, M., Sankaran, M., Bhalla, R., Badiger, S., 2015. Non-stationary and non-linear influence of ENSO and Indian ocean dipole on the variability of Indian monsoon rainfall and extreme rain events. *Climate Dynamics*, 45(1–2): 175–184.

Kroon, D., Steens, T., Troelstra, S. R., 1991. Onset of Monsoonal related upwelling in the western Arabian Sea as revealed by planktonic foraminifers. In: Prell, W., Niitsuma, N. (eds.), *Proceedings of the Ocean Drilling Program, Scientific Results*. Ocean Drilling Program, College Station, TX, pp.257–263.

Kühn, T., Partanen, A. I., Laakso, A., Lu, Z., Bergman, T., Mikkonen, S., Kokkola, H., Korhonen, H., Räisänen, P., Streets, D.G., Romakkaniemi, S., Laaksonen, A., 2014. Climate impacts of changing aerosol emissions since 1996. *Geophysical Research Letters*, 41(13): 4711–4718. DOI:10.1002/2014GL060349.

Kulp, S. A., Strauss, B. H., 2019. New elevation data triple estimates of global vulnerability to sea-level rise and coastal flooding. *Nature Communications*, 10(1): 4844. DOI:10.1038/s41467-019-12808-z.

Kumar, K. K., Rajagopalan, B., Cane, M. A., 1999. On the weakening relationship between the Indian monsoon and ENSO. *Science*, 284(5423): 2156–2159. DOI:10.1126/science.284.5423.2156.

Kumar, K. K., Rajagopalan, B., Hoerling, M., Bates, G., Cane, M., 2006. Unraveling the mystery of Indian monsoon failure during El Niño. *Science*, 314(5796): 115–119 DOI:10.1126/science.1131152.

Kumar, R., Mishra, V., Buzan, J., Kumar, R., Shindell, D., Huber, M., 2017. Dominant control of agriculture and irrigation on urban heat island in India. *Scientific Reports*, 7 (1): 14054. DOI:10.1038/s41598-017-14213-2.

Kummerow, C., Barnes, W., Kozu, T., Shiue, J., Simpson, J., 1998. The tropical rainfall measuring mission (TRMM) sensor package. *Journal of Atmospherc and Oceanic Technology*, 15: 809–817.

Kurz, J., 2011. *China's Southern Tang Dynasty, 937–976*. Routledge, London, p. 160.

Kutzbach, J. E., 1981. Monsoon climate of the early Holocene – climate experiment with the Earths orbital parameters for 9000 years ago. *Science*, 214(4516): 59–61.

Kuzmin, Y., 2006a. Chronology of the earliest pottery in East Asia: progress and pitfalls. *Antiquity*, 80(2006): 362–371.

Kuzmin, Y. V., 2006b. Chronology of the earliest pottery in East Asia: progress and pitfalls. *Antiquity*, 80(308): 362–371.

Kuzmin, Y. V., 2010. The origin of pottery in East Asia and its relationship to environmental changes in the late glacial. *Radiocarbon*, 52(2–3): 415–420.

Lansing, J. S., Fox, K. M., 2011. Niche construction on Bali: the gods of the countryside. *Philosophical Transactions of the Royal Society B: Biological Sciences*, 366(1566): 927–934. DOI:10.1098/rstb.2010.0308.

Lambeck, K., Rouby, H., Purcell, A., Sun, Y., Sambridge, M., 2014. Sea level and global ice volumes from the last glacial maximum to the holocene. *Proceedings of the National Academy of Sciences*, 111(43): 15296–15303. 10.1073/pnas.1411762111.

Lansing, J. S., Kremer, J. N., 2011. Rice, fish, and the planet. *Proceedings of the National Academy of Sciences*, 108(50): 19841–19842.

Lau, K. M., Kim, K. M., 2006. Observational relationships between aerosol and Asian monsoon rainfall, and circulation. *Geophysical Research Letters*, 33: L21810. DOI:10.1029/2006GL027546.

Lau, N. C., Wang, B., 2005. Monsoon–ENSO interactions. *The Global Monsoon System: Research and Forecast. WMO Technical Document 1266 and TMRP Report*, 70: 299–309.

Lau, W. K. M., Kim, K. M., 2010. Fingerprinting the impacts of aerosols on long-term trends of the Indian summer monsoon regional rainfall. *Geophysical Research Letters*, 37(L16705). DOI:10.1029/2010GL043255.

Laufer, B., 1907. *The Introduction of Maize into Eastern Asia*. Dussault & Proulx, Québec City.

Lawson, P., 2014. *The East India Company: A History*. Routledge, London, 200 pp.

Lear, C. H., Rosenthal, Y., Slowey, N., 2002. Benthic foraminiferal Mg/Ca-paleothermometry: a revised core-top calibration. *Geochimica et Cosmochimica Acta*, 66(19): 3375–3387. DOI:10.1016/S0016-7037(02)00941-9.

Lee, G. -A., 2011. The transition from foraging to farming in prehistoric korea. *Current Anthropology*, 52(S4): S307–S329. DOI:10.1086/658488.

Lee, G. -A., Crawford, G. W., Liu, L., Sasaki, Y., Chen, X., 2011. Archaeological soybean (Glycine max) in East Asia: does size matter? *PLOS ONE*, 6(11): e26720. 10.1371/journal.pone.0026720.

Lee, J., 1982. Food supply and population growth in southwest China, 1250–1850. *The Journal of Asian Studies*, 41(4): 711–747. DOI:10.2307/2055447.

Lee, J., Feng, W., 1999. Malthusian models and Chinese realities: The Chinese demographic system 1700–2000. *Population and Development Review*, 25(1): 33–65.

Lemordant, L., Gentine, P., Swann, A. S., Cook, B. I., Scheff, J., 2018. Critical impact of vegetation physiology on the continental hydrologic cycle in response to increasing CO2. *Proceedings of the National Academy of Sciences*, 115(16): 4093–4098. DOI:10.1073/pnas.1720712115.

Leng, M. J., Barker, P. A., 2006. A review of the oxygen isotope composition of lacustrine diatom silica for palaeoclimate reconstruction. *Earth-Science Reviews*, 75(1–4): 5–27.

Leshnik, L. S., 1973. Land use and ecological factors in prehistoric Northwest India. In: Hammond, N. (ed.), *South Asian Archaeology*. Duckworth, London, pp.67–84.

Leung, Y. K., Wu, M. C., Yeung, K. K., 2007. *Recent Decline in Typhoon Activity in the South China Sea, International Conference on Climate Change*. Hong Kong Observatory, Hong Kong, China.

Levine, R. C., Turner, A. G., 2012. Dependence of Indian monsoon rainfall on moisture fluxes across the Arabian Sea and the impact of coupled model sea surface temperature biases. *Climate Dynamics*, 38: 2167–2190.

Lewandowsky, S., Risbey, J. S., Oreskes, N., 2016. The "pause" in global warming: Turning a routine fluctuation into a problem for science. *Bulletin of the American Meteorological Society*, 97(5): 723–733.

Li, C., McLinden, C., Fioletov, V., Krotkov, N., Carn, S., Joiner, J., Streets, D., He, H., Ren, X., Li, Z., Dickerson, R. R., 2017a. India is overtaking China as the world's largest emitter of anthropogenic sulfur dioxide. *Scientific Reports*, 7(1): 14304. DOI:10.1038/s41598-017-14639-8.

Li, H. M., Dai, A. G., Zhou, T. J., Lu, J., 2010. Responses of East Asian summer monsoon to historical SST and atmospheric forcing during 1950–2000. *Climate Dynamics*, 34: 501–514.

Li, J., Liu, Z., He, C., Yue, H., Gou, S., 2017b. Water shortages raised a legitimate concern over the sustainable development of the drylands of northern China: Evidence from the water stress index. *Science of The Total Environment*, 590–591: 739–750. DOI:10.1016/j.scitotenv.2017.03.037.

Li, K., Xu, Z., 2006. Overview of Dujiangyan irrigation scheme of ancient China with current theory. *Irrigation and Drainage*, 55(3): 291–298. DOI:10.1002/ird.234.

Li, X., Dodson, J., Zhou, J., Zhou, X., 2009. Increases of population and expansion of rice agriculture in Asia, and anthropogenic methane emissions since 5000 BP. *Quaternary International*, 202(1–2): 41–50.

Li, X., Ting, M., Li, C., Henderson, N., 2015. Mechanisms of Asian summer monsoon changes in response to anthropogenic forcing in CMIP5 models. *Journal of Climate*, 28(10): 4107–4125. 10.1175/jcli-d-14–00559.1.

Li, X., Zhou, Y., Asrar, G. R., Imhoff, M., Li, X., 2017c. The surface urban heat island response to urban expansion: A panel analysis for the conterminous United States. *Science of the Total Environment*, 605: 426–435.

Li, Z., Lau, W. K. -M., Ramanathan, V., Wu, G., Ding, Y., Manoj, M. G., Liu, J., Qian, Y., Li, J., Zhou, T., Fan, J., Rosenfeld, D., Ming, Y., Wang, Y., Huang, J., Wang, B.,

Xu, X., Lee, S.-S., Cribb, M., Zhang, F., Yang, X., Zhao, C., Takemura, T., Wang, K., Xia, X., Yin, Y., Zhang, H., Guo, J., Zhai, P. M., Sugimoto, N., Babu, S. S., Brasseur, G. P., 2016. Aerosol and monsoon climate interactions over Asia. *Reviews of Geophysics*, 54(4): 866–929. DOI:10.1002/2015rg000500.

Liang, X., Wood, E. F., Lettenmaier, D. P., 1996. Surface soil moisture parameterization of the VIC-2L model: Evaluation and modification. *Global and Planetary Change*, 13 (1): 195–206. DOI:10.1016/0921-8181(95)00046-1.

Licht, A., Cappelle, M. v., Abels, H. A., Ladant, J. -B., Trabucho-Alexandre, J., France-Lanord, C., Donnadieu, Y., Vandenberghe, J., Rigaudier, T., Lecuyer, C., Terry, D., Adriaens, R., Boura, A., Guo, Z., Soe, A. N., Quade, J., Dupont-Nivet, G., Jaeger, J. -J., 2014. Asian monsoons in a late Eocene greenhouse world. *Nature*, 513: 501–506. DOI:10.1038/nature13704.

Lieberman, V., 2003. *Strange Parallels: Southeast Asia in Global Context, C. 800–1830, vol. 1, Integration on the Mainland*. Cambridge University Press, Cambridge.

Linwang, X. Z. Y., 2012. Fluctuation characteristics of Holocene sea-level change and its environmental implications. *Quaternary Sciences*, 6: 004.

Liu, L., Chen, X., Shi, J., 2014. Shanxi Wuxiang Niupiziwan shimopan, mobang de weihen yu canliuwu fenxi (Usewear and residue analyses of the grinding stones from Niupiziwan in Wuxiang county, Shanxi province). *Kaogu yu Wenwu*, 3: 109–119.

Liu, B., Xu, M., Henderson, M., Qi, Y., 2005a. Observed trends of precipitation amount, frequency, and intensity in China, 1960–2000. *Journal of Geophysical Research*, 110 (D08103). DOI:10.1029/2004JD004864.

Liu, J., Xu, K., Li, A., Milliman, J., Velozzi, D., Xiao, S., Yang, Z., 2007a. Flux and fate of Yangtze River sediment delivered to the East China Sea. *Geomorphology*, 85(3–4): 208–224.

Liu, K. S., Chan, J. C. L., 2019. Inter-decadal variability of the location of maximum intensity of category 4–5 typhoons and its implication on landfall intensity in East Asia. *International Journal of Climatology*, 39(4): 1839–1852. DOI:10.1002/joc.5919.

Liu, L., Chen, X., Zhao, H., 2013. Henan Mengjin Zhaigen, Bangou chutu Peiligang wanqi shimopan gongneng fenxi (Functional analysis of grinding slabs of the late Peiligang culture from Zhaigen and Bangou in Mengjin, Henan). *Zhongyuan Wenwu*, 5: 76–86.

Liu, L., Field, J., Fullagar, R., Chaohong, Z., Chen, X., Yu, J., 2010. A functional analysis of grinding stones from an early holocene site at Donghulin, North China. *Journal of Archaeological Science*, 37(2010): 2630–2639.

Liu, L., Ge, W., Bestel, S., Jones, D., Shi, J., Song, Y., Chen, X., 2011. Plant exploitation of the last foragers at Shizitan in the middle yellow river valley China: evidence from grinding stones. *Journal of Archaeological Science*, 38(12): 3524–3532.

Liu, L., Lee, G. -A., Jiang, L., Zhang, J., 2007b. Evidence for the early beginning (c. 9000 cal. BP) of rice domestication in China: a response. *The Holocene*, 17(8): 1059–1068.

Liu, W., Huang, Y., 2005. Compound specific D/H ratios and molecular distributions of higher plant leaf waxes as novel paleoenvironmental indicators in the Chinese Loess Plateau. *Organic Geochemistry*, 36(6): 851–860. DOI:10.1016/j.orggeochem.2005.01.006.

Liu, W., Huang, Y., An, Z., Clemens, S. C., Li, L., Prell, W. L., Ning, Y., 2005b. Summer monsoon intensity controls C4/C3 plant abundance during the last 35 ka in the Chinese Loess Plateau; carbon isotope evidence from bulk organic matter and individual leaf waxes. *Palaeogeography, Palaeoclimatology, Palaeoecology*, 220(3–4): 243–254.

Liu, X., 1998. The success or failure of ancient cities and paleoclimate in Chengdu plain. *Sichuan Wenwu*, 4: 34–37.

Liu, Y., Jiang, D., 2016. Last glacial maximum permafrost in China from CMIP5 simulations. *Palaeogeography, Palaeoclimatology, Palaeoecology*, 447: 12–21. DOI:10.1016/j.palaeo.2016.01.042.

Liu, Z., Otto-Bliesner, B., Kutzbach, J., Li, L., Shields, C., 2003. Coupled climate simulation of the evolution of global monsoons in the Holocene. *Journal of Climatology*, 16: 2472–2490.

Lobell, D. B., Burke, M. B., 2008. Why are agricultural impacts of climate change so uncertain? The importance of temperature relative to precipitation. *Environmental Research Letters*, 3(034007), DOI:10.1088/1748-9326/3/3/034007.

Lobell, D. B., Field, C. B., 2007. Global scale climate–crop yield relationships and the impacts ofrecent warming. *Environmental Research Letters*, 2(014002), DOI:10.1088/1748-9326/2/1/014002.

Lockwood, M., Bell, C., Woollings, T., Harrison, R. G., Gray, L. J., Haigh, J. D., 2010. Top-down solar modulation of climate: evidence for centennial-scale change. *Environmental Research Letters*, 5(3): 034008. 10.1088/1748–9326/5/3/034008.

Londo, J. P., Chiang, Y. -C., Hung, K. -H., Chiang, T. -Y., Schaal, B. A., 2006. Phylogeography of Asian wild rice, Oryza rufipogon, reveals multiple independent domestications of cultivated rice, Oryza sativa. *Proceedings of the National Academy of Sciences* 103(25): 9578–9583.

Looney, R., 2012. Economic impacts of the floods in Pakistan. *Contemporary South Asia*, 20(2): 225–241. 10.1080/09584935.2012.670203.

Lu, H., Liu, Z., Wu, N., BernÉ, S., Saito, Y., Liu, B., Wang, L. U. O., 2008. Rice domestication and climatic change: phytolith evidence from East China. *Boreas*, 31 (4): 378–385. 10.1111/j.1502–3885.2002.tb01081.x.

Lu, H., Zhang, J., Liu, K. -b., Wu, N., Li, Y., Zhou, K., Ye, M., Zhang, T., Zhang, H., Yang, X., Shen, L., Xu, D., Li, Q., 2009. Earliest domestication of common millet (Panicum miliaceum) in East Asia extended to 10,000 years ago. *Proceedings of the National Academy of Sciences*, 106(18): 7367–7372. 10.1073/pnas.0900158106.

Lu, T., 2006. The occurrence of cereal cultivation in China. *Asian Perspectives*, 45(2): 129–158.

Lutz, A. F., Immerzeel, W. W., Shrestha, A. B., Bierkens, M. F. P., 2014. Consistent increase in High Asia's runoff due to increasing glacier melt and precipitation. *Nature Climate Change*, 4: 587–592. DOI 10.1038/NCLIMATE2237.

Mabbett, I. W., 1964. The date of the arthaśāstra. *Journal of the American Oriental Society*, 84(2): 162–169.

MacNeish, R. S., Cunnar, G., Zhao, Z., Libby, J. G., 1998. Re-revised Second Annual Report of the Sino-American Jiangxi (PRC) Origin of Rice Project SAJOR. Andover Foundation for Archaeological Research, Andover, MA.

Madden, R. A., Julian, P. R., 1972. Description of global-scale circulation cells in tropics with a 40–50 day period. *Journal of Atmospheric Science*, 29: 1109–1123.

Madden, R. A., Julian, P. R., 1994. Observations of the 40–50-day tropical oscillation—a review. *Monthly Weather Review*, 122: 814–837.

Madella, M., 2003. Investigating agriculture and environment in South Asia: present and future contributions from opal phytoliths In: Weber SA, Belcher WM, (eds.), *Indus Ethnobiology. New Perspectives from the Field*. Altamira Press, Lanham, MD, pp. 201–250.

Madella, M., Fuller, D. Q., 2006. Palaeoecology and the Harappan civilisation of south Asia: a reconsideration. *Quaternary Science Reviews*, 25: 1283–1301. DOI:10.1016/j.quascirev.2005.10.012.

Magee, J. W., Miller, G. H., Spooner, N. A., Questiaux, D., 2004. Continuous 150 k.y. monsoon record from Lake Eyre, Australia: Insolation-forcing implications and unexpected Holocene failure. *Geology*, 32: 885–888.

Manabe, S., Terpstra, T. B., 1974. The effects of mountains on the general circulation of the atmosphere as identified by numerical experiments. *Journal of Atmospheric Science*, 31: 3–42.

Manghnani, V., Morrison, J. M., Hopkins, T. S., Böhm, E., 1998. Advection of upwelled waters in the form of plumes off Oman during the Southwest Monsoon. *Deep Sea Research Part II: Topical Studies in Oceanography*, 45(10–11): 2027–2052.

Mann, H., 1946. Millets in the middle east. *Empire Journal of Experimental Agriculture*, 14: 208–16.

Mann, M. E., Zhang, Z., Rutherford, S., Bradley, R. S., Hughes, M. K., Shindell, D., Ammann, C., Faluvegi, G., Ni, F., 2009. Global signatures and dynamical origins of the little ice age and medieval climate anomaly. *Science*, 326(5957): 1256–1260.

Mao, L., Zhang, Y., Bi, H., 2006. Modern pollen deposits in coastal mangrove swamps from northern Hainan Island, China. *Journal of Coastal Research*, 226: 1423–1436. DOI:/10.2112/05-0516.1

Marcott, S. A., Shakun, J. D., Clark, P. U., Mix, A. C., 2013. A reconstruction of regional and global temperature for the past 11,300 years. *Science*, 339(6124): 1198–1201. 10.1126/science.1228026.

Marks, R., 1998. *Tigers, Rice, Silk, and Silt: Environment and Economy in Late Imperial South China*. Cambridge University Press, Cambridge.

Marshall, J. H., 1931. *Mohenjo-daro and the Indus Civilisation*. Arthur Probsthain, London.

Marshall, M., 2010. Frozen jet stream links Pakistan floods, Russian fires, New Scientist. www.newscientist.com/article/mg20727730-101-frozen-jet-stream-links-pakistan-floods-russian-fires/.

Marston, J. M., 2011. Archaeological markers of agricultural risk management. *Journal of Anthropological Archaeology*, 30(2): 190–205.

Masson-Delmotte, V., Schulz, M., Abe-Ouchi, A., Beer, J., Ganopolski, A., Rouco, J. F. G., Jansen, E., Lambeck, K., Luterbacher, J., Naish, T., Osborn, T., Otto-Bliesner, B., Quinn, T., Ramesh, R., Rojas, M., Shao, X., Timmermann, A., 2013. Information from paleoclimate archives. In: Stocker, T.F. et al. (eds.), *Climate Change 2013: The Physical Science Basis. Contribution of Working Group I to the Fifth Assessment Report of the Intergovernmental Panel on Climate Change*. Cambridge University Press, Cambridge, United Kingdom and New York, NY, USA.

Matsui, A., 1996. Archaeological investigations of andromous salminoid fishing in Japan. *World Archaeology*, 27(3): 444–460.

Matthew, R., 2013. Climate change and water security in the Himalayan region. *Asia Policy*, 16: 39–44.

May, T. M., 2012. *The Mongols Conquests in World History*. Reaktion Books, London, p. 313.

McGlade, J., 1995. Archaeology and the ecodynamics of human-modified landscapes. *Antiquity*, 68: 113–132.

McInerney, F. A., Helliker, B. R., Freeman, K. H., 2011. Hydrogen isotope ratios of leaf wax n-alkanes in grasses are insensitive to transpiration. *Geochimica et Cosmochimica Acta*, 75(2): 541–554. DOI:10.1016/j.gca.2010.10.022.

McIntosh, R. J., Tainter, J. A., McIntosh, S. K., 2000. *The Way the Wind Blows: Climate, History, and Human Action*. Columbia University Press, New York.

McManus, J., Francois, R., Gherardi, J., Keigwin, L., Brown-Leger, S., 2004. Collapse and rapid resumption of Atlantic meridional circulation linked to deglacial climate changes. *Nature*, 428: 834–837.

Meadow, R., 1984. Notes on the faunal remains from Mehrgarh, with a focus on cattle (Bos). In: Allchin, B. (ed.), *South Asian Archaeology*. Cambridge University Press, Cambridge, pp. 34–40.

Meehl, G. A., Arblaster, J. M., Lawrence, D. M., Seth, A., Schneider, E. K., Kirtman, B. P., Min, D., 2006. Monsoon regimes in the CCSM3. *Journal of Climate*, 19: 2482–2495.

Memon, A. A., 2004. *Evaluation of Impacts on the Lower Indus River Basin Due to Upstream Water Storage and Diversion, World Water & Environmental Resources Congress*. American Society of Civil Engineers, Salt Lake City, Utah.

Meybeck, M., 1987. Global chemical weathering of surficial rocks estimates from river dissolved loads. *American Journal of Science*, 287: 401–428.

Meyers, P. A., 1997. Organic geochemical proxies of paleoceanographic, paleolimnologic, and paleoclimatic processes. *Organic Geochemistry*, 27(5): 213–250. DOI:10.1016/S0146-6380(97)00049-1.

Middleton, G. D., 2017. The show must go on: Collapse, resilience, and transformation in 21st-century archaeology. *Reviews in Anthropology*, 46(2–3): 78–105.

Miller, H., 2006a. Water supply, labor requirements, and land ownership in Indus floodplain agricultural systems. In: Marcus, J., Stanish, C. (eds.), *Agricultural Strategies*, Cotsen Institute of Archaeology, UCLA, pp. 92–127.

Miller, M. -L. H., 2006b. Water supply, Labor organization and land ownership in Indus floodplain agricultural systems. In: Stanish, C., Marcus, J. (eds.), *Agriculture and Irrigation in Archaeology*. Cotsen Institute of Archaeology Press, Los Angeles, pp.92–128.

Milliman, J. D., Farnsworth, K. L., Jones, P. D., Xu, K. H., Smith, L. C., 2008. Climatic and anthropogenic factors affecting river discharge to the global ocean, 1951–2000. *Planet. Change. Global and Planetary Change*, 62: 187–194.

Milliman, J. D., Syvitski, J. P. M., 1992. Geomorphic/tectonic control of sediment discharge to the ocean; the importance of small mountainous rivers. *Journal of Geology*, 100: 525–544.

Milne, G. A., Mitrovica, J. X., 2008. Searching for eustasy in deglacial sea-level histories. *Quaternary Science Reviews*, 27(25): 2292–2302. DOI:10.1016/j.quascirev.2008.08.018.

Minnis, P. E., 1999. Sustainability: The long view from archaeology. *New Mexico Journal of Science*, 39: 23–41.

Mishra, A. K., Singh, V. P., 2010. A review of drought concepts. *Journal of Hydrology*, 391 (1): 202–216. DOI:10.1016/j.jhydrol.2010.07.012.

Mishra, V., Shah, R., Thrasher, B., 2014. Soil moisture droughts under the retrospective and projected climate in India. *Journal of Hydrometeorology*, 15(6): 2267–2292. https://doi.org/10.1175/JHM-D-13-0177.1

Mishra, V., Smoliak, B. V., Lettenmaier, D. P., Wallace, J. M., 2012. A prominent pattern of year-to-year variability in Indian summer monsoon rainfall. *Proceedings of the National Academy of Sciences*, 109(19): 7213–7217. DOI:10.1073/pnas.1119150109.

Mishra, V., Tiwari, A. D., Aadhar, S., Shah, R., Xiao, M., Pai, D. S., Lettenmaier, D., 2019. Drought and famine in India, 1870–2016. *Geophysical Research Letters*, 46(4): 2075–2083. DOI:10.1029/2018GL081477.

Miteva, D. A., Murray, B. C., Pattanayak, S. K., 2015. Do protected areas reduce blue carbon emissions? A quasi-experimental evaluation of mangroves in Indonesia. *Ecological Economics*, 119: 127–135. DOI:10.1016/j.ecolecon.2015.08.005.

Mitrovica, J. X., Tamisiea, M. E., Davis, J. L., Milne, G. A., 2001. Recent mass balance of polar ice sheets inferred from patterns of global sea-level change. *Nature*, 409(6823): 1026–1029. DOI:10.1038/35059054.

Mohammed, A. R., Tarpley, L., 2009. High nighttime temperatures affect rice productivity through altered pollen germination and spikelet fertility. *Agricultural and Forest Meteorology*, 149(6): 999–1008. DOI:10.1016/j.agrformet.2008.12.003.

Mohtadi, M., Oppo, D. W., Steinke, S., Stuut, J. -B. W., Pol-Holz, R. D., Hebbeln, D., Lückge, A., 2011. Glacial to Holocene swings of the Australian-Indonesian monsoon. *Nature Geoscience*, 4: 540–544.

Molnar, P., England, P., Martinod, J., 1993. Mantle dynamics, uplift of the Tibetan Plateau, and the Indian monsoon. *Reviews of Geophysics*, 31(4): 357–396.

Molnar, P. H., Rajagopalan, B., 2012. Late Miocene upward and outward growth of eastern Tibet and decreasing monsoon rainfall over the northwestern Indian subcontinent since ~10 Ma. *Geophysical Research Letters*, 39: L09702. DOI:10.1029/2012GL051305.

Morisaki, K., 2012. The evolution of lithic technology and human behavior from MIS 3 to MIS 2 in the Japanese Upper Paleolithic. *Quaternary International*, 248: 56–69. DOI:10.1016/j.quaint.2010.11.011.

Morrill, C., Anderson, D. M., Bauer, B. A., Buckner, R., Gille, E. P., Gross, W. S., Hartman, M., Shah, A., 2013a. Proxy benchmarks for intercomparison of 8.2 ka simulations. *Clim. Past*, 9(1): 423–432. DOI:10.5194/cp-9-423-2013.

Morrill, C., LeGrande, A. N., Renssen, H., Bakker, P., Otto-Bliesner, B. L., 2013b. Model sensitivity to North Atlantic freshwater forcing at 8.2 ka. *Clim. Past*, 9(2): 955–968. DOI:10.5194/cp-9-955-2013.

Morrill, C., Wagner, A., Otto-Bliesner, B., Rosenbloom, N., 2011. Evidence for significant climate impacts in monsoonal Asia at 8.2 ka from multiple proxies and model simulations. *Journal of Earth Environment*, 2: 426–441.

Morrill, C., Wagner, A. J., Otto-Bliesner, B. L., Rosenbloom, N., 2014. Evidence for significant climate impacts in monsoonal Asia at 8.2 ka from multiple proxies and model simulations. *Journal of Earth Environment*, 2: 426–441.

Moulherat, C., Tengberg, M., Haquet, J. -F., Mille, B. t., 2002. First evidence of cotton at Neolithic Mehrgarh, Pakistan: analysis of mineralized fibres from a copper bead. *Journal of Archaeological Science*, 29(12): 1393–1401.

Mughal, M. R., 1997. *Ancient Cholistan: Archaeology and Architecture*. Ferozsons Pvt. Ltd, Rawalpindi.

Mughal, R., 1990. The protohistoric settlement patterns in the Cholistan desert, Pakistan. In: Teddei, M. (ed.), *South Asian Archaeology 1987*. Instituto Italiano per il Medio ed Estremo Oriente, Rome, pp.143–156.

Mughal, R., 1992. The geographical extent of the Indus civilization during the early, mature and late Harappan times. In: Possehl, G. (ed.), *South Asian Archaeology Studies*. Oxford & IBH Publishing Co. Pvt. Ltd, New Delhi, pp.123–143.

Muir, J., 1873. *Original Sanskrit Texts on the Origin and History of the People of India, Their Religions and Institutions*. Trübner & Company.

Murakami, H., Vecchi, G. A., Underwood, S., 2017. Increasing frequency of extremely severe cyclonic storms over the Arabian Sea. *Nature Climate Change*, 7(12): 885–889. DOI:10.1038/s41558-017-0008-6.

Murra, J., 1985. The limits and limitations of the "vertical archipelago" in the andes. In: Masuda, S., Shimada, I., Morris, C. (eds.), *Andean Ecology and Civilization: An Interdisciplinary Perspective on Andean Ecological Complementarity*. University of Tokyo, Tokyo, pp.15–20.

Mustafa, D., Wrathall, D., 2011. Indus basin floods of 2010: souring of a Faustian bargain? *Water Alternatives*, 4(1): 72–85.

Naidu, P. D., Malmgren, B. A., 1995. A 2,200 years periodicity in the Asian monsoon system. *Geophysical Research Letters*, 22: 2361–2364.

Nakajima, T., Hudson, M. J., Uchiyama, J., Makibayashi, K., Zhang, J. 2019. Common carp aquaculture in Neolithic China dates back 8,000 years. *Nature Ecology & Evolution*, 3 (10): 1415–1418, DOI:10.1038/s41559-019-0974-3.

Naqvi, S. A., 2012. *Indus Waters and Social Change: The Evolution and Transition of Agrarian Society in Pakistan*.Oxford University Press, Pakistan, p. 541.

Neena, J. M., Suhas, E., Goswami, B. N., 2011. Leading role of internal dynamics in the 2009 Indian summer monsoon drought. *Journal of Geophysical Research: Atmospheres*, 116(D13). DOI:10.1029/2010jd015328.

Nesbitt, H. W., Markovics, G., Price, R. C., 1980. Chemical processes affecting alkalis and alkaline earths during continental weathering. *Geochimica et Cosmochimica Acta*, 44: 1659–1666.

Netting, R. M., 1993. *Smallholders, Householders: Farm Families and the Ecology of Intensive, Sustainable Agriculture*. Stanford University Press, Palo Alto, CA, p. 446.

Neumann, B., Vafeidis, A. T., Zimmermann, J., Nicholls, R. J., 2015a. Future coastal population growth and exposure to sea-level rise and coastal flooding – a global assessment. *PLOS ONE*, 10(3): e0118571. DOI;10.1371/journal.pone.0118571.

Neumann, J. E., Emanuel, K. A., Ravela, S., Ludwig, L. C., Verly, C., 2015b. Risks of coastal storm surge and the effect of sea level rise in the red river delta, vietnam. *Sustainability*, 7: 6553–6572. DOI;10.3390/su7066553.

Ni, J., Ge, Y., Harrison, S., Prentice, C. I., 2010. Palaeovegetation in China during the late Quaternary: Biome reconstructions based on a global scheme of plant functional types. *Palaeogeography, Palaeoclimatology, Palaeoecology*, 289(2010): 44–61.

Nie, J., Stevens, T., Rittner, M., Stockli, D., Garzanti, E., Limonta, M., Bird, A., Ando, S., Vermeesch, P., Saylor, J., Lu, H., Breecker, D., Hu, X., Liu, S., Resentini, A., Vezzoli, G., Peng, W., Carter, A., Ji, S., Pan, B., 2015. Loess Plateau storage of Northeastern Tibetan Plateau-derived Yellow River sediment. *Nature Communications*, 6: 8511. DOI:10.1038/ncomms9511.

North Greenland Ice Core Project members, 2004. High-resolution record of Northern Hemisphere climate extending into the last interglacial period. *Nature*, 431: 147–151.

Norton, C., Kim, B., Bae, K., 1999. Differential processing of fish during the Korean Neolithic: Konam-Ri. *Artic Anthropology*, 36(1/2): 151–165.

O'Brien, P. J., 1972. The sweet potato: its origin and dispersal1. *American Anthropologist*, 74(3): 342–365. DOI:10.1525/aa.1972.74.3.02a00070.

Oberlies, T., 1998. Die Religion des Rgveda: Ester Teil: das religiöse system des Rgveda, 26. *Publications of the De Nobili Research library, Vienna, Austria*, 632 pp.

Oka, H. I., 1958. Photoperiodic adaption to latitude in rice varieties. *Phyton*, 11: 153–160.

Oram, P. A., De Haan, C., 1995. *Technologies for Rainfed Agriculture in Mediterranean Climates: A Review of World Bank Experiences*. The World Bank, New York.

Ortiz, R., Sayre, K. D., Govaerts, B., Gupta, R., Subbarao, G. V., Ban, T., Hodson, D., Dixon, J. M., Iván Ortiz-Monasterio, J., Reynolds, M., 2008. Climate change: Can wheat beat the heat? *Agriculture, Ecosystems & Environment*, 126(1): 46–58. DOI:10.1016/j.agee.2008.01.019.

Ozturk, T., Turp, M. T., Türkeş, M., Kurnaz, M. L., 2017. Projected changes in temperature and precipitation climatology of Central Asia CORDEX Region 8 by using RegCM4.3.5. *Atmospheric Research*, 183: 296–307. DOI:10.1016/j.atmosres.2016.09.008.

Padma Kumari, B., Londhe, A. L., Daniel, S., Jadhav, D. B., 2007. Observational evidence of solar dimming: Offsetting surface warming over India. *Geophysical Research Letters*, 34(21): L21810. 10.1029/2007GL031133.

Pal, J. S., Eltahir, E. A. B., 2015. Future temperature in southwest Asia projected to exceed a threshold for human adaptability. *Nature Climate Change*, 6: 197. DOI:10.1038/nclimate2833.

Palmer, W. C., 1965. *Meteorological Drought*, Dept. of Commerce, Washington, D.C.

Palmieri, A., Shah, F., Dinar, A., 2001. Economics of reservoir sedimentation and sustainable management of dams. *Journal of Environmental Management*, 61(2): 149–163. DOI:10.1006/jema.2000.0392.

Pant, G. B., Parthasarathy, B., 1981. Some aspects of an association between the southern oscillation and Indian summer monsoon. *Arch. Meteor. Geophys. Bioklimatol., Archiv für Meteorologie, Geophysik und Bioklimatologie, Serie B*, 1329: 245–252.

Panwar, N., Kaushik, S., Kothari, S., 2011. Role of renewable energy sources in environmental protection: a review. *Renewable and Sustainable Energy Reviews*, 15(3): 1513–1524.

Park, C. E., Jeong, S. J., Ho, C. H., Park, H., Piao, S., Kim, J., Feng, S., 2017. Dominance of climate warming effects on recent drying trends over wet monsoon regions. *Atmos. Chem. Phys.*, 17(17): 10467–10476. 10.5194/acp-17-10467-2017.

Parsons, J. B., 1970. *The Peasant Rebellions of the Late Ming Dynasty*. University of Arizona Press, Tuscon, 292 pp.

Parthasarathy, B., Kumar, K. R., Munot, A. A., 1992. Surface pressure and summer monsoon rainfall over India. *Advances in Atmospheric Sciences*, 9(3): 359–366. 10.1007/bf02656946.

Parthasarathy, B., Munot, A., Kothawale, D., 1994. All-India monthly and seasonal rainfall series: 1871–1993. *Theoretical and Applied Climatology*, 49(4): 217–224.

Parthasarathy, B., Munot, A. A., Kothawale, D. R., 1995. *Monthly and seasonal rainfall series for All-India homogeneous regions and meteorological subdivisions: 1871–1994. RR-065*, Contributions from Indian Institute of Tropical Meteorology, Pune 411 008 India.

Partin, J. W., Cobb, K. M., Adkins, J. F., Clark, B., Fernandez, D. P., 2007. Millennial-scale trends in west Pacific warm pool hydrology since the last glacial maximum. *Nature*, 449(7161): 452–455. DOI:10.1038/nature06164.

Patil, N., Venkataraman, C., Muduchuru, K., Ghosh, S., Mondal, A., 2019. Disentangling sea-surface temperature and anthropogenic aerosol influences on recent trends in South Asian monsoon rainfall. *Climate Dynamics*, 52(3): 2287–2302. DOI:10.1007/s00382-018-4251-y.

Pattanaik, D., 2000. *Devi, the Mother-Goddess: An Introduction*. Vakils, Feffer, and Simons Limited.

Patwardhan, S., Kulkarni, A., Rao, K. K., 2018. Projected changes in rainfall and temperature over homogeneous regions of India. *Theoretical & Applied Climatology*, 131 (1–2): 581–592. DOI:10.1007/s00704-016-1999-z.

Pearson, R., 2006. Jomon hot spot: increasing sedentism in south- western Japan in the Incipient Jomon (14,000–9250 cal. bc) and Earliest Jomon (9250–5300 cal. bc) periods.*World Archaeology*, 38(2): 239–258.

Pederson, N., Hessl, A. E., Baatarbileg, N., Anchukaitis, K. J., Di Cosmo, N., 2014. Pluvials, droughts, the Mongol Empire, and modern Mongolia. *Proceedings of the National Academy of Sciences*, 111(12): 4375–4379. DOI:10.1073/pnas.1318677111.

Pei, Q., Zhang, D. D., Li, J., Lee, H. F., 2017. Proxy-based Northern Hemisphere temperature reconstruction for the mid-to-late Holocene. *Theoretical and Applied Climatology*, 130(3): 1043–1053. 10.1007/s00704-016-1932-5.

Peltier, W. R., 1994. Ice age paleotopography. *Science*, 265: 195–201.

Peltier, W. R., 2002. On eustatic sea level history: Last Glacial Maximum to Holocene. *Quaternary Science Reviews*, 21(1–3): 377–396.

Peng, X., 1987. Demographic consequences of the great leap forward in China's provinces. *Population and Development Review*, 13(4): 639–670. DOI:10.2307/1973026.

Perkins, D. H., 1969. *Agricultural Development in China 1368–1968*. Aldine Publishing Co., Chicago.

Perkins, J. H., 1997. *Geopolitics and the Green Revolution: Wheat, Genes, and the Cold War*. Oxford University Press, New York, 336 pp.

Peterson, C. E., Shelach, G., 2012. Jiangzhai: Social and economic organization of a Middle Neolithic Chinese village. *Journal of Anthropological Archaeology*, 31(2012): 265–301.

Peterson, L. C., Murray, D. W., Ehrmann, W. U., Hempel, P., 1992. Cenozoic carbonate accumulation and compensation depth changes in the Indian Ocean. In: Duncan, R. A., Rea, D. K., Kidd, R. B., von Rad, U., Weissel, J. K. (eds.), *Synthesis of Results from Scientific Drilling in the Indian Ocean. Geophysical Monograph*. American Geophysical Union, Washington, DC, pp.311–333.

Petrie, C. A., Bates, J., 2017. 'Multi-cropping', *Intercropping and Adaptation to Variable Environments in Indus South Asia. Journal of World Prehistory*, 30(2): 81–130.

Phadtare, N. R., 2000. Sharp decrease in summer monsoon strength 4000 –3500 cal yr B. P. in the central higher himalaya of India based on pollen evidence from alpine peat. *Quaternary Research*, 53: 122–129.

Pingali, P. L., 2012. Green revolution: Impacts, limits, and the path ahead. *Proceedings of the National Academy of Sciences*, 109(31): 12302–12308. DOI:10.1073/pnas.0912953109.

Ponte, R. M., Quinn, K. J., Piecuch, C. G., 2018. Accounting for gravitational attraction and loading effects from land ice on absolute sea level. *Journal of Atmospheric and Oceanic Technology*, 35(2): 405–410. DOI:10.1175/JTECH-D-17-0092.1.

Ponton, C., Giosan, L., Eglinton, T. I., Fuller, D., Johnson, J. E., Kumar, P., Collet, T. S., 2012. Holocene Aridification of India. *Geophysical Research Letters*, 39(L03704). DOI:10.1029/2011GL050722.

Porter, J., Xie, L., Challinor, A., Cochrane, K., Howden, S., Iqbal, M., Lobell, D., Travasso, M., 2014. Food security and food production systems. In: Field, C. B. et al. (eds.), *Climate Change 2014: Impacts, Adaptation, and Vulnerability. Part A: Global and Sectoral Aspects. Contribution of Working Group II to the Fifth Assessment Report of the Intergovernmental Panel on Climate Change*. Cambridge University Press, Cambridge, pp. 485–533.

Prasad, S., Enzel, Y., 2006. Holocene paleoclimates of India. *Quaternary Research*, 66: 442–453.

Prasad, S., Kusumgar, S., Gupta, S. K., 1997. A mid–late Holocene record of palaeoclimatic changes from Nal Sarovar—A palaeodesert margin lake in western India. *Journal of Quaternary Science*, 12(2): 153–159.

Prasanna, V., 2014. Impact of monsoon rainfall on the total foodgrain yield over India. *Journal of Earth System Science*, 123(5): 1129–1145. 10.1007/s12040-014–0444-x.

Prell, W. L., Kutzbach, J. E., 1992. Sensitivity of the Indian monsoon to forcing parameters and implications for its evolution. *Nature*, 360(6405): 647–652.

Prell, W. L., Murray, D. W., Clemens, S. C., Anderson, D. M., 1992. Evolution and variability of the Indian ocean summer monsoon: evidence from the western Arabian sea drilling program. In: Duncan, R. A., Rea, D. K., Kidd, R. B., von Rad, U., Weissel, J. K. (eds.), *Synthesis of Results from Scientific Drilling in the Indian Ocean. Geophysical Monograph*. American Geophysical Union, Washington, DC, pp.447–469.

Prendergast, M. E., Yuan, J., Bar-Yosef, O., 2009. Resource intensification in the Late Upper Paleolithic: a view from southern China. *Journal of Archaeological Science*, 36 (2009): 1027–1037.

Pretzsch, H., 2009. *Forest dynamics, growth, and yield, Forest Dynamics, Growth and Yield.* Springer, pp. 1–39.

Pritchard, H. D., 2017. Asia's glaciers are a regionally important buffer against drought. *Nature*, 545: 169. DOI:10.1038/nature22062.

Purugganan, M. D., Fuller, D. Q., 2009. The nature of selection during plant domestication. *Nature*, 457(7231): 843–848.

Qian, W. -h., Huang, J., 2018. Impact of different climatic flows on typhoon tracks. *Meteorology and Atmospheric Physics*, 130(2): 137–152. DOI:10.1007/s00703-017-0515-z.

Qinghai Sheng Huangyuan Xian Bowuguan, Qinghai Sheng Wenwu Kaogudui, 1985. Qinghai Huangyuan Xian Dahuazhongzhuang Kayue Wenhua Mudi Fajue Jianbao. *Kaogu yu Wenwu*, 1985(5): 11–34.

Qu, T., Bar-Yosef, O., Wang, Y., Wu, X., 2013. The chinese upper paleolithic: geography, chronology and techno-typology. *Journal of Archaeological Research*, 21 (2013): 1–73.

Quade, J., Cerling, T. E., Bowman, J. R., 1989. Development of Asian monsoon revealed by marked ecological shift during the latest Miocene in northern Pakistan. *Nature*, 342 (6246): 163–166.

Radić, V., Bliss, A., Beedlow, A. C., Hock, R., Miles, E., Cogley, J. G., 2013. Regional and global projections of twenty-first century glacier mass changes in response to climate scenarios from global climate models. *Climate Dynamics*, 42: 37–58. DOI:10.1007/s00382-013-1719-7.

Rajeevan, M., Bhate, I., Kale, J. D., Lal, B., 2006. High resolution daily gridded rainfall data for the Indian region: Analysis of break and active monsoon spells. *Current Science*, 91: 296–306.

Rajeevan, M., Gadgil, S., Bhate, J., 2010. Active and break spells of the Indian summer monsoon. *Journal of Earth System Science*, 119(3): 229–247. DOI:10.1007/s12040-010-0019-4.

Ramanathan, V., Carmichael, G., 2008. Global and regional climate changes due to black carbon. *Nature Geoscience*, 1: 221. 10.1038/ngeo156.

Ramanathan, V., Chung, C., Kim, D., Bettge, T., Buja, L., Kiehl, J. T., Washington, W. M., Fu, Q., Sikka, D. R., Wild, M., 2005. Atmospheric brown clouds: Impacts on South Asian climate and hydrological cycle. *Proceedings of the National Academy of Sciences*, 102(15): 5326–5333. DOI:10.1073/pnas.0500656102.

Ramankutty, N., Foley, J. A., 1999. Estimating historical changes in global land cover: Croplands from 1700 to 1992,. *Global Biogeochemical Cycles*, 13(4): 997–1027. DOI:10.1029/1999GB900046.

Ramaswamy, V., Chanin, M. L., Angell, J., Barnett, J., Gaffen, D., Gelman, M., Keckhut, P., Koshelkov, Y., Labitzke, K., Lin, J. J. R., O'Neill, A., Nash, J., Randel, W., Rood, R., Shine, K., Shiotani, M., Swinbank, R., 2001. Stratospheric temperature trends: Observations and model simulations. *Reviews of Geophysics*, 39(1): 71–122. 10.1029/1999RG000065.

Ramstein, G., Fluteau, F., Besse, J., Joussaume, S., 1997. Effect of orogeny, plate motion and land-sea distribution on Eurasian climate change over the past 30 million years. *Nature*, 386: 788–795.

Rao, S. R., 1973. *Lothal and the Indus Civilisation.* Asia Publishing House, London.

Rao, V. B., Ferreira, C. C., Franchito, S. H., Ramakrishna, S. S. V. S., 2008. In a changing climate weakening tropical easterly jet induces more violent tropical storms over the north. *Geophysical Research Letters*, 35(L15710), DOI:10.1029/2008GL034729.

Rasmusson, E. M., Carpenter, T. H., 1982. Variations in tropical sea surface temperature and surface wind fields associated with the Southern oscillation/El Niño. *Monthly Weather Review*, 110: 354–384.

Rasul, G., Mahmood, A., Sadiq, A., Khan, S. I., 2012. Vulnerability of the Indus delta to climate change in Pakistan. *Pakistan Journal of Meteorology*, 8(16): 89–107.

Raven, P. H., 2009. *Biology of Plants W H*. Freeman & Company.

Rea, D. K., 1994. The paleoclimatic record provided by eolian deposition in the deep sea; the geologic history of wind. *Reviews in Geophysics*, 32: 159–195.

Reddy, S.N., 1997. If the threshing floor could talk: Integration of agriculture and pastoralism during the late Harappan in Gujarat, India. *Journal of Anthropological Archaeology*, 16(2): 162–187. DOI:10.1006/jaar.1997.0308.

Reddy, S. N., 2003. Food and fodder: Plant usage and changing socio-cultural landscapes during the Harappan phase in Gujarat, India. In: Weber, S., Belcher, W. R. (eds.), *Indus Ethnobiology*. New perspectives from the field. Lexington Books, Lanham, pp.327–342.

Redman, C., 2005. Resilience theory in archaeology. *American Anthropologist*, 107(1): 70–77.

Redman, C. L., Kinzig, A. P., 2003. Resilience of past landscapes resilience theory, society, and the longue duree. *Conservation Ecology*, 7(1): 14. DOI:10.5751/ES-00510-070114.

Regattieri, E., Giaccio, B., Galli, P., Nomade, S., Peronace, E., Messina, P., Sposato, A., Boschi, C., Gemelli, M., 2016. A multi-proxy record of MIS 11–12 deglaciation and glacial MIS 12 instability from the Sulmona basin (central Italy). *Quaternary Science Reviews*, 132: 129–145. DOI:10.1016/j.quascirev.2015.11.015.

Ren, M. -E., Shi, Y. -L., 1986. Sediment discharge of the Yellow River (China) and its effect on the sedimentation of the Bohai and the Yellow Sea Original Research. *Continental Shelf Research*, 6(6): 785–810. DOI:10.1016/0278-4343(86)90037-3.

Revadekar, J. V., Kothawale, D. R., Patwardhan, S. K., Pant, G. B., Rupa Kumar, K., 2012. About the observed and future changes in temperature extremes over India. *Natural Hazards*, 60(3): 1133–1155. DOI:10.1007/s11069-011-9895-4.

Rispoli, F., 2007. The incised & impressed pottery style of mainland Southeast Asia: following the paths of Neolithization. *East and West*, 57(1/4): 235–304.

Rispoli, F., Ciarla, R., Pigott, V. C., 2013. Establishing the prehistoric cultural sequence for the Lopburi region, Central Thailand. *Journal of World Prehistory*, 26(2): 101–171. DOI:10.1007/s10963-013-9064-7.

Roberts, P., Boivin, N., Petraglia, M., Masser, P., Meece, S., Weisskopf, A., Silva, F., Korisettar, R., Fuller, D. Q., 2016. Local diversity in settlement, demography and subsistence across the southern Indian Neolithic-Iron age transition: site growth and abandonment at Sanganakallu-Kupgal. *Archaeological and Anthropological Sciences*, 8(3): 575–599.

Rolett, B. V., Zheng, Z., Yue, Y., 2011. Holocene sea-level change and the emergence of Neolithic seafaring in the Fuzhou Basin (Fujian, China). *Quaternary Science Reviews*, 30(7): 788–797. DOI:10.1016/j.quascirev.2011.01.015.

Ropelewski, C. F., Halpert, M. S., 1987. Global and regional scale precipitation patterns associated with the El Niño/Southern Oscillation. *Monthly Weather Review*, 115(8): 1606–1626.

Rosen, A., 2008. The impact of environmental change and human land use on alluvial valleys in the loess plateau of China during the middle holocene. *Geomorphology*, 101 (2008): 298–307.

Rosen, A., Macphail, R., Liu, L., Chen, X., Weisskopf, A., 2017. Rising social complexity, agricultural intensification, and the earliest rice paddies on the Loess Plateau of northern China. *Quaternary International*, 437, Part B: 50–59. DOI:10.1016/j. quaint.2015.10.013.

Rosenfeld, D., Dai, J., Yu, X., Yao, Z., Xu, X., Yang, X., Du, C., 2007. Inverse relations between amounts of air pollution and orographic precipitation. *Science*, 315(5817): 1396. DOI:10.1126/science.1137949.

Rowley, D. B., Currie, B. S., 2006. Palaeo-altimetry of the late Eocene to Miocene Lunpola basin, central Tibet. *Nature*, 439: 677–681.

Roy, T. N., 1984. The concept, provenance and chronology of painted grey ware. *East and West*, 34(1/3): 127–137.

Royden, L. H., Burchfiel, B. C., Van Der Hilst, R. D., 2008. The geological evolution of the Tibetan plateau. *Science*, 321(5892): 1054–1058. DOI:10.1126/science.1155371.

Rozanski, K., Araguás-Araguás, L., Gonfiantini, R., 2013. Isotopic patterns in modern global precipitation. In: Swart, P.K., Lohmann, K.C., Mckenzie, J., Savin, S. (eds.), *Climate Change in Continental Isotopic Records. Geophysical Monograph*. American Geophysical Union, Washington DC, pp. 1–36.

Ruddiman, W. F., 2003. The anthropogenic greenhouse era began thousands of years ago. *Climatic Change*, 61(3): 261–293. DOI:10.1023/B:CLIM.0000004577.17928.fa.

Ryan, W. B. F., Carbotte, S. M., Coplan, J. O., O'Hara, S., Melkonian, A., Arko, R., Weissel, R.A., Ferrini, V., Goodwillie, A., Nitsche, F., Bonczkowski, J., Zemsky, R., 2009. Global multi-resolution topography synthesis. *Geochemistry Geophysics Geosystems*, 10(Q03014). DOI:10.1029/2008GC002332.

Sagawa, T., Kuwae, M., Tsuruoka, K., Nakamura, Y., Ikehara, M., Murayama, M., 2014. Solar forcing of centennial-scale East Asian winter monsoon variability in the mid- to late Holocene. *Earth and Planetary Science Letters*, 395(1): 124–135.

Sage, S. F., 1992. *Ancient Sichuan and the Unification of China*. SUNY Press, New York.

Sahni, H. K., 2006. The politics of water in South Asia: the case of the Indus waters treaty. *SAIS Review of International Affairs*, 26(2): 153–165. DOI:10.1353/sais.2006.0043.

Saini, H. S., Tandon, S. K., Mujtaba, S. A. I., Pant, N. C., Khorana, R. K., 2009. Reconstruction of buried channel-floodplain systems of the northwestern Haryana Plains and their relation to the 'Vedic' Saraswati. *Current Science*, 97(11): 1634–1643.

Saith, N., Slingo, J. M., 2006. The role of the Madden Julian oscillation in the Indian drought of 2002. *International Journal of Climatology*, 26(10): 1361–1378. DOI:10.1002/joc.1317.

Sankaran, A., 1999. Saraswati–the ancient river lost in the desert. *Current Science*, 77(8): 1054–1060.

Sarkar, J., 2011. Chapter 4 drought, its impacts and management: scenario in India. In: Shaw, R., Nguyen, H. (eds.), *Droughts in Asian Monsoon Region (Community, Environment and Disaster Risk Management, Volume 8)*. Emerald Group Publishing Limited, pp.67–85.

Sarthi, P. P., Agrawal, A., Rana, A., 2014. Possible future changes in cyclonic storms in the Bay of Bengal, India under warmer climate. *International Journal of Climatology*, 35: 1267–1277. DOI:10.1002/joc.4053.

Saseendran, S. A., Nielsen, D. C., Lyon, D. J., Ma, L., Felter, D. G., Baltensperger, D. D., Hoogenboom, G., Ahuja, L. R., 2009. Modeling responses of dryland spring triticale, proso millet and foxtail millet to initial soil water in the high plains. *Field Crops Research*, 113: 48–63.

Sato, T., 2009. Influences of subtropical jet and Tibetan Plateau on precipitation pattern in Asia: Insights from regional climate modeling. *Quaternary International*, 194(1): 148–158. DOI:10.1016/j.quaint.2008.07.008.

Sato, Y. -I., Yamanaka, S., Takahashi, M., 2003. Evidence for Jomon Plant Cultivation based on DNA analysis of chestnut remains. *Senri Ethnological Studies*, 63: 187–198.

Schewe, J., Levermann, A., 2012. A statistically predictive model for future monsoon failure in India. *Environmental Research Letters*, 7(4): 044023. DOI:10.1088/1748-9326/7/4/044023.

Schouten, S., Hopmans, E. C., Schefuß, E., Sinninghe Damsté, J. S., 2002. Distributional variations in marine crenarchaeotal membrane lipids: a new tool for reconstructing ancient sea water temperatures? *Earth and Planetary Science Letters*, 204(1): 265–274. DOI:10.1016/S0012-821X(02)00979-2.

Schuldenrein, J., Wright, R., Khan, M. A., 2007. *Harappan geoarchaeology reconsidered: Holocene landscapes and environments of the Greater Indus Plain. Settlement and Society: Essays Dedicated to Robert McCormick Adams, Cotsen Institute of Archaeology*, University of California, Los Angeles: 83–116.

Schulz, H., von Rad, U., Ittekkot, V., 2002. Planktic Foraminifera, particle flux and oceanic productivity off Pakistan, NE Arabian Sea; modern analogues and application to the palaeoclimatic record. In: The Tectonic and Climatic Evolution of the Arabian Sea region. In: Clift, P. D., Kroon, D., Gaedicke, C., Craig, J. (eds.), *Special Publication*. Geological Society, London, pp. 499–516.

Scott, J. C., 1998. *Seeing Like a State: How Certain Schemes to Improve the Human Condition Have Failed*. Yale University Press, New Haven, 460 pp.

Scussolini, P., Tran, T. V. T., Koks, E., Diaz-Loaiza, A., Ho, P. L., Lasage, R., 2017. Adaptation to sea level rise: A multidisciplinary analysis for Ho Chi Minh City, Vietnam. *Water Resources Research*, 53(12): 10841–10857. DOI:10.1002/2017WR021344.

Sen, A., 1977. Starvation and exchange entitlements: a general approach and its application to the great Bengal famine. *Cambridge Journal of Economics*, 1(1): 33–59. DOI:10.1093/oxfordjournals.cje.a035349.

Shakir, A. S., Khan, N. M., Qureshi, M. M., 2010. Canal water management: Case study of upper Chenab Canal in Pakistan. *Irrigation and Drainage*, 59(1): 76–91. 10.1002/ird.556.

Shanahan, T. M., Overpeck, J. T., Anchukaitis, K. J., Beck, J. W., Cole, J. E., Dettman, D. L., Peck, J. A., Scholz, C. A., King, J. W., 2009. Atlantic forcing of persistent drought in West Africa. *Science*, 324: 377–380. DOI:10.1126/science.1166352.

Shao, X., Wang, Y., Cheng, H., Kong, X., Wu, J., Edwards, R. L., 2006. Long-term trend and abrupt events of the Holocene Asian monsoon inferred from a stalagmite δ10O record from Shennongjia in Central China. *Chinese Science Bulletin*, 51(2): 222—228. DOI:10.1007/s11434-005-0882-6.

Shapiro, J., 2001. *Mao's War Against Nature: Politics and the Environment in Revolutionary China*. Cambridge University Press, Cambridge, 276 pp.

Sharma, A., Sharma, D., Panda, S. K., Dubey, S. K., Pradhan, R. K., 2018. Investigation of temperature and its indices under climate change scenarios over different regions of Rajasthan state in India. *Global and Planetary Change*, 161: 82–96. DOI:10.1016/j.gloplacha.2017.12.008.

Sharmila, S., Joseph, S., Sahai, A. K., Abhilash, S., Chattopadhyay, R., 2015. Future projection of Indian summer monsoon variability under climate change scenario: An assessment from CMIP5 climate models. *Global and Planetary Change*, 124: 62–78. DOI:10.1016/j.gloplacha.2014.11.004.

Sheffield, J., Goteti, G., Wen, F., Wood, E. F., 2004. A simulated soil moisture based droughtanalysis for the United States. *Journal of Geophysical Research: Atmospheres*, 109: 1984–2012.

Shelach, G., 2006. Economic adaptation, community structure, and sharing strategies of households at early sedentary communities in northeast China. *Journal of Anthropological Archaeology*, 25(2006): 318–345.

Shelach, G., 2009. *Prehistoric Socities on the Northern Frontiers of China*. Equinox Publishing, London.

Shelach, G., 2012. On the invention of pottery. *Science*, 336(6089): 1644–1645. 10.1126/science.1224119.

Shelach-Lavi, G., Teng, M., Goldsmith, Y., Wachtel, I., Stevens, C. J., Marder, O., Wan, X., Wu, X., Tu, D., Shavit, R., Polissar, P., Xu, H., Fuller, D. Q., 2019. Sedentism and plant cultivation in northeast China emerged during affluent conditions. *PLOS ONE*, 14(7): e0218751. 10.1371/journal.pone.0218751.

Sheng, E., Yu, K., Xu, H., Lan, J., Liu, B., Che, S., 2015. Late Holocene Indian summer monsoon precipitation history at Lake Lugu, northwestern Yunnan Province, southwestern China. *Palaeogeography, Palaeoclimatology, Palaeoecology*, 438: 24–33. DOI:10.1016/j.palaeo.2015.07.026.

Sheng, M., Wang, X., Zhang, S., Chu, G., Su, Y., Yang, Z., 2017. A 20,000-year high-resolution pollen record from Huguangyan Maar Lake in tropical–subtropical South China. *Palaeogeography, Palaeoclimatology, Palaeoecology*, 472: 83–92.

Shi, S., 1956. 氾勝之書今釋：初稿 / 石声漢著 *Fan Shengzhi shu jin shi : chu gao Shi Shenghan zhu (The Book of Fan Shengzhi)* Kexue Chubanshe, Beijing, China.

Shi, S., 1974. *On "Fan Sheng-chih shu" : An Agriculturist Book of China written by Fan Sheng-chih in the first century B.C.* Kexue Chubanshe, Beijing, China.

Shi, Y. F., 2002. Characteristics of late quaternary monsoonal glaciation on the Tibetan Plateau and in East Asia. *Quaternary International*, 97–98: 79–91.

Shimpei, A., 2007. Agricultural technologies of terraced rice cultivation in the Ailao Mountains, Yunnan, China.*Asian and African Area Studies*, 6(2): 173–196.

Shinde, V., Deshpande, S. S., Osada, T., Uno, T., 2006. Basic issues in Harappan archaeology: some thoughts. *Ancient Asia*, 1: 63–72. DOI:10.5334/aa.06107.

Shiva, V., 2016. *The Violence of the Green Revolution: Third World Agriculture, Ecology, and Politics*. University Press of Kentucky, 264.

Shoda, S., Lucquin, A., Yanshina, O., Kuzmin, Y., Shevkomud, I., Medvedev, V., Derevianko, E., Lapshina, Z., Craig, O. E., Jordan, P., 2020. Late Glacial hunter-gatherer pottery in the Russian far east: Indications of diversity in origins and use. *Quaternary Science Reviews*, 229: 106124. DOI:10.1016/j.quascirev.2019.106124.

Siegel, B. R., 2018. *Hungry Nation: Food, Famine, and The Making of Modern India*. Cambridge University Press, Cambridge, 279 pp.

Simoons, F. J., 1991. *Food in China: A Cultural and Historical Inquiry*. CRC Press, Boca Raton, FL.

Singapore Red Cross, 2010. Pakistan Floods:The Deluge of Disaster – Facts & Figures as of 15 September 2010. ReliefWeb, www.reliefweb.int/rw/rwb.nsf/db900SID/LSGZ89GD7W?OpenDocument.

Singh, A., Thomsen, K. J., Sinha, R., Buylaert, J. -P., Carter, A., Mark, D. F., Mason, P. J., Densmore, A. L., Murray, A. S., Jain, M., Paul, D., Gupta, S., 2017. Counter-intuitive influence of Himalayan river morphodynamics on Indus Civilisation urban settlements. *Nature Communications*, 8(1): 1617. 10.1038/s41467-017–01643-9.

Singhvi, A. K., Williams, M. A. J., Rajaguru, S. N., Misra, V. N., Chawla, S., Stokes, S., Chauhan, N., Francis, T., Ganjoo, R. K., Humphreys, G. S., 2010. A 200 ka record of climatic change and dune activity in the Thar Desert, India. *Quaternary Science Reviews*, 29(23–24): 3095–3105. DOI:10.1016/j.quascirev.2010.08.003.

Sinha, A., Cannariato, K. G., Stott, L. D., Li, H. -C., You, C. -F., Cheng, H., Edwards, R. L., Singh, I. B., 2005. Variability of southwest Indian summer monsoon precipitation during the Bolling-Allerod. *Geology*, 33(10): 813–816.

Sivakumar, M. V. K., Singh, P., Williams, J. S., 1983. *Agroclimatic Aspects in Planning for Improved Productivity in Alfisols', Alfisols in the Semiarid Tropics: A Consultant's Workshop*, ICRISAT Centre, India.

Slingerland, R., Smith, N. D., 2004. River avulsions and their deposits. *Annu. Rev. Earth Planet. Sci.*, 32: 257–285.

Smith, B. D., 2015. A comparison of niche construction theory and diet breadth models as explanatory frameworks for the initial domestication of plants and animals. *Journal of Archaeological Research*, 23(3): 215–262.

Smith, B. N., Epstein, S., 1971. Two categories of 13C/12C ratios for higher plants. *Plant Physiology*, 47: 380–384.

Smith, S. J., Edmonds, J., Hartin, C. A., Mundra, A., Calvin, K., 2015. Near-term acceleration in the rate of temperature change. *Nature Climate Change*, 5: 333. DOI:10.1038/nclimate2552.

Solomon, S., Plattner, G. -K., Knutti, R., Friedlingstein, P., 2009. Irreversible climate change due to carbon dioxide emissions. *Proceedings of the National Academy of Sciences*, 106(6): 1704–1709.

Sorrel, P., Eymard, I., Leloup, P. -H., Maheo, G., Olivier, N., Sterb, M., Gourbet, L., Wang, G., Jing, W., Lu, H., Li, H., Yadong, X., Zhang, K., Cao, K., Chevalier, M. -L., Replumaz, A., 2017. Wet tropical climate in SE Tibet during the Late Eocene. *Scientific Reports*, 7(1): 7809. 10.1038/s41598-017–07766-9.

Sperber, K. R., Annamalai, H., Kang, I. -S., Kitoh, A., Moise, A., Turner, A., Wang, B., Zhou, T., 2013. The Asian summer monsoon: an intercomparison of CMIP5 vs. CMIP3 simulations of the late 20th century. *Climate Dynamics*, 41: 2711–2744. DOI 10.1007/s00382-012–1607-6.

Spicer, R. A., Valdes, P. J., Spicer, T. E. V., Craggs, H. J., Srivastava, G., Mehrotra, R. C., Yang, J., 2009. New developments in CLAMP: Calibration using global gridded meteorological data. *Palaeogeography, Palaeoclimatology, Palaeoecology*, 283(1): 91–98. DOI:10.1016/j.palaeo.2009.09.009.

Spignesi, S. J., 2004. *Catastrophe!: The 100 Greatest Disasters of All Time*. Citadel Press, New York, p. 37.

Stager, J., Mayewski, P. A., 1997. Abrupt early to mid-Holocene climatic transition registered at the equator and the poles. *Science*, 276(5320): 1834–1836.

Staubwasser, M., Sirocko, F., Grootes, P. M., Segl, M., 2003. Climate change at the 4.2 ka BP termination of the Indus valley civilization and Holocene south Asian monsoon variability. *Geophysical Research Letters*, 30: 1425. DOI:10.1029/2002GL016822.

Staubwasser, M., Weiss, H., 2006. Holocene climate and cultural evolution in late prehistoric–early historic West Asia. *Quaternary Research*, 66(3): 372–387. DOI:10.1016/j.yqres.2006.09.001.

Steffen, W., Rockström, J., Richardson, K., Lenton, T. M., Folke, C., Liverman, D., Summerhayes, C. P., Barnosky, A. D., Cornell, S. E., Crucifix, M., 2018. Trajectories of the earth system in the anthropocene. *Proceedings of the National Academy of Sciences*, 115(33): 8252–8259. DOI:10.1073/pnas.1810141115.

Stein, M. A., 1942. A survey of ancient sites along the 'Lost' Saraswati river. *Geographical Journal*, 99: 173–182.

Steinke, S., Groeneveld, J., Johnstone, H., Rendle-Bühring, R., 2010. East Asian summer monsoon weakening after 7.5 Ma: Evidence from combined planktonic foraminifera

Mg/Ca and δ18O (ODP Site 1146; northern South China Sea). *Palaeogeography, Palaeoclimatology, Palaeoecology*, 289(1–4): 33–43. DOI:10.1016/j.palaeo.2010.02.007.

Stocker, T. F., Qin, D., Plattner, G. -K., Alexander, L. V., Allen, S. K., Bindoff, N. L., Bréon, F. -M., Church, J. A., Cubasch, U., Emori, S., Forster, P., Friedlingstein, P., Gillett, N., Gregory, J. M., Hartmann, D. L., Jansen, E., Kirtman, B., Knutti, R., Kumar, K. K., Lemke, P., Marotzke, J., Masson-Delmotte, V., Meehl, G. A., Mokhov, I. I., Piao, S., Ramaswamy, V., Randall, D., Rhein, M., Rojas, M., Sabine, C., Shindell, D., Talley, L. D., Vaughan, D. G., Xie, S. -P., 2013. Technical summary. In: Stocker, T. F. et al. (eds.), *Climate Change 2013: The Physical Science Basis. Contribution of Working Group I to the Fifth Assessment Report of the Intergovernmental Panel on Climate Change.* Cambridge University Press, Cambridge, United Kingdom and New York, NY.

Stokes, E., 1975. Agrarian society and the pax britannica in Northern India in the early nineteenth century. *Modern Asian Studies*, 9(4): 505–528.

Stone, Glenn D., 2002. Both sides now: Fallacies in the genetic-modification wars, implications for developing countries, and anthropological perspectives. *Current Anthropology*, 43(4): 611–630. DOI;10.1086/341532.

Stone, G. D., 2019. Commentary: New histories of the Indian green revolution. *The Geographical Journal*, 185(2): 243–250.

Stone, R., 2008. Three Gorges Dam: Into the unknown. *Science*, 321(5889): 628–632. 10.1126/science.321.5889.628.

Stott, L., Cannariato, K., Thunell, R., Haug, G. H., Koutavas, A., Lund, S., 2004. Decline of surface temperature and salinity in the western tropical Pacific Ocean in the Holocene epoch. *Nature*, 431: 56–59.

Stuiver, M., Grootes, P. M., 2000. GISP2 oxygen isotope ratios. *Quaternary Research (New York)*, 53(3): 277–284.

Stuiver, M., Reimer, P. J., 1993. Extended C-14 data-base and revised calib 3.0 C-14 age calibration program. *Radiocarbon*, 35(1): 215–230.

Subramanian, K., 2015. *Revisiting the Green Revolution: Irrigation and food production in twentieth-century India.* PhD Thesis, King's College London, p. 259.

Sugiura, T., Sumida, H., Yokoyama, S., Ono, H., 2012. Overview of recent effects of global warming on agricultural production in Japan. *Japan Agricultural Research Quarterly: JARQ*, 46(1): 7–13. DOI:10.6090/jarq.46.7.

Sun, J., Oey, L., Xu, F. H., Lin, Y. C., 2017. Sea level rise, surface warming, and the weakened buffering ability of South China Sea to strong typhoons in recent decades. *Scientific Reports*, 7(1): 7418. DOI:10.1038/s41598-017-07572-3.

Sun, X., Wang, P., 2005. How old is the Asian monsoon system? Palaeobotanical records from China. *Palaeogeography, Palaeoclimatology, Palaeoecology*, 222(3–4): 181–222.

Sun, X., Xu, L., Luo, Y., Chen, X., 2000. The vegetation and climate at the last glaciation on the emerged continental shelf of the South China Sea. *Palaeogeography, Palaeoclimatology, Palaeoecology*, 160(2000): 301–316.

Sweeney, M. T., McCouch, S. R., 2007. The complex history of the domestication of rice. *Annals of Botany*, 100: 951–957.

Syvitski, J. P., Brakenridge, G. R., 2013. Causation and avoidance of catastrophic flooding along the Indus River, Pakistan. *GSA Today*, 23(1): 4–10.

Syvitski, J. P., Kettner, A. J., Overeem, I., Giosan, L., Brakenridge, G. R., Hannon, M., Bilham, R., 2013. Anthropocene metamorphosis of the Indus Delta and lower floodplain. *Anthropocene*, 3: 24–35. DOI:10.1016/j.ancene.2014.02.003.

Syvitski, J. P. M., C., V., Kettner, A. J., Green, P., 2005. Impact of humans on the flux of terrestrial sediment to the global coastal ocean. *Science*, 308: 376–380.

Syvitski, J. P. M., Kettner, A. J., 2011. Sediment flux and the Anthropocene. *Philosophical Transactions of the Royal Society of London, Series A: Mathematical and Physical Sciences*, 369: 957–975. DOI:10.1098/rsta.2010.0329.

Tada, R., Zheng, H., Clift, P. D., 2016. Evolution and variability of the Asian monsoon and its potential linkage with uplift of the Himalaya and Tibetan Plateau. *Progress in Earth and Planetary Science*, 3(4): 1–26. DOI 10.1186/s40645-016-0080-y.

Tainter, J., A, 1988. *The Collapse of Complex Societies*. Cambridge University Press, Cambridge.

Taiz, L., Zeiger, E., 2002. *Plant Physiology*. Sinauer Associates, Sunderland MA, 623 pp.

Takata, K., Saito, K., Yasunari, T., 2009. Changes in the Asian monsoon climate during 1700 –1850 induced by preindustrial cultivation. *Proceedings of the National Academy of Sciences*, 106(24): 9586–9589.

Talbot, M., 1990. A review of the palaeohydrological interpretation of carbon and oxygen isotopic ratios in primary lacustrine carbonates. *Chemical Geology: Isotope Geoscience Section*, 80(4): 261–279.

Tan, L., Cai, Y., Cheng, H., An, Z., Edwards, R. L., 2009. Summer monsoon precipitation variations in central China over the past 750 years derived from a high-resolution absolute-dated stalagmite. *Palaeogeography, Palaeoclimatology, Palaeoecology*, 280 (3–4): 432–439.

Tan, L. C., An, Z. S., Cai, Y. J., 2008. The hydrological exhibition of 4.2 ka BP event in China and its global linkages. *Geological Reviews*, 54: 94–104.

Tan, Z., Huang, C., Pang, J., 2011. Holocene wildfires related to climate and land-use change over the Weihe River Basin, China. *Quaternary International*, 234: 167–173.

Tao, J. -s., 1976. *The Jurchen in Twelfth Century China*. University of Washington Press, Seattle.

Tao, S., Eglinton, T. I., Montluçon, D. B., McIntyre, C., Zhao, M., 2015. Pre-aged soil organic carbon as a major component of the Yellow River suspended load: Regional significance and global relevance. *Earth and Planetary Science Letters*, 414: 77–86. DOI:10.1016/j.epsl.2015.01.004.

Tariq, M. A. U. R., van de Giesen, N., 2012. Floods and flood management in Pakistan. *Physics and Chemistry of the Earth, Parts A/B/C*, 47–48: 11–20. DOI:10.1016/j.pce.2011.08.014.

Tate, E. L., Farquharson, F. A. K., 2000. Simulating reservoir management under the threat of sedimentation: the case of Tarbela dam on the river Indus. *Water Resources Management*, 14(3): 191–208. 10.1023/a:1026579230560.

Thomas, E. R., Wolff, E. W., Mulvaney, R., Steffensen, J. P., Johnsen, S. J., Arrowsmith, C., White, J. W. C., Vaughn, B., Popp, T., 2007. The 8.2ka event from Greenland ice cores. *Quaternary Science Reviews*, 26: 70–81.

Thompson, L. G., Yao, T., Mosley-Thompson, E., Davis, M. E., Henderson, K. A., Lin, P. -N., 2000. A high-resolution millennial record of the South Asian Monsoon from Himalayan ice cores. *Science*, 289: 1916–1919.

Thornalley, D. J. R., Barker, S., Broecker, W. S., Elderfield, H., I. N. McCave, 2011. The deglacial evolution of North Atlantic deep convection. *Science*, 331: 202–205.

Tian, X., Matsui, T., Li, S., Yoshimoto, M., Kobayasi, K., Hasegawa, T., 2010. Heat-induced floret sterility of hybrid rice (Oryza sativa L.) cultivars under humid and low wind conditionsin the field of Jianghan basin, China. *Plant Production Science*, 13(3): 243–251. DOI:10.1626/pps.13.243.

Tipple, B. J., Pagani, M., 2010. A 35 Myr North American leaf-wax compound-specific carbon and hydrogen isotope record: Implications for C4 grasslands and hydrologic cycle dynamics. *Earth and Planetary Science Letters*, 299: 250–262. DOI:10.1016/j.epsl.2010.09.006.

Tiwari, M., Ramesh, R., Somayajulu, B. L. K., Jull, A. J. T., Burr2, G. S., 2005. Solar control of southwest monsoon on centennial timescales. *Current Science*, 89(9): 1583–1588.

Tjallingii, R., Stattegger, K., Stocchi, P., Saito, Y., Wetzel, A., 2014. Rapid flooding of the southern Vietnam shelf during the early to mid-Holocene. *Journal of Quaternary Science*, 29(6): 581–588. DOI:10.1002/jqs.2731.

Tong, W., 1984. 磁山遗址的原始农业依存及其相关的问题 Cishan Yizhi de Yuanshi Nongye Yicun Jiqi Xianguan de Wenti (The Agricultural Remains from the Site of Cishan and related Problems. *Nongye Kaogu*, 1984(1): 194–207.

Trentesaux, A., Liu, Z., Colin, C. J. G., Boulay, S., Wang, P., Arnold, E. M., Buehring, C. J., Chen, M. -P., Clift, P. D., Colin, C. J. G., Farrell, J. W., Higginson, M. J., Jian, Z., Kuhnt, W., Laj, C. E., Lauer-Leredde, C., Leventhal, J. S., Li, A., Li, Q., Lin, J., McIntyre, K., Miranda, C. R., Nathan, S. A., Shyu, J. -P., Solheid, P. A., Su, X., Tamburini, F., Trentesaux, A., Wang, L., 2006. Pleistocene paleoclimatic cyclicity of southern China; clay mineral evidence recorded in the South China Sea (ODP Site 1146). *Proceedings of the Ocean Drilling Program, Scientific Results (CD-ROM)*, 184: 10.

Truschke, A., 2017. *Aurangzeb: The Life and Legacy of India's Most Controversial King*. Stanford University Press, Palo Alto, CA, p. 152.

Tsuboki, K., Yoshioka, M. K., Shinoda, T., Kato, M., Kanada, S., Kitoh, A., 2015. Future increase of supertyphoon intensity associated with climate change. *Geophysical Research Letters*, 42(2): 646–652. DOI:10.1002/2014gl061793.

Tu, J. -Y., Chou, C., Chu, P. -S., 2009. The abrupt shift of typhoon activity in the vicinity of Taiwan and its association with Western North Pacific–East Asian climate change. *Journal of Climate*, 22(13): 3617–3628. DOI:10.1175/2009JCLI2411.1.

Turner, A. G., Annamalai, H., 2012. Climate change and the south Asian summer monsoon. *Nature Climate Change*, 2: 1–9.

Twitchett, D. C., Fairbank, J. K., Feuerwerker, A., Peterson, W. J., Liuv, K. -C., MacFarquhar, R., 1978. *The Cambridge History of China, 1991*. Cambridge University Press, Cambridge.

Twitchett, D. C., Fairbank, J. K., Franke, H., 1994. *The Cambridge History of China: Volume 6, Alien Regimes and Border States, 907–1368*, 6. Cambridge University Press, Cambridge.

Tyndall, J., 1861. On the absorption and radiation of heat by gases and vapours. *Philosophical Magazine, Series*, 4, 22: 273–285.

UN Office for the Coordination of Humanitarian Affairs, 2009. Nepal: OCHA Koshi Flood Response Update 13 May 2009. https://reliefweb.int/report/nepal/nepal-ocha-koshi-flood-response-update-13-may-2009.

Upadhyaya, H. D., Vetriventhan, M., Dwivedi, S. L., Pattanashetti, S. K., Singh, S. K., 2016. 8 – Proso, barnyard, little, and kodo millets. In: Singh, M., Upadhyaya, H. D. (eds.), *Genetic and Genomic Resources for Grain Cereals Improvement*. Academic Press, San Diego, pp. 321–343.

Valdiya, K. S., 2002. *Saraswati: The River that Disappeared*. 1st. University Press (India) Limited, Hyderabad, India, 116 pp.

Van Buren, M., 1996. Rethinking the vertical archipelago. *American Anthropologist*, 98(2): 338–351. 10.1525/aa.1996.98.2.02a00100.

Van de Noort, R., 2011. Conceptualizing climate change archaeology. *Antiquity*, 85: 1039–1048.

Van der Leeuw, S., Redman, C., 2002. Placing archaeology at the center of socionatural studies. *American Antiquity*, 67(4): 597–605.

van Vliet, M. T. H., Sheffield, J., Wiberg, D., Wood, E. F., 2016. Impacts of recent drought and warm years on water resources and electricity supply worldwide. *Environmental Research Letters*, 11(12): 124021.

Vanwalleghem, T., Gómez, J., Amate, J. I., de Molina, M. G., Vanderlinden, K., Guzmán, G., Laguna, A., Giráldez, J.V., 2017. Impact of historical land use and soil management change on soil erosion and agricultural sustainability during the Anthropocene. *Anthropocene*, 17: 13–29.

Vaughan, D., Lu, B. R., Tomooka, N., 2008. The evolving story of rice evolution. *Plant Science*, 174(2008): 394–408.

Vitousek, S., Barnard, P. L., Fletcher, C. H., Frazer, N., Erikson, L., Storlazzi, C. D., 2017. Doubling of coastal flooding frequency within decades due to sea-level rise. *Scientific Reports*, 7(1): 1399. DOI:10.1038/s41598-017-01362-7.

von Rad, U., Schaaf, M., Michels, K. H., Schulz, H., Berger, W. H., Sirocko, F., 1999. A 5,000 year record of climate change in varved sediments from the Oxygen Minimum Zone off Pakistan (Northeastern Arabian Sea). *Quaternary Research*, 51: 39–53.

Wainer, I., Webster, P. J., 1996. Monsoon/El Niño-Southern oscillation relationships in a simple coupled ocean-atmosphere model. *Journal of Geophysical Research: Oceans*, 101(C11): 25599–25614. 10.1029/96JC00670.

Walsh, D., 2010. *Pakistan Floods: The Indus Delta*, The Guardian, London.

Wan, H. L., Huang, C. C., Pang, J. L., 2010a. Holocene extreme floods of the Baoji Gorges of the Weihe River. *Quaternary Sciences*, 2010(30): 430–440.

Wan, S., Clift, P. D., Li, A., Li, T., Yin, X., 2010b. Geochemical records in the South China Sea: implications for East Asian summer monsoon evolution over the last 20 Ma. In: Clift, P. D., Tada, R., Zheng, H. (eds.), *Monsoon Evolution and Tectonics–Climate Linkage in Asia*. Special Publication. Geological Society, London, pp.245–263.

Wang, A., Lettenmaier, D. P., Sheffield, J., 2011a. Soil moisture drought in China, 1950–2006. *Journal of Climate*, 24(13): 3257–3271. DOI:10.1175/2011jcli3733.1.

Wang, B. (ed.), 2006. *The Asian Monsoon*. Springer-Verlag, Berlin, 795 pp.

Wang, B., Ding, Q. H., 2008. Global monsoon: Dominant mode of annual variation in the tropics. *Dynamics of Atmospheres and Oceans*, 44: 165–183.

Wang, B., Kim, H. J., Kikuchi, K., Kitoh, A., 2011b. Diagnostic metrics for evaluation of annual and diurnal cycles. *Climate Dynamics*, 37: 941–955.

Wang, B., Wu, R., Li, T., 2003. Atmosphere–warm ocean interaction and its impacts on Asian–Australian monsoon variation. *Journal of Climate*, 16(8): 1195–1211. DOI:10.1175/1520-0442(2003)16<195:aoiaii>2.0.co;2.

Wang, B., Xu, S., Wu, L., 2012a. Intensified Arabian sea tropical storms. *Nature*, 489: E1. DOI:10.1038/nature11470.

Wang, E., Kirby, E., Furlong, K. P., Soest, M. v., Xu, G., Shi, X., Kamp, P. J. J., Hodges, K. V., 2012b. Two-phase growth of high topography in eastern Tibet during the Cenozoic. *Nature Geoscience*, 5: 640–645. DOI:10.1038/ngeo1538.

Wang, G., Su, J., Ding, Y., Chen, D., 2007a. Tropical cyclone genesis over the South China Sea. *Journal of Marine Systems*, 68(3): 318–326. DOI:10.1016/j.jmarsys.2006.12.002.

Wang, H., Yang, Z., Saito, Y., Liu, J.P., Sun, X., Wang, Y., 2007b. Stepwise decreases of the Huanghe (Yellow River) sediment load (1950–2005): Impacts of climate change and human activities. *Global and Planetary Change*, 57(3): 331–354. DOI:10.1016/j.gloplacha.2007.01.003.

Wang, J., 1978. Xiachun Wenhua (The Xiachuan Culture). *Kaogu Xuebao*, 3: 259–288.

Wang, L., Chen, W., 2014. A CMIP5 multimodel projection of future temperature, precipitation, and climatological drought in China. *International Journal of Climatology*, 34(6): 2059–2078. DOI:10.1002/joc.3822.

Wang, Q. -S., Pan, C. -H., Zhang, G. -Z., 2018. Impact of and adaptation strategies for sea-level rise on Yangtze River Delta. *Advances in Climate Change Research*, 9(2): 154–160. DOI:10.1016/j.accre.2018.05.005.

Wang, R., Wu, L., Wang, C., 2011c. Typhoon track changes associated with global warming. *Journal of Climate*, 24: 3748–3752. DOI:10.1175/JCLI-D-11-00074.1.

Wang, S., Fu, B., Piao, S., Lü, Y., Ciais, P., Feng, X., Wang, Y., 2015. Reduced sediment transport in the Yellow River due to anthropogenic changes. *Nature Geoscience*, 9: 38. DOI:10.1038/ngeo2602.

Wang, S. S. Y., Kim, H., Coumou, D., Yoon, J. -H., Zhao, L., Gillies, R. R., 2019. Consecutive extreme flooding and heat wave in Japan: Are they becoming a norm? *Atmospheric Science Letters*, 20(10): e933. DOI:10.1002/asl.933.

Wang, Y., Cheng, H., Edwards, R. L., He, Y., Kong, X., An, Z., Wu, J., Kelly, M. J., Dykoski, C. A., Li, X. D., 2005. The Holocene Asian monsoon; links to solar changes and North Atlantic climate. *Science*, 308: 854–857.

Wang, Y., Cheng, H., Edwards, R. L., Kong, X., Shao, X., Chen, S., Wu, J., Jiang, X., Wang, X., An, Z., 2008. Millennial- and orbital-scale changes in the East Asian monsoon over the past 224,000 years. *Nature*, 451: 1090–1093. DOI:10.1038/nature06692.

Wang, Y., Wan, Q., Meng, W., Liao, F., Tan, H., Zhang, R., 2011d. Long-term impacts of aerosols on precipitation and lightning over the Pearl River Delta megacity area in China. *Atmospheric Chemistry and Physics*, 11: 12421–12436. DOI:10.5194/acp-11-12421-2011.

Wang, Y. J., Cheng, H., Edwards, R. L., An, Z. S., Wu, J. Y., Shen, C. -C., Dorale, J. A., 2001. A high-resolution absolute-dated late Pleistocene Monsoon record from Hulu Cave, China. *Science*, 294: 2345–2348.

Wang, Z., Zhan, Q., Long, H., Saito, Y., Gao, X., Wu, X., Li, L. I. N., Zhao, Y., 2013. Early to mid-Holocene rapid sea-level rise and coastal response on the southern Yangtze delta plain, China. *Journal of Quaternary Science*, 28(7): 659–672. 10.1002/jqs.2662.

Wanner, H., Beer, J., Bütikofer, J., Crowley, T. J., Cubasch, U., Flückiger, J., Goosse, H., Grosjean, M., Joos, F., Kaplan, J. O., 2008. Mid-to late holocene climate change: an overview. *Quaternary Science Reviews*, 27(19–20): 1791–1828.

Wasson, R. J., Smith, G. I., Agrawal, D. P., 1984. Late quaternary sediments, minerals, and inferred geochemical history of Didwana lake, Thar desert India. *Palaeogeography, Palaeoclimatology, Palaeoecology*, 46: 345–372.

Weaver, A. J., Saenko, O. A., Clark, P. U., Mitrovica, J. X., 2003. Meltwater pulse 1A from Antarctica as a trigger of the Bølling-Allerød warm interval. *Science*, 299: 1709–1713.

Weber, S., 1998. Out of Africa: the initial impact of millets in South Asia. *Current Anthropology*, 39(2): 267–274. 10.1086/204725.

Weber, S., 1999. Seeds of urbanism: palaeoethnobotany and the Indus Civilization. *Antiquity*, 73: 813–826.

Weber, S., 2003. Archaeobotany at Harappa: Indications for Change. In: Weber, S., Belcher, B. (eds.), *Indus Ethnobiology: New Perspectives from the Field*, Lexington Books, Lanham, MD, pp.175–198.

Weber, S., Kashyap, A., Harriman, D., 2010a. Does size matter: the role and significance of cereal grains in the Indus civilization. *Archaeological and Anthropological Sciences*, 2: 35–43.

Weber, S., Lehman, H., Barela, T., Hawks, S., Harriman, D., 2010b. Rice or millets: Early farming strategies in prehistoric central Thailand. *Archaeological and Anthropological Sciences*, 2(2): 79–88.

Weber, S. A., 1989. *Plants and harappan Subsistence: An Example of Stability and Change from Rojdi*. PhD Thesis, University of Pennsylvania.

Weber, S. A., Barela, T., Lehman, H., 2010c. Ecological continuity: An explanation for agricultural diversity in the Indus Civilization and beyond. *Man and Environment*, 35 (1): 62–75.

Webster, P. J., 1987. The elementary monsoon. In: Fein, J. S., Stephens, P. L. (eds.), *Monsoons*. John Wiley, New York, pp. 3–32.

Webster, P. J., Magana, V. O., Palmer, T. N., Shukla, J., R. A., Tomas, M., Yanai, Y., Yasunari, T., 1998. Monsoons: Processes, predictability, and the prospects for prediction, in the TOGA decade. *Journal of Geophysical Research*, 103: 14,451–14,510.

Wei, K. Y., Lee, M. Y., Duan, W. W., Chen, C. Y., Wang, C. H., 1998. Palaeoceanographic change in the northeastern South China Sea during the last 15,000 years. *Journal of Quaternary Science*, 13: 55–64.

Weinkle, J., Maue, R., Pielke, R., 2012. Historical global tropical cyclone landfalls. *Journal of Climate*, 25(13): 4729–4735. 10.1175/JCLI-D-11–00719.1.

Weisman, S. R., 1987. *India's Drought Is Worst in Decades*, New York Times, New York, pp. 1001016.

Weisskopf, A., 2018. Elusive wild foods in South East Asian subsistence: Modern ethnography and archaeological phytoliths. *Quaternary International*, 489: 80–90. DOI:10.1016/j.quaint.2016.09.028.

Wells, N. A., Dorr, J. A., 1987. Shifting of the Kosi River, northern India. *Geology*, 15(3): 204–207. 10.1130/0091–7613(1987)15<204:sotkrn>2.0.CO;2.

Wen, X., Bai, S., Na, Z., Chamberlain, C. P., Wang, C., Huang, C., Zhang, Q., 2012. Interruptions of the ancient shu civilization: Triggered by climate change or natural disaster? *International Journal of Earth Sciences*: 1–15.

Wescoat, J. L., Halvorson, S. J., Mustafa, D., 2000. Water management in the Indus basin of Pakistan: A half-century perspective. *International Journal of Water Resources Development*, 16(3): 391–406. 10.1080/713672507.

West, A. J., Galy, A., Bickle, M. J., 2005. Tectonic and climatic controls on silicate weathering. *Earth and Planetary Science Letters*, 235: 211–228. DOI:10.1016/j.epsl.2005.03.020.

West, B. J., Han, W., Li, Y., 2018. The role of Oceanic processes in the initiation of Indian summer monsoon intraseasonal oscillations over the Indian Ocean. *Journal of Geophysical Research: Oceans*, 123(5): 3685–3704. DOI:10.1029/2017jc013564.

White, S., 2005. Sediment yield prediction and modelling. *Hydrological Processes*, 19(15): 3053–3057. 10.1002/hyp.6003.

Wiersma, A., Roche, D., Renssen, H., 2011. Fingerprinting the 8.2 ka event climate response in a coupled climate model. *Journal of Quaternary Science*, 26: 118–127.

Wilks, D. S., 1995. *Statistical Methods in the Atmospheric Sciences*, 59. Academic Press, San Diego, CA, USA, p. 467.

Williams, A., 2008. Turning the tide: Recognizing climate change refugees in International Law. *Law & Policy*, 30(4): 502–529. DOI:10.1111/j.1467-9930.2008.00290.x.

Willmott, W. E., 1989. Dujiangyan: Irrigation and society in Sichuan, China. *The Australian Journal of Chinese Affairs*, 22: 143–153. 10.2307/2158849.

Wilson, E., 1911. The use of maize. *Agricultural Journal of the Union of South Africa*, 1 (4): 386.

Wilson, H. H., 1868. *The Vishnu Purana: A System of Hindu Mythology and Tradition*, 9. Trübner & co., London, p. 342.

Wilson, R., Anchukaitis, K., Briffa, K. R., Büntgen, U., Cook, E., D'Arrigo, R., Davi, N., Esper, J., Frank, D., Gunnarson, B., Hegerl, G., Helama, S., Klesse, S., Krusic, P. J., Linderholm, H. W., Myglan, V., Osborn, T. J., Rydval, M., Schneider, L., Schurer, A., Wiles, G., Zhang, P., Zorita, E., 2016. Last millennium northern hemisphere summer temperatures from tree rings: Part I: The long term context. *Quaternary Science Reviews*, 134: 1–18. DOI:10.1016/j.quascirev.2015.12.005.

Wing, S. L., Greenwood, D. R., Allen, J. R. L., Hoskins, B. J., Sellwood, B. W., Spicer, R. A., Valdes, P. J., 1993. Fossils and fossil climate: the case for equable continental interiors in the Eocene. Philosophical transactions of the royal society of London. *Series B: Biological Sciences*, 341(1297): 243–252. DOI:10.1098/rstb.1993.0109.

Wright, R., 2010. *The Ancient Indus: Urbanism, Economy and Society*. Cambridge University Press, Cambridge.

Wright, R. P., Bryson, R. A., Schuldenrein, J., 2008. Water supply and history: Harappa and the Beas regional survey. *Antiquity*, 82: 37–48.

Wu, L., Wang, B., Geng, S., 2005. Growing typhoon influence on east Asia. *Geophysical Research Letters*, 32(L18703). DOI:10.1029/2005GL022937.

Wu, L., Zong, H., Liang, J., 2012a. Observational Analysis of Tropical Cyclone Formation Associated with Monsoon Gyres. *Journal of the Atmospheric Sciences*, 70(4): 1023–1034. DOI:10.1175/JAS-D-12-0117.1.

Wu, Q., Zhao, Z., Liu, L., Granger, D. E., Wang, H., Cohen, D. J., Wu, X., Ye, M., Bar-Yosef, O., Lu, B., Zhang, J., Zhang, P., Yuan, D., Qi, W., Cai, L., Bai, S., 2016a. Outburst flood at 1920 BCE supports historicity of China's Great Flood and the Xia dynasty. *Science*, 353(6299): 579.

Wu, W., Liu, T., 2004. Possible role of the "Holocene Event 3" on the collapse of Neolithic Cultures around the Central Plain of China. *Quarternary International*, 117(2004): 153–166.

Wu, X., Menzel, W. P., Wade, G. S., 1999. Estimation of sea surface temperatures using GOES-8/9 radiance measurements. *Bulletin of the American Meteorological Society*, 80(6): 1127–1138.

Wu, X., Zhang, C., Goldberg, P., Cohen, D., Pan, Y., Arpin, T., Bar-Yosef, O., 2012b. Early Pottery at 20,000 Years Ago in Xianrendong Cave, China. *Science*, 336(6089): 1696–1700. DOI;10.1126/science.1218643.

Wu, Y., Wu, S. -Y., Wen, J., Tagle, F., Xu, M., Tan, J., 2016b. Future Changes in Mean and Extreme Monsoon Precipitation in the Middle and Lower Yangtze River Basin, China, in the CMIP5 Models. *Journal of Hydrometeorology*, 17(11): 2785–2797. DOI:10.1175/JHM-D-16-0033.1.

Wu, Z., Li, J., Jiang, Z., He, J., Zhu, X., 2012c. Possible effects of the North Atlantic Oscillation on the strengthening relationship between the East Asian summer monsoon and ENSO. *International Journal of Climatology*, 32(5): 794–800.

Wuchter, C., Schouten, S., Coolen, M. J. L., Sinninghe Damsté, J. S., 2004. Temperature-dependent variation in the distribution of tetraether membrane lipids of marine Crenarchaeota: Implications for TEX86 paleothermometry. *Paleoceanography*, 19 (4). DOI:10.1029/2004PA001041.

Wünnemann, B., Demske, D., Tarasov, P., Kotlia, B. S., Reinhardt, C., Bloemendal, J., Diekmann, B., Hartmann, K., Krois, J., Riedel, F., Arya, N., 2010. Hydrological evolution during the last 15 kyr in the Tso Kar lake basin (Ladakh, India), derived from geomorphological, sedimentological and palynological records. *Quaternary Science Reviews*, 29: 1138–1155.

Xie, J., Hu, L., Tang, J., Wu, X., Li, N., Yuan, Y., Yang, H., Zhang, J., Luo, S., Chen, X., 2011. Ecological mechanisms underlying the sustainability of the agricultural heritage rice–fish coculture system. *Proceedings of the National Academy of Sciences*, 108 (50): E1381–E1387.

Xie, P., Arkin, P. A., 1997. Global precipitation: A 17-year monthly analysis based on gauge observations, satellite esti- mates and numerical model outputs. *Bulletin of the American Meteorological Society*, 78: 2539–2558.

Xu, J., 2018. A cave δ18O based 1800-year reconstruction of sediment load and streamflow: The Yellow River source area. *CATENA*, 161: 137–147. DOI:10.1016/j.catena.2017.09.028.

Xu, K., Milliman, J. D., 2009. Seasonal variations of sediment discharge from the Yangtze River before and after impoundment of the Three Gorges Dam. *Geomorphology*, 104 (3): 276–283. DOI:10.1016/j.geomorph.2008.09.004.

Xu, Q., 2001. Abrupt change of the mid-summer climate in central east China by the influence of atmospheric pollution. *Atmospheric Environment*, 35(30): 5029–5040. DOI:10.1016/S1352-2310(01)00315-6.

Xue, Y., 2010. 云南剑川海门口遗址植物遗存初步研究*Yunnan Jianchuan Haimenkou Yizhi Zhiwu Yicun Chubu Yanjiu (A Preliminary Investigation on the Archaeobotanical Material from the Site of Haimenkou in Jianchuan County, Yunnan)*. MA Thesis, Peking University.

Yan, W., 1984. Lun Zhongguo de tongshi bingyong de dai. *Shiqian yanjiu*, 1: 36–44.

Yancheva, G., Nowaczyk, N. R., Mingram, J., Dulski, P., Schettler, G., Negendank, J. F. W., Liu, J., Sigman, D. M., Peterson, L. C., Haug, G. H., 2007. Influence of the inter-tropical convergence zone on the East Asian monsoon. *Nature*, 445: 74–77. DOI:10.1038/nature05431.

Yang, B., Shi, Y., Braeuning, A., 2004. Evidence for a warm humid climate in arid north-western China during 40–30 ka BP. *Quaternary Science Reviews*, 23: 2537–2548.

Yang, J., 2012. *Tombstone: The Great Chinese Famine, 1958–1962*. Macmillan, New York, p. 629.

Yang, S., Ding, Z., Li, Y., Wang, X., Jiang, W., Huang, X., 2015a. Warming-induced northwestward migration of the East Asian monsoon rain belt from the Last Glacial Maximum to the mid-Holocene. *Proceedings of the National Academy of Sciences*, 112(43): 13178–13183.

Yang, S.L., Li, M., Dai, S.B., Liu, Z., Zhang, J., Ding, P. X., 2006a. Drastic decrease in sediment supply from the Yangtze River and its challenge to coastal wetland management. *Geophysical Research Letters*, 33(L06408). DOI:10.1029/2005GL025507.

Yang, S. L., Xu, K. H., Milliman, J. D., Yang, H. F., Wu, C. S., 2015b. Decline of Yangtze River water and sediment discharge: Impact from natural and anthropogenic changes. *Scientific Reports*, 5: 12581. DOI:10.1038/srep12581.

Yang, X., Ma, Z., Li, J., Yu, J., Stevens, C., Zhuang, Y., 2015c. Comparing subsistence strategies in different landscapes of North China 10,000 years ago. *The Holocene*, 25 (12): 1957–1964. DOI:10.1177/0959683615596833.

Yang, X., Wang, W., Zhuang, Y., Li, Z., Ma, Z., Ma, Y., Cui, Y., Wei, J., Fuller, D. Q., 2016. New radiocarbon evidence on early rice consumption and farming in South China. *The Holocene*, 27(7): 1045–1051. DOI:10.1177/0959683616678465.

Yang, X., Zhou, L., Zhao, C., Yang, J., 2018. Impact of aerosols on tropical cyclone-induced precipitation over the mainland of China. *Climatic Change*, 148(1): 173–185. DOI:10.1007/s10584-018-2175-5.

Yang, Z. -S., Wang, H. -J., Saito, Y., Milliman, J., Xu, K., Qiao, S., Shi, G., 2006b. Dam impacts on the Changjiang (Yangtze) river sediment discharge to the sea: The past 55 years and after the three Gorges dam. *Water Resources Research*, 42(4): W04407. DOI:10.1029/2005WR003970.

Yao, T., Thompson, L., Yang, W., Yu, W., Gao, Y., Guo, X., Yang, X., Duan, K., Zhao, H., Xu, B., 2012. Different glacier status with atmospheric circulations in Tibetan Plateau and surroundings. *Nature Climate Change*, 2(9): 663.

Yasuda, Y., 2002. Origins of pottery and agriculture in East Asia. In: Yasuda, Y. (ed.), *The Origins of Pottery and Agriculture*. International Research Center for Japanese Studies, Kyoto, pp. 119–142.

Yasuda, Y., Fujiki, T., Nasu, H., Kato, M., Morita, Y., Mori, Y., Kanehara, M., Toyama, S., Yano, A., Okuno, M., Jiejun, H., Ishihara, S., Kitagawa, H., Fukusawa, H., Naruse, T., 2004. Environmental archaeology at the Chengtoushan site, Hunan Province, China, and implications for environmental change and the rise and fall of the Yangtze river civilization. *Quaternary International*, 123–125: 149–158. DOI:10.1016/j.quaint.2004.02.016.

Ye, D., 1981. Some characteristics of the summer circulation over the Qinghai-Xizang (Tibet) plateau and its neighborhood. *Bulletin of the American Meteorological Society*, 62(1): 14–19. 10.1175/1520–0477(1981)062<0014:scotsc>2.0.Co;2.

Ye, Q. C., 1989. Landforms system of the great plain of North China and its tendency of environmental evolution. *Geographical Research*, 8(3): 10–20.

Yi, M., Barton, L., Morgan, C., Liu, D., Chen, F., Zhang, Y., Pei, S., Guan, Y., Wang, H., Gao, X., Bettinger, R. L., 2013. Microblade technology and the rise of serial specialists in North-Central China. *Journal of Anthropological Archaeology*, 32(2013): 212–223.

Yihui, D., 1992. Effects of the Qinghai-Xizang (Tibetan) plateau on the circulation features over the plateau and its surrounding areas. *Advances in Atmospheric Sciences*, 9(1): 112–130. 10.1007/bf02656935.

Yin, H., Li, C., 2001a. Human impact on floods and flood disasters on the Yangtze River. *Geomorphology*, 41(2): 105–109. DOI:10.1016/S0169-555X(01)00108-8.

Yin, H., Li, C., 2001b. Human impact on floods and flood disasters on the Yangtze River. *Geomorphology*, 41(2–3): 105–109. DOI:10.1016/S0169-555X(01)00108-8.

Yin, Y., Ma, D., Wu, S., Pan, T., 2015. Projections of aridity and its regional variability over China in the mid-21st century. *International Journal of Climatology*, 35(14): 4387–4398. DOI:10.1002/joc.4295.

Yokoyama, Y., Naruse, T., Ogawa, N.O., Tada, R., Kitazato, H., Ohkouchi, N., 2006. Dust influx reconstruction during the last 26,000 years inferred from a sedimentary leaf wax record from the Japan Sea. *Global and Planetary Change*, 54(3–4): 239–250.

Yoshida, S., 1981. *Fundamentals of Rice Crop Science*. In: IRRI (Editor). IRRI, Los Banos, Phillipines.

You, Y., 2019. *Climate Change, Agriculture and Human Adaptation at the Indus site of Harappa*. Masters Thesis, Washington State University.

Yu, F., Chen, Z., Ren, X., Yang, G., 2009a. Analysis of historical floods on the Yangtze River, China: Characteristics and explanations. *Geomorphology*, 113: 210–216. DOI:10.1016/j.geomorph.2009.03.008.

Yu, F., Chen, Z., Ren, X., Yang, G., 2009b. Analysis of historical floods on the Yangtze River, China: Characteristics and explanations. *Geomorphology*, 113(3): 210–216. DOI:10.1016/j.geomorph.2009.03.008.

Yu, G., Chen, X., Ni, J., Cheddadi, R., Guiot, J., Han, H., Harrison, S. P., Huang, C., Ke, M., Kong, Z., Li, S., Li, W., Liew, P., Liu, G., Liu, J., Liu, K. -B., Prentice, I. C., Qui, W., Ren, G., Song, C., Sugita, S., Sun, X., Tang, L., van Campo, E., Xia, Y., Xu, Q., Yan, S., Yang, X., Zhao, J., Zheng, Z., 2000. Palaeovegetation of China: a pollen data-based synthesis for the mid-Holocene and last glacial maximum. *Journal of Biogeography*, 27: 635–664.

Yu, R., Wang, B., Zhou, T., 2004. Tropospheric cooling and summer monsoon weakening trend over East Asia. *Geophysical Research Letters*, 31(22): L22212. DOI:10.1029/2004gl021270.

Yu, X., 2005. *Da Yuejin Kurezi*. Shidai Chaoliu Chubanshe, Hong Kong, 124 pp.

Yun, K. S., Timmermann, A., 2018. Decadal monsoon-ENSO relationships reexamined. *Geophysical Research Letters*, 45(4): 2014–2021.

Zawahri, N. A., 2009. India, Pakistan and cooperation along the Indus River system. *Water Policy*, 11(1): 1–20. 10.2166/wp.2009.010.

Zeder, M. A., 2015. Core questions in domestication research. *Proceedings of the National Academy of Sciences*, 112(11): 3191–3198.

Zeebe, R. E., Ridgwell, A., Zachos, J. C., 2016. Anthropogenic carbon release rate unprecedented during the past 66 million years. *Nature Geoscience*, 9(4): 325.

Zeng, X., 1998. 中国历史上的黄穋稻 Zhongguo Lishi Shang Huangludao (Huang Lu Rice in Chinese History). *Nongye Kaogu*, 1998(1): 292–307.

Zhang, C., Hung, H. -c., 2010. The emergence of agriculture in southern China. *Antiquity*, 84: 11–25.

Zhang, C., Hung, H. -c., 2012. Later hunter-gatherers in southern China 18000–3000 BC. *Antiquity*, 86(331): 11–29.

Zhang, C., Hung, H. -c., 2013. Jiahu 1: earliest farmers beyond the Yangtze. *Antiquity*, 87 (2013): 46–63.

Zhang, D., Cervantes, J., Huamán, Z., Carey, E., Ghislain, M., 2000. Assessing genetic diversity of sweet potato (Ipomoea batatas (L.) Lam.) cultivars from tropical America using AFLP. *Genetic Resources and Crop Evolution*, 47(6): 659–665. DOI:10.1023/a:1026520507223.

Zhang, L., Li, J., Wei, G., Liao, W., Wang, Q., Xiang, C., 2017. Analysis of the relationship between water level fluctuation and seismicity in the Three Gorges Reservoir (China). *Geodesy and Geodynamics*, 8(2): 96–102. DOI:10.1016/j.geog.2017.02.004.

Zhang, L., Wang, Q., Liu, Q., 2009. Sweet potato in China. In: Loebenstein, G., Thottappilly, G. (eds.), *The Sweet Potato*. Springer, Dortrecht, pp.325–358.

Zhang, P., Cheng, H., Edwards, R. L., Chen, F., Wang, Y., Yang, X., Liu, J., Tan, M., Wang, X., Liu, J., An, C., Dai, Z., Zhou, J., Zhang, D., Jia, J., Jin, L., Johnson, K. R., 2008. A test of climate, sun, and culture relationships from an 1810-year Chinese cave record. *Science*, 322: 940–942. DOI:10.1126/science.1163965.

Zhang, R., Delworth, T. L., 2006. Impact of Atlantic multidecadal oscillations on India/Sahel rainfall and Atlantic hurricanes. *Geophysical Research Letters*, 33(L17712). DOI:10.1029/2006GL026267.

Zhang, S., Yi, Y., Liu, Y., Wang, X., 2013. Hydraulic principles of the 2,268-year-old Dujiangyan project in China. *Journal of Hydraulic Engineering*, 139(5): 538–546. DOI:10.1061/(ASCE)HY.1943-7900.0000675.

Zhang, W., Yuan, J., 1998. A preliminary study of ancient excavated rice from Yuchanyan site, Dao County, Hunan Province. *Acta Agronomica Sinica*, 24(4): 416–420.

Zhang, X., Wang, J., 2003. 古人类食物结构研究 Gurenlei Shiwu Jiegou Yanjiu (Study on the Diet of ancient populations). 考古 *Kaogu (Archaeology)*, 2003(2): 158–171.

Zhang, X., Zhong, S., Wu, Z., Li, Y., 2018a. Seasonal prediction of the typhoon genesis frequency over the Western North Pacific with a Poisson regression model. *Climate Dynamics*, 51(11): 4585–4600. DOI:10.1007/s00382-017-3654-5.

Zhang, X., Zhou, A., Zhang, C., Hao, S., Zhao, Y., An, C., 2016a. High-resolution records of climate change in arid eastern central Asia during MIS 3 (51 600–25 300 cal a BP) from Wulungu Lake, north-western China. *Journal of Quaternary Science*, 31(6): 577–586. DOI:10.1002/jqs.2881.

Zhang, Y. G., Zheng, W. J., Wang, Y. J., Zhang, D. L., Tian, Y. T., Wang, M., Zhang, Z. Q., Zhang, P. Z., 2018b. Contemporary deformation of the North China plain from global positioning system data. *Geophysical Research Letters*, 45(4): 1851–1859. DOI:10.1002/2017gl076599.

Zhang, Z., Jin, Q., Chen, X., Xu, C. -Y., Jiang, S., 2016. On the linkage between the extreme drought and pluvial patterns in China and the large-scale atmospheric circulation. *Advances in Meteorology*, 2016: 8010638. DOI:10.1155/2016/8010638.

Zhang, Z., Wang, H., Guo, Z., Jiang, D., 2007. What triggers the transition of palaeoenvironmental patterns in China, the Tibetan Plateau uplift or the Paratethys Sea retreat? *Palaeogeography, Palaeoclimatology, Palaeoecology*, 245: 317–331.

Zhang, Z., Zhao, M., Eglinton, G., Lu, H., Huang, C. -Y., 2006. Leaf wax lipids as paleovegetational and paleoenvironmental proxies for the Chinese Loess Plateau over the last 170 kyr. *Quaternary Science Reviews*, 25(5–6): 575–594.

Zhao, J., An, C. -B., Huang, Y., Morrill, C., Chen, F. -H., 2017a. Contrasting early Holocene temperature variations between monsoonal East Asia and westerly dominated Central Asia. *Quaternary Science Reviews*, 178: 14–23.

Zhao, K., Tung, C. -W., Eizenga, G. C., Wright, M. H., Ali, M. L., Price, A. H., Norton, G. J., Islam, M. R., Reynolds, A., Mezey, J., McClung, A. M., Bustamante, C. D., McCouch, S. R., 2011. Genome-wide association mapping reveals a rich genetic architecture of complex traits in Oryza sativa. *Nature Communications*, 2: 467.

Zhao, S., Chang, L., 1999. 论两汉时代冬小麦在我国北方的推广普及 Lun Lianghan Shidai Dong Xiaomai Zai Wo Guo Beifang Tuiguang Puji (A Discussion on Han Dynasty Winter Wheat in our Country and its Distribution in Northern China). *Zhongguo Lishi Dili Luncong*, 1999(2): 37–46.

Zhao, Y., An, C. -B., Duan, F., Zhao, J., Mao, L., Zhou, A., Cao, Z., Chen, F., 2017b. Consistent vegetation and climate deterioration from early to late MIS3 revealed by multi-proxies (mainly pollen data) in north-west China. *Review of Palaeobotany and Palynology*, 244: 43–53. DOI:10.1016/j.revpalbo.2017.04.010.

Zhao, Y., Yu, Z., Herzschuh, U., Yang, B., Zhao, H., Fang, K., Li, H., Li, Q., 2014. Vegetation and climate change during marine isotope stage 3 in China. *Chinese Science Bulletin*, 59(33): 4444–4455.

Zhao, Z., 1998. The middle Yangtze region in China is one place where rice was domesticated: Phytolith evidence from the diaotonghuan cave, Northern Jiangxi. *Antiquity*, 72(278): 885–897.

Zhao, Z., 2003. 浮选与植碳化植物依存的研究 Fuxuan Yu Tanhua Zhiwu Yicun de Yanjiu (Flotation and Carbonized Plant remains). In: Zhongguo Shehui Kexueyuan Kaogu Yanjiusuo, Guangzi Zhuangzu Zizhiqu Wenwu Gongzuodui, Guilin Zengpiyan Yizhi Bowuguan, Guilin Shi Wenwu Gongzuodui (eds.), 桂林甑皮岩 *Guilin Zengpiyan (The site of Zengpiyan in Guilin)*. Wenwu Chubanshe, Beijing, pp. 93–97.

Zhao, Z., 2005. 从兴隆沟遗址浮选结果谈中国北方旱作农业起源问题 Cong Xinglonggou Yizhi fuxuan jieguo tan Zhongguo Beifang Zaozuo Nongye Qiyuan Wenti(Discussion on Chinese dry crop agriculture from flotation results in

Xinglonggou Site). In: Wang, R., Tang, H. (eds.), 东亚古物 *Dongya Guwu (Antiquities of Eastern Asia)*. Wenwu Chubanshe Beijing, pp. 188–199.

Zhao, Z., Zhang, J., 2009. 贾湖遗址2001年度浮选结果分析报告 Jiahu Yizhi 2001 Niandu Fuxuan Jieguo Fenxi Baogao (Report on the Analysis of the Results of the 2001 Floatation of the Jiahu Site). *Kaogu*, 2009(8): 84–93.

Zhejiang Sheng Kaogu Yanjiusuo, 2004. 浙江跨湖桥 *Zhejiang Kuahuqiao (The Site of Kuahuqiao in Zhejiang)*. Wenwu Chubanshe, Beijing.

Zhen, J., 2000. Rice-wheat cropping system in China. In: Hobbs, P. R., Gupta, R. K. (eds.), *Soil and crop management practices for enhanced productivity of the rice-wheat cropping system in the Sichuan province of China*. Rice-Wheat Consortium Paper Series. Rice-Wheat Consortium for the Indo-Gangetic Plains, New Delhi, India, pp. 1–10.

Zheng, H., Clift, P. D., Tada, R., Jia, J. T., He, M. Y., Wang, P., 2013. A Pre-Miocene birth to the Yangtze River. *Proceedings of the National Academy of Sciences*: 1–6. DOI:10.1073/pnas.1216241110.

Zheng, H., Wei, X., Tada, R., Clift, P. D., Wang, B., Jourdan, F., Wang, P., He, M., 2015. Late oligocene–early miocene birth of the Taklimakan Desert. *Proceedings of the National Academy of Sciences*, 110(19): 7556–7561. DOI:10.1073/pnas.1424487112.

Zheng, J., Hao, Z., Ge, Q., 2005. Variation of precipitation for the last 300 years over the middle and lower reaches of the yellow river. *Science in China Series D- Earth Science*, 48: 2182. DOI:10.1360/03yd0392.

Zheng, J., Wang, W. -C., Ge, Q., Man, Z., Zhang, P., 2006. Precipitation variability and extreme events in Eastern China during the past 1500 years. *Terrestrial, Atmospheric and Oceanic Sciences*, 17(3): 579–592.

Zheng, Z., Lei, Z. Q., 1999. A 400,000 year record of vegetaitonal and climatic changes from a volcanic basin, Leizhou Peininsula, Southern China. *Palaeogeogr Palaeoclimatol Palaeoecol*, 145: 339–362.

Zhongguo Shehui Kexueyuan Kaoguyanjiusuo, 2003. 桂林甑皮岩 *Guilin Zengpiyan (The site of Zengpiyan in Guilin Province)*. Wenwu Chubanshe, Beijing.

Zhongyuan, C., Stanley, D. J., 1998. Sea-level rise on Eastern China's Yangtze Delta. *Journal of Coastal Research*, 14(1): 360–366.

Zhou, T., Bronnimann, S., Griesser, T., Fischer, A. M., Zou, L., 2010. A reconstructed dynamic Indian monsoon index extended back to 1880. *Climate Dynamics*, 34: 573–585. DOI 10.1007/s00382-009–0552-5.

Zhou, T. J., Zhang, L. X., Li, H. M., 2008. Changes in global land monsoon area and total rainfall accumulation over the last half century. *Geophysical Research Letters*, 35 (16): L16707. DOI:10.1029/2008GL034881.

Zhou, W., Lu, X., Wu, Z., Deng, L., Jull, A. J. T., Donahue, D., Beck, W., 2002. Peat record reflecting Holocene climatic change in the Zoige Plateau and AMS radiocarbon dating. *Chinese Science Bulletin*, 47(1): 66–70.

Zhou, Y., Jeppesen, E., Li, J., Zhang, Y., Zhang, X., Li, X., 2016. Impacts of three gorges reservoir on the sedimentation regimes in the downstream-linked two largest Chinese freshwater lakes. *Scientific Reports*, 6: 35396. DOI:10.1038/srep35396.

Zhu, C., Zheng, C., Ma, C., Yang, X., Gao, X., Wang, H., Shao, J., 2003. On the Holocene sea-level highstand along the Yangtze delta and Ningshao plain, east China. *Chinese Science Bulletin*, 48(24): 2672–2683.

Zhu, S., Ge, F., Sein, D., Remedio, A., Sielmann, F., Fraedrich, K., Zhi, X., 2019. Projected changes in surface air temperature over the Indochina Peninsula from the regionally coupled model ROM. *Geophysical Research Abstracts*, 21: EGU2019–3848.

Zhu, X. F., Huang, C. C., Pang, J. L., 2010. Palaeo-hydrological studies of the Holocene extreme floods in the Tianshui Gorges of the Weihe River. *Progress in Geography*, 2010(29): 840–846.

Zhuang, C., 2009. 雞肋編／莊季裕撰 *Ji le bian / Zhuang Jiyu zhuan (The 12th Century Ji le Bian)*. Beijing Tushuguan Chubanshe, Beijing.

Zhuang, Y., Kidder, T. R., 2014. Archaeology of the anthropocene in the yellow river region, China, 8000–2000 cal. *BP. The Holocene*, 24(11): 1602–1623. DOI:10.1177/0959683614544058.

Zohary, D., Hopf, M., Weiss, E., 2012. *Domestication of Plants in the Old World: The origin and spread of domesticated plants in Southwest Asia, Europe, and the Mediterranean Basin*. Oxford University Press on Demand, Oxford.

Zong, Y., 2004. Mid-Holocene sea-level highstand along the Southeast Coast of China. *Quaternary International*, 117(1): 55–67. DOI:10.1016/S1040-6182(03)00116-2.

Zong, Y., Chen, X., 2000. The 1998 Flood on the Yangtze, China. *Natural Hazards*, 22(2): 165–184. 10.1023/a:1008119805106.

Zong, Y., Chen, Z., Innes, J.B., Chen, C., Wang, Z., Wang, H., 2007. Fire and flood management of coastal swamp enabled first rice paddy cultivation in East China. *Nature*, 449(7161): 459–462.

Zuo, X., Lu, H., Jiang, L., Zhang, J., Yang, X., Huan, X., He, K., Wang, C., Wu, N., 2017. Dating rice remains through phytolith carbon-14 study reveals domestication at the beginning of the Holocene. *Proceedings of the National Academy of Sciences*, 114 (25): 6486–6491.

Index

Printed in the United States
By Bookmasters